AF344605

Oscillation Theory of
Differential Equations with
Deviating Arguments

PURE AND APPLIED MATHEMATICS

A Program of Monographs, Textbooks, and Lecture Notes

EXECUTIVE EDITORS

Earl J. Taft
Rutgers University
New Brunswick, New Jersey

Zuhair Nashed
University of Delaware
Newark, Delaware

CHAIRMEN OF THE EDITORIAL BOARD

S. Kobayashi
University of California, Berkeley
Berkeley, California

Edwin Hewitt
University of Washington
Seattle, Washington

EDITORIAL BOARD

M. S. Baouendi
Purdue University

Donald Passman
University of Wisconsin-Madison

Jack K. Hale
Brown University

Fred S. Roberts
Rutgers University

Marvin Marcus
University of California, Santa Barbara

Gian-Carlo Rota
Massachusetts Institute of Technology

W. S. Massey
Yale University

David Russell
University of Wisconsin-Madison

Leopoldo Nachbin
Centro Brasileiro de Pesquisas Físicas
and University of Rochester

Jane Cronin Scanlon
Rutgers University

Anil Nerode
Cornell University

Walter Schempp
Universität Siegen

Mark Teply
University of Wisconsin-Milwaukee

MONOGRAPHS AND TEXTBOOKS IN PURE AND APPLIED MATHEMATICS

1. *K. Yano,* Integral Formulas in Riemannian Geometry (1970) *(out of print)*
2. *S. Kobayashi,* Hyperbolic Manifolds and Holomorphic Mappings (1970) *(out of print)*
3. *V. S. Vladimirov,* Equations of Mathematical Physics (A. Jeffrey, editor; A. Littlewood, translator) (1970) *(out of print)*
4. *B. N. Pshenichnyi,* Necessary Conditions for an Extremum (L. Neustadt, translation editor; K. Makowski, translator) (1971)
5. *L. Narici, E. Beckenstein, and G. Bachman,* Functional Analysis and Valuation Theory (1971)
6. *S. S. Passman,* Infinite Group Rings (1971)
7. *L. Dornhoff,* Group Representation Theory (in two parts). Part A: Ordinary Representation Theory. Part B: Modular Representation Theory (1971, 1972)
8. *W. Boothby and G. L. Weiss (eds.),* Symmetric Spaces: Short Courses Presented at Washington University (1972)
9. *Y. Matsushima,* Differentiable Manifolds (E. T. Kobayashi, translator) (1972)
10. *L. E. Ward, Jr.,* Topology: An Outline for a First Course (1972) *(out of print)*
11. *A. Babakhanian,* Cohomological Methods in Group Theory (1972)
12. *R. Gilmer,* Multiplicative Ideal Theory (1972)
13. *J. Yeh,* Stochastic Processes and the Wiener Integral (1973) *(out of print)*
14. *J. Barros-Neto,* Introduction to the Theory of Distributions (1973) *(out of print)*
15. *R. Larsen,* Functional Analysis: An Introduction (1973) *(out of print)*
16. *K. Yano and S. Ishihara,* Tangent and Cotangent Bundles: Differential Geometry (1973) *(out of print)*
17. *C. Procesi,* Rings with Polynomial Identities (1973)
18. *R. Hermann,* Geometry, Physics, and Systems (1973)
19. *N. R. Wallach,* Harmonic Analysis on Homogeneous Spaces (1973) *(out of print)*
20. *J. Dieudonné,* Introduction to the Theory of Formal Groups (1973)
21. *I. Vaisman,* Cohomology and Differential Forms (1973)
22. *B. -Y. Chen,* Geometry of Submanifolds (1973)
23. *M. Marcus,* Finite Dimensional Multilinear Algebra (in two parts) (1973, 1975)
24. *R. Larsen,* Banach Algebras: An Introduction (1973)
25. *R. O. Kujala and A. L. Vitter (eds.),* Value Distribution Theory: Part A; Part B: Deficit and Bezout Estimates by Wilhelm Stoll (1973)
26. *K. B. Stolarsky,* Algebraic Numbers and Diophantine Approximation (1974)
27. *A. R. Magid,* The Separable Galois Theory of Commutative Rings (1974)
28. *B. R. McDonald,* Finite Rings with Identity (1974)
29. *J. Satake,* Linear Algebra (S. Koh, T. A. Akiba, and S. Ihara, translators) (1975)

30. *J. S. Golan*, Localization of Noncommutative Rings (1975)
31. *G. Klambauer*, Mathematical Analysis (1975)
32. *M. K. Agoston*, Algebraic Topology: A First Course (1976)
33. *K. R. Goodearl*, Ring Theory: Nonsingular Rings and Modules (1976)
34. *L. E. Mansfield*, Linear Algebra with Geometric Applications: Selected Topics (1976)
35. *N. J. Pullman*, Matrix Theory and Its Applications (1976)
36. *B. R. McDonald*, Geometric Algebra Over Local Rings (1976)
37. *C. W. Groetsch*, Generalized Inverses of Linear Operators: Representation and Approximation (1977)
38. *J. E. Kuczkowski and J. L. Gersting*, Abstract Algebra: A First Look (1977)
39. *C. O. Christenson and W. L. Voxman*, Aspects of Topology (1977)
40. *M. Nagata*, Field Theory (1977)
41. *R. L. Long*, Algebraic Number Theory (1977)
42. *W. F. Pfeffer*, Integrals and Measures (1977)
43. *R. L. Wheeden and A. Zygmund*, Measure and Integral: An Introduction to Real Analysis (1977)
44. *J. H. Curtiss*, Introduction to Functions of a Complex Variable (1978)
45. *K. Hrbacek and T. Jech*, Introduction to Set Theory (1978)
46. *W. S. Massey*, Homology and Cohomology Theory (1978)
47. *M. Marcus*, Introduction to Modern Algebra (1978)
48. *E. C. Young*, Vector and Tensor Analysis (1978)
49. *S. B. Nadler, Jr.*, Hyperspaces of Sets (1978)
50. *S. K. Segal*, Topics in Group Rings (1978)
51. *A. C. M. van Rooij*, Non-Archimedean Functional Analysis (1978)
52. *L. Corwin and R. Szczarba*, Calculus in Vector Spaces (1979)
53. *C. Sadosky*, Interpolation of Operators and Singular Integrals: An Introduction to Harmonic Analysis (1979)
54. *J. Cronin*, Differential Equations: Introduction and Quantitative Theory (1980)
55. *C. W. Groetsch*, Elements of Applicable Functional Analysis (1980)
56. *I. Vaisman*, Foundations of Three-Dimensional Euclidean Geometry (1980)
57. *H. I. Freedman*, Deterministic Mathematical Models in Population Ecology (1980)
58. *S. B. Chae*, Lebesgue Integration (1980)
59. *C. S. Rees, S. M. Shah, and C. V. Stanojević*, Theory and Applications of Fourier Analysis (1981)
60. *L. Nachbin*, Introduction to Functional Analysis: Banach Spaces and Differential Calculus (R. M. Aron, translator) (1981)
61. *G. Orzech and M. Orzech*, Plane Algebraic Curves: An Introduction Via Valuations (1981)
62. *R. Johnsonbaugh and W. E. Pfaffenberger*, Foundations of Mathematical Analysis (1981)
63. *W. L. Voxman and R. H. Goetschel*, Advanced Calculus: An Introduction to Modern Analysis (1981)
64. *L. J. Corwin and R. H. Szcarba*, Multivariable Calculus (1982)
65. *V. I. Istrătescu*, Introduction to Linear Operator Theory (1981)
66. *R. D. Järvinen*, Finite and Infinite Dimensional Linear Spaces: A Comparative Study in Algebraic and Analytic Settings (1981)

Other Volumes in Preparation

Oscillation Theory of Differential Equations with Deviating Arguments

G. S. LADDE
V. LAKSHMIKANTHAM
The University of Texas
Arlington, Texas

B. G. ZHANG
Shandong College of Oceanography
Qingdao, Shandong
People's Republic of China

DISCARDED

LIBRARY
Appalachian State University
Boone. North Carolina
DISCARDED

MARCEL DEKKER, INC.   New York and Basel

Library of Congress Cataloging in Publication Data

Ladde, G. S.
   Oscillation theory of differential equations with
deviating arguments.

   (Pure and applied mathematics ; 110)
   Bibliography: p.
   Includes index.
   1. Differential equations. 2. Oscillations.
I. Lakshmikantham, V. II. Zhang, B. G., (date).
III. Title. IV. Series: Monographs and textbooks in pure
and applied mathematics ; v. 110.
QA372.L19   1987   515.3'5   87-9232
ISBN 0-8247-7738-7

COPYRIGHT © 1987 by MARCEL DEKKER, INC. ALL RIGHTS RESERVED

Neither this book nor any part may be reproduced or transmitted
in any form or by any means, electronic or mechanical, including
photocopying, microfilming, and recording, or by any information
storage and retrieval system, without permission in writing from
the publisher.

MARCEL DEKKER, INC.
270 Madison Avenue, New York, New York 10016

Current printing (last digit):
10 9 8 7 6 5 4 3 2 1

PRINTED IN THE UNITED STATES OF AMERICA

Preface

The mathematical modeling of several real-world problems leads to
differential equations that depend on the past history rather than only the
current state. The models may have discrete time lags as well as distributed lags or delays.

Most of the work in the theory of oscillations is centered around second
or higher order <u>ordinary differential equations</u> (ODE) because of the fact
that the first order scalar ODEs do not possess oscillatory behavior.
Bernoulli (1728) studied the problem of sound vibrating in a tube of finite
length and investigated the properties of first order <u>ordinary differential
equations with deviating arguments</u> (ODEWDA). Myskis investigated several
oscillation problems of first order ODEWDA, which are recorded in his
book. Since 1950 oscillation theory of ODEWDA has received the attention
of several applied mathematicians as well as other scientists around the
world. In light of this, it is essential to present an up-to-date account in
a systematic way.

This book offers a systematic treatment of oscillation and nonoscillation
theory of differential equations with deviating arguments. The book is
divided into six chapters. The first chapter consists of preliminary material
that is essential for the rest of the book. Chapters 2 and 3 deal with first
order linear and nonlinear differential equations with deviating arguments.
Chapter 4 is devoted to second order equations. Chapter 5 extends the study
to higher order equations. In Chapter 6, we introduce the oscillation theory
to systems of differential equations with deviating arguments.

Some important features of the book are the following:

(i) This is the first English language book that offers a systematic study
of the theory of oscillation of ordinary differential equations with deviating arguments. It complements the book by Swanson [260] which discusses the oscillation theory of linear ordinary differential equations
only.

(ii) This book stresses the importance of deviating arguments in the sense
that their presence causes or destroys the oscillation phenomena,

and does not merely preserve the oscillatory behavior of equations
without deviating arguments.

(iii)   This book contains recent results and consequently is a valuable source
for research workers in the field.

We wish to express our thanks to Ms. Vickie Kearn and the staff
of Marcel Dekker, Inc., for their cooperation during the publication of the
book. We also wish to thank Mrs. Sandra Weber for her excellent typing of
the manuscript.

G. S. Ladde

V. Lakshmikantham

B. G. Zhang

Contents

1
Preliminaries

1.0 INTRODUCTION

This chapter is essentially introductory in nature. Its main purpose is to introduce some basic concepts from the theory of differential equations with deviating arguments, to sketch some important results from the theory of oscillation of ordinary differential equations, and to demonstrate some new problems in oscillation theory caused by deviating arguments.

Section 1.1 is concerned with the statement of the basic initial value problems and classification of equations with deviating arguments. Section 1.2 provides definitions of oscillation of solutions with or without deviating arguments. Section 1.3 sketches necessary important results and summarizes certain main topics in oscillation theory. In Section 1.4, we offer some examples to illustrate new oscillation phenomena caused by deviating arguments. Finally, we introduce some fixed point theorems which are important tools in oscillation theory, especially, when one proves the existence of nonoscillatory solutions.

1.1 INITIAL VALUE PROBLEMS

Let us consider the <u>ordinary differential equation</u> (ODE)

$$x' = f(t, x) \tag{1.1.1}$$

together with the initial condition

$$x(t_0) = x_0 \tag{1.1.2}$$

It is well known that under certain assumptions with respect to f the initial value problem (1.1.1) and (1.1.2) has a unique solution and it is equivalent to the integral equation

$$x(t) = x(t_0) + \int_{t_0}^{t} f(s, x(s))ds, \quad t \geq t_0 \tag{1.1.3}$$

Next, we consider the differential equation of the form

$$\frac{dx(t)}{dt} = f(t, \, x(t), \, x(t - \tau)), \qquad \tau > 0, \; t \geq t_0 \qquad (1.1.4)$$

in which the right-hand side depends not only on the instantaneous position $x(t)$, but also on $x(t - \tau)$, the position at τ units back, that is to say, the equation has past memory. Such an equation is called an ODE with a _delay_ or _retarded argument_. Whenever necessary, we shall consider the integral equation

$$x(t) = x(t_0) + \int_{t_0}^{t} f(s, \, x(s), \, x(s - \tau)) \, ds \qquad (1.1.5)$$

that is equivalent to (1.1.4). In order to define a solution of (1.1.4), we need to have a known function $\phi(t)$ on $[t_0 - \tau, \, t_0]$, instead of just the initial condition $x(t_0) = x_0$.

The basic initial value problem for an ordinary differential equation with delay argument is posed as follows: On the interval $[t_0, T]$, $T \leq +\infty$, we seek a continuous function x that satisfies (1.1.4) and an initial condition

$$x(t) = \phi(t), \qquad t \in E_{t_0} \qquad (1.1.6)$$

where t_0 is an initial point, $E_{t_0} = [t_0 - \tau, \, t_0]$ is the initial set; the known function $\phi(t)$ on E_{t_0} is called the _initial function_. Usually, it is assumed that $x(t_0 + 0) = \phi(t_0)$. We always mean a one-sided derivative when we speak of the derivative at an endpoint of an interval.

Under general assumptions, the existence and uniqueness of solutions to the initial value problem (1.1.4) and (1.1.6) can be established. The solution sometimes is denoted by $x(t, \phi)$. In the case of a variable delay $\tau = \tau(t) > 0$ in Eq. (1.1.4), it is also required to find a solution of this equation for $t > t_0$ such that on the initial set $E_{t_0} = t_0 \cup \{t - \tau(t): t - \tau(t) < t_0, \, t \geq t_0\}$, $x(t)$ coincides with the given initial function $\phi(t)$. If it is required to determine the solution on the interval $[t_0, T]$, then the initial set $E_{t_0, T}$ is $= \{t_0\} \cup \{t - \tau(t): t - \tau(t) < t_0, \, t_0 \leq t \leq T\}$.

EXAMPLE 1.1.1 For the equation

$$x'(t) = f(t, \, x(t), \, x(t - \cos^2 t)) \qquad (1.1.7)$$

$t_0 = 0$, $E_0 = [-1, 0]$, and the initial function $\phi(t)$ must be given on the interval $[-1, 0]$.

The initial set E_{t_0} depends on the initial point t_0. This statement can be justified by the following example.

EXAMPLE 1.1.2 For the equation

$$x'(t) = ax\left(\frac{t}{2}\right) \tag{1.1.8}$$

$E_0 = \{0\}$ and $E_1 = [\frac{1}{2}, 1]$.

Now we consider the differential equation of nth order with ℓ deviating arguments, of the form

$$x^{(m_0)}(t) = f(t, x(t), \ldots, x^{(m_0-1)}(t),$$
$$x(t - \tau_1(t)), \ldots, x^{(m_1-1)}(t - \tau_1(t)), \ldots,$$
$$x(t - \tau_\ell(t)), \ldots, x^{(m_\ell-1)}(t - \tau_\ell(t))) \tag{1.1.9}$$

where the deviations $\tau_i(t) > 0$, and $\max_{0 \le i \le \ell} m_i = n$.

In order to formulate the initial value problem with respect to (1.1.9), we require the following notation. Let the given initial point be t_0. Each deviation $\tau_i(t)$ defines the initial set $E_{t_0}^{(i)}$ given by

$$E_{t_0}^{(i)} = \{t_0\} \cup \{t - \tau_i(t) : t - \tau_i(t) < t_0 , \ t \ge t_0\}$$

We denote $E_{t_0} = \cup_{i=1}^{\ell} E_{t_0}^{(i)}$, and on E_{t_0} we must be given continuous functions $\phi_k(t)$, $k = 0, 1, \ldots, \mu$, with $\mu = \max_{1 \le i \le \ell} m_i$. In applications, it is most natural to consider the case where on E_{t_0},

$$\phi_k(t) = \phi_0^{(k)}(t), \quad k = 0, 1, 2, \ldots, \mu$$

but it is not generally necessary.

The nth order differential equation should be given initial values $x_0^{(k)}$, $k = 0, 1, 2, \ldots, n - 1$. Now let $x_0^{(k)} = \phi_k(t_0)$, $k = 0, 1, \ldots, \mu$. If $\mu < n - 1$, then, in addition, the numbers $x_0^{(\mu+1)}, \ldots, x_0^{(n-1)}$ are given. If the point t_0 is an isolated point of E_{t_0}, then $x_0^{(0)}, \ldots, x_0^{(n-1)}$ are also given.

For equation (1.1.9), the basic initial value problem consists of the determination of an (n - 1) times continuously differentiable function x that satisfies Eq. (1.1.9) for $t > t_0$ and the conditions

$$x^{(k)}(t_0 + 0) = x_0^{(k)}, \quad k = 0, 1, \ldots, n - 1$$

$$x^{(k)}(t - \tau_i(t)) = \phi_k(t - \tau_i(t)), \quad \text{if } t - \tau_i(t) < t_0$$

$$k = 0, 1, \ldots, \mu; \, i = 1, 2, \ldots, \ell$$

At a point $t_0 + (k - 1)\tau$ the derivative $x^{(k)}(t)$, generally speaking, is discontinuous, but the derivatives of lower order are continuous.

EXAMPLE 1.1.3 Consider

$$x''(t) = f\left(t, \, x(t), \, x'(t), \, x(t - \cos^2 t), \, x\left(\frac{t}{2}\right)\right) \tag{1.1.10}$$

For $t_0 = 0$, we have $n = 2$, $\ell = 2$, $\mu = 0$, the initial set $E_0^{(1)}$ as an interval $-1 \le t \le 0$, $E_0^{(2)} = \{0\}$, and $E_0 = [-1, 0]$, on which is given the initial function $\phi_0(t) : x_0^{(0)} = \phi_0(0)$ and $x_0^{(1)}$ is any given number.

For equation (1.1.9) a classification method was proposed by G. A. Kamenskii [105]. We let $\lambda = m_0 - \mu$. If $\lambda > 0$, (1.1.9) is called an <u>equation with retarded arguments</u> or <u>with delay</u>. If $\lambda = 0$, it is called an <u>equation of neutral type</u>. If $\lambda < 0$, it is called an <u>equation of advanced type</u>.

EXAMPLE 1.1.4 The equations

$$x'(t) + a(t)x(t - \tau) = 0, \quad \tau > 0 \tag{1.1.11}$$

$$x'(t) + a(t)x(t + \tau) = 0, \quad \tau > 0 \tag{1.1.12}$$

and

$$x'(t) + a(t)x(t) + b(t)x'(t - \tau) = 0, \quad \tau > 0 \tag{1.1.13}$$

are of retarded type ($\lambda = 1$), advanced type ($\lambda = -1$), and neutral type ($\lambda = 0$), respectively.

In applications, the equation with retarded arguments is most important; the theory of such equations has been developed extensively. In this book we study mainly the equations of retarded type.

The deviating arguments can be very complex; therefore, the above classification is incomplete. In later chapters, we will discuss also some ODE with complex deviating arguments.

1.2 DEFINITION OF OSCILLATION

Before we define oscillation of solutions, let us consider some simple examples.

EXAMPLE 1.2.1 The equation

$$y'' + y = 0$$

has periodic solutions $x(t) = \cos t$, $y(t) = \sin t$.

EXAMPLE 1.2.2 Consider

$$y''(t) - \frac{1}{t}y'(t) + 4t^2 y(t) = 0 \tag{1.2.2}$$

whose solution is $y(t) = \sin t^2$. This solution is not periodic but has an oscillatory property.

EXAMPLE 1.2.3 Consider

$$y''(t) + \frac{1}{2}y(t) - \frac{1}{2}y(t - \pi) = 0, \quad t \geq 0 \tag{1.2.3}$$

whose solution $y(t) = 1 - \sin t$ has an infinite sequence of multiple zeros. This solution also has an oscillatory property.

EXAMPLE 1.2.4 Consider

$$y''(t) - y(-t) = 0 \tag{1.2.4}$$

which has an oscillatory solution $y_1(t) = \sin t$ and a nonoscillatory solution $y_2(t) = e^t + e^{-t}$.

Let us now restrict our discussion to those solutions $y(t)$ of the equation

$$y''(t) + a(t)y(t - \tau(t)) = 0 \tag{1.2.5}$$

which exist on some ray $[T_y, \infty)$ and satisfy $\sup\{|y(t)| : t \geq T\} > 0$ for every $T \geq T_y$. In other words, $|y(t)| \not\equiv 0$ on any infinite interval $[T, \infty)$. Such a solution sometimes is said to be a <u>regular solution</u>.

We usually assume that $a(t) \geq 0$ or $a(t) \leq 0$ in (1.2.5), and in doing so we mean to imply that $a(t) \not\equiv 0$ on any infinite interval $[T, \infty)$.

There are various definitions for the oscillation of solutions of ODE (with or without deviating arguments). In this section, we give two definitions of oscillation, which are used in the rest of the book; these are the ones most frequently used in the literature.

As we see from the above examples, the definition of oscillation of regular solutions can have two different forms.

DEFINITION 1.2.1 A nontrivial solution $y(t)$ (implying a regular solution always) is said to be <u>oscillatory</u> iff it has arbitrarily large zeros for $t \geq t_0$, that is, there exists a sequence of zeros $\{t_n\}$ $(y(t_n) = 0)$ of $y(t)$ such that $\lim_{n \to \infty} t_n = +\infty$. Otherwise, $y(t)$ is said to be <u>nonoscillatory</u>.

For nonoscillatory solutions there exists a t_1 such that $y(t) \neq 0$, for all $t \geq t_1$.

DEFINITION 1.2.2 A nontrivial solution $y(t)$ is said to be <u>oscillatory</u> if it changes sign on (T, ∞), where T is any number.

When $\tau(t) \equiv 0$ and $a(t)$ is continuous in (1.2.5), the two definitions given above are equivalent. This is because of the fact that the uniqueness of the solution makes multiple zeros impossible. However, as Example 1.2.3 suggests, a differential equation with deviating arguments can have solutions with multiple zeros. These two definitions are different, especially for higher order ordinary differential equations which may have solutions with multiple zeros.

Definition 1.2.1 is more general than Definition 1.2.2. The solution $y(t) = 1 - \sin t$ of Eq. (1.2.3) is oscillatory according to Definition 1.2.1 and is nonoscillatory according to Definition 1.2.2.

In Example 1.2.3, the possibility of multiple zeros of nontrivial solution is a consequence of the retardation, since if $\tau(t) \equiv 0$, the corresponding equation has no solutions with multiple zeros.

For the system of first order equations with deviating arguments

$$\begin{cases} x'(t) = f_1(t,\ x(t),\ x(\tau_1(t)),\ y(t),\ y(\tau_2(t))) \\ y'(t) = f_2(t,\ x(t),\ x(\tau_1(t)),\ y(t),\ y(\tau_2(t))) \end{cases} \tag{1.2.6}$$

the solution $(x(t), y(t))$ is said to be <u>strongly (weakly) oscillatory</u> if each (at least one) of its components is oscillatory.

1.3 REVIEW OF THE OSCILLATION THEORY OF ODE

Since Sturm (1836) introduced the concept of oscillation when he studied the problem of heat transmission, oscillation theory has been an important area of research in the qualitative theory of ODE. Oscillation theory of <u>ordinary differential equations with deviating arguments</u> (ODEWDA) is a natural extension of ODE generated from oscillation theory of ODE, while certain known results in oscillation theory for ODE carry over to ODEWDA some-

what. Therefore some background in oscillation theory for ODE is essential
for understanding oscillation theory of ODEWDA.

We shall recall only some facts concerning oscillation theory of ODE
that are useful for our discussion.

We consider a second order linear ODE

$$y''(t) + a(t)y(t) = 0 \qquad\qquad (1.3.1)$$

Sturm's comparison theorem for Eq. (1.3.1) is a very important result
[165] in oscillation theory. Using this comparison theorem, it is easy to
see the following conclusions:

(a) For the linear differential equation (1.3.1), solutions are either all
     oscillatory or all nonoscillatory. Eq. (1.3.1) is said to be oscillatory
     if every solution of (1.3.1) is oscillatory and it is said to be nonoscil-
     latory otherwise.

(b) We consider another second order linear ODE

$$y''(t) + b(t)y(t) = 0 \qquad\qquad (1.3.2)$$

   If $a(t) \leq b(t)$ for all $t \geq t_0$, and (1.3.1) is oscillatory, then so is (1.3.2).
Moreover, from (a), if (1.3.2) is oscillatory then so is (1.3.1).

Using Sturm's comparison theorem, we can obtain the oscillatory property of an ODE from some other ODE with known oscillatory behavior. In
fact, many good oscillation criteria have been obtained from Sturm's comparison theorem. For example, consider the Euler equation

$$y''(t) + \frac{a}{t^2}\, y(t) = 0 \qquad\qquad (1.3.3)$$

It is well known that (1.3.3) is nonoscillatory when $a = 1/4$, and (1.3.3) is
oscillatory when $a = (1 + \epsilon)/4$, $\epsilon > 0$. According to (b) we obtain the following
oscillation criteria: $t^2 a(t) \leq 1/4$ implies (1.3.1) is nonoscillatory, and
$t^2 a(t) > (1 + \epsilon)/4$, $\epsilon > 0$, implies (1.3.1) is oscillatory.

(c) Assume that $a(t) \leq 0$. Then Eq. (1.3.1) is nonoscillatory. This follows
     from the conclusions in (b).

The comparison method is one of the important methods in oscillation
theory of second order linear ODE. There is much literature dealing with
extensions of the comparison method to nonlinear ODE and higher order ODE.

Now we consider a second order nonlinear ODE

$$y''(t) + q(t)f(y(t)) = 0 \qquad\qquad (1.3.4)$$

Interest in nonlinear oscillation problems for equations of this type began
with the publication of the pioneering work by Atkinson [7]. We would like
to point out the fact that the nonlinearity in (1.3.4) may generate both oscillatory and nonoscillatory solutions.

A special case of (1.3.4) is

$$y''(t) + a(t)y^{\alpha}(t) = 0 \tag{1.3.5}$$

Equation (1.3.5) is said to be <u>superlinear</u> if $\alpha > 1$, and <u>sublinear</u> if $\alpha < 1$. We usually need to distinguish between these cases in our study because of the difference in the type of results that are known. For example, consider

$$y''(t) + a(t)|y(t)|^{\alpha} \text{ sgn } y(t) = 0 \tag{1.3.6}$$

where $a(t) \in C(R_+)$ and $a(t) \geq 0$. Then

For $\alpha > 1$ (superlinear), (1.3.6) is oscillatory iff

$$\int^{\infty} sa(s) \, ds = \infty$$

For $\alpha < 1$ (sublinear), (1.3.6) is oscillatory iff

$$\int^{\infty} s \, a(s) \, ds = \infty$$

Finally we would like to summarize the main topics discussed extensively in the literature of oscillation theory of ODE:

(1) Establishing criteria for oscillation or nonoscillation of all solutions
(2) Obtaining conditions such that an ODE has an oscillatory solution or a nonoscillatory solution with some asymptotic property
(3) Discussing the distribution of zeros and the variability of amplitude of the oscillatory solutions
(4) Investigating the oscillation and asymptotic property of nonoscillatory solutions of ODE with a forcing term
(5) Finding the relation between oscillation and other qualitative properties, such as boundedness, convergence to zero.

1.4 SOME OSCILLATORY AND NONOSCILLATORY PHENOMENA CAUSED BY DEVIATING ARGUMENTS

We shall present some examples to show that the oscillation theory of differential equations with deviating arguments is complex.

EXAMPLE 1.4.1 Consider the equation with delay

$$y'(t) + y\left(t - \frac{\pi}{2}\right) = 0 \tag{1.4.1}$$

It has oscillatory solutions $y = \sin t$ and $y = \cos t$. The equation

$$y'(t) + y\left(t + \frac{\pi}{2}\right) = 0 \tag{1.4.2}$$

also has oscillatory solutions $y = \sin t$ and $y = \cos t$. However, the equation without delay

$$y'(t) + y(t) = 0 \tag{1.4.3}$$

has no oscillatory solutions.

This example illustrates the need to study oscillation of first order ODEWDA. This forms the content of Chapters 2 and 3.

EXAMPLE 1.4.2 Consider the second order equation with delay

$$y''(t) + y(\pi - t) = 0 \tag{1.4.4}$$

It has both an oscillatory solution $y_1 = \sin t$ and a nonoscillatory solution $y_2 = e^t - e^{\pi-t}$.

As we mentioned in Section 1.3, for second order linear ODE either all solutions oscillate or all solutions are nonoscillatory. Thus we see that second order equations with delay create some new problems in oscillation theory. For example, consider

$$y''(t) + p(t)y(\tau(t)) = 0 \tag{1.4.5}$$

We need to establish various sets of conditions under which either: (a) all solutions are oscillatory; (b) all solutions are nonoscillatory; (c) the equation has a nonoscillatory solution; (d) the equation has an oscillatory solution; or (e) the equation has both oscillatory and nonoscillatory solutions.

EXAMPLE 1.4.3 The equation with delay given by

$$y''(t) - y(t - \pi) = 0 \tag{1.4.6}$$

has the oscillatory solutions $y = \sin t$ and $y = \cos t$. But

$$y''(t) - y(t) = 0$$

has no oscillatory solution.

This example suggests that we need to find conditions for oscillatory solutions of (1.4.5), whenever $p(t) \leq 0$.

EXAMPLE 1.4.4 Consider the system

$$\begin{cases} x'(t) = 2x(t) - y(t) \\ y'(t) = x(t) + y(t) \end{cases} \tag{1.4.7}$$

Every solution $(x(t), y(t))$ oscillates. But the system with delay

$$
\begin{cases}
x'(t) = 2x(t) - y\!\left(t - \dfrac{1}{3}\ln 4\right) \\[2ex]
y'(t) = x\!\left(t - \dfrac{1}{3}\ln 4\right) + y(t)
\end{cases}
\qquad (1.4.8)
$$

has the nonoscillatory solution $x(t) = \exp((3/2)t)$, $y(t) = \exp((3/2)t)$.

Therefore we need to study the effect of deviating arguments on the oscillation of systems.

Deviating arguments can occur in many complex forms. For example, we will consider equations where the deviating argument depends on the solution itself

$$y''(t) + p(t)y(t - \tau(y(t))) = 0 \qquad (1.4.9)$$

and the equation with deviating argument of distributed type

$$y'(t) = \int_{0}^{\sigma(t)} y(t - s)\, dr(t,s) \qquad t \geq 0 \qquad (1.4.10)$$

Since the oscillation theory of ODEWDA presents some new problems that are not relevant for the corresponding ODE, a study of the oscillation and nonoscillation caused by deviating arguments is most interesting.

1.5 SOME FIXED POINT THEOREMS

Fixed point theorems are important tools in proving the existence of nonoscillatory solutions. In this section we state some fixed point theorems that we need later. Let us begin with the following notation:

Let S be any fixed set, and C_S be the relation of strict inclusion on subsets of S:

$$C_S = \{\langle A, B\rangle \mid A \subseteq B \subseteq S \text{ and } A \neq B\}$$

We write $A \subset_S B$ in place of the notation $\langle A, B\rangle \in C_S$.

For the set of real numbers, we have the usual ordering relation $<$. For any distinct real numbers x and y, either $x < y$ or $y < x$.

DEFINITION 1.5.1 A partial ordering is a relation R satisfying the following two conditions

(a) R is a transitive relation: xRy and $yRz \Rightarrow xRz$

(b) R is antisymmetric: xRy and $yRx \Rightarrow x = y$.

If $<$ is such a relation, then we can define

$$x \leq y \text{ iff either } x < y \text{ or } x = y$$

It is easy to see that

$$x \leq y < z \Rightarrow x < z$$

LEMMA 1.5.1 Assume that $<$ is a partial ordering relation. Then for any x, y, and z,

(a) At most one of the three alternatives:

$$x < y, \quad x = y, \quad y < x$$

can hold.

(b) $x \leq y \leq x \Rightarrow x = y$.

DEFINITION 1.5.2 Suppose that $<$ is a partial ordering relation on A, and consider a subset C of A. An <u>upper bound</u> of C is an element $b \in A$ such that $x \leq b$ for all $x \in C$. Here b may or may not belong to C. If it belongs to C, then it is clearly the greatest element of C. If b is the least element of the set of all upper bounds for C, then b is the <u>least upper bound</u> (or <u>supremum</u>) of C.

EXAMPLE 1.5.1 Consider a fixed set S. The set consisting of all subsets of S is denoted by P(S). Let the partial ordering be $\subset_S$ on S. For A and B in P(S), the set $\{A, B\}$ has a least upper bound (w.r.t. $\subset_S$), namely, $A \cup B$.

THEOREM 1.5.1 Let $\leq$ be a partial ordering relative to a field A, and suppose that every $B \subseteq A$ has a least upper bound. Suppose that F maps A into A in such a way that for all x, y in A, $x \leq y$ implies that $Fx \leq Fy$. Then $Fx = x$ for some $x \in A$.

DEFINITION 1.5.3 A subset S of a normed space X is called <u>bounded</u> if there is a number M such that $\|x\| \leq M$ for all $x \in S$.

DEFINITION 1.5.4 A set S in a vector space X is called <u>convex</u> if, for any $x, y \subset S$, $ax + (1 - a)y \in S$ for all $a \in [0, 1]$.

DEFINITION 1.5.5 Let N, M be normed linear spaces, and X be a subset of N. An operator $T: X \to M$ is <u>continuous at a point</u> $x \in X$ iff for any $\epsilon > 0$ there is a $\delta > 0$ such that $\|Tx - Ty\| < \epsilon$ for all $y \in X$ such that $\|x - y\| < \delta$. T is continuous on X, or simply <u>continuous</u>, if it is continuous at all points of X.

THEOREM 1.5.2 Every continuous mapping of a closed bounded convex set in R^n into itself has a fixed point.

DEFINITION 1.5.6   A subset S of a normed space B is compact iff every infinite sequence of elements of S has a subsequence which converges to an element of S.

We can prove that compact sets are closed and bounded, but not vice versa, in general.

LEMMA 1.5.2   Continuous mappings take compact sets into compact sets. In other words, if M, N are normed linear spaces, $X \subset M$ is compact, and T: $X \to N$ is continuous, the set T(x) = $\{Tx: x \in X\}$, the image of X under T, is compact.

DEFINITION 1.5.7   A subset S of a normed linear space N is <u>relatively compact</u> iff every sequence in S has a subsequence converging to an element of N.

It is obvious that every subset of a compact or relatively compact set is relatively compact.

LEMMA 1.5.3   The closure of a relatively compact set is compact, and a closed and relatively compact set is compact.

DEFINITION 1.5.8   A function f: $R \to C$ is <u>bounded</u> on an interval $I \subset R$ iff there is a positive real M such that $|f(x)| \leq M$ for all $x \in I$. A family F of functions is <u>uniformly bounded</u> on I if there is an M such that $|f(x)| \leq M$ for all $x \in I$ and all $f \in F$.

LEMMA 1.5.4   A continuous mapping of a compact set is uniformly continuous.

DEFINITION 1.5.9   A family F of functions is <u>equicontinuous</u> on an interval $I \subset R$ iff for every $\epsilon > 0$ there is a $\delta > 0$ such that for all $f \in F$, $|f(x) - f(y)| < \epsilon$ whenever $|x - y| < \delta$, $x, y \in I$.

LEMMA 1.5.5   (Arzela-Ascoli)   A set of functions in C[a,b] with

$$\|f\| = \sup_{x \in [a,b]} |f(x)|$$

is relatively compact if and only if it is uniformly bounded and equicontinuous on [a,b].

THEOREM 1.5.3   (Schauder's first theorem)   If S is a convex, compact subset of a normed linear space, then every continuous mapping of S into itself has a fixed point.

THEOREM 1.5.4 (Schauder's second theorem) If S is a convex closed subset of a normed linear space and R a relatively compact subset of S, then every continuous mapping of S into R has a fixed point.

Theorem 1.5.4 is the more useful form for the theory of ODE.

REMARK 1.5.1 We should point out that we need to use Lemma 1.5.5 carefully, because we usually discuss problems on the infinite interval $[t_0, +\infty)$ in the qualitative theory of ODE. That is, we want usually to prove that the family of functions is uniformly bounded and equicontinuous on $[t_0, +\infty)$. Levitan's result [167] provides a correct formulation. According to his result, the family of functions is equicontinuous on $[t_0, +\infty)$ if for any given $\epsilon > 0$, the interval $[t_0, +\infty)$ can be decomposed into a finite number of subintervals in such a way that on each subinterval all functions of the family have oscillations less than ϵ.

DEFINITION 1.5.10 A real-valued function $\rho(x)$ defined on a linear space X is called a <u>seminorm</u> on X iff the following conditions are satisfied:

$$\rho(x + y) \leq \rho(x) + \rho(y)$$

$$\rho(\alpha x) = |\alpha| \rho(x), \quad \alpha \text{ any scalar.}$$

From this definition, we can prove that a seminorm $\rho(x)$ satisfies $\rho(0) = 0$, $\rho(x_1 - x_2) \geq |\rho(x_1) - \rho(x_2)|$. In particular, $\rho(x) \geq 0$. However, it may happen that $\rho(x) = 0$ for $x \neq 0$.

DEFINITION 1.5.11 A family P of seminorms on X is said to be <u>separating</u> iff to each $x \neq 0$ there corresponds at least one $\rho \in P$ with $\rho(x) \neq 0$.

For a separating seminorm family P, if $\rho(x) = 0$ for every $\rho \in P$, then $x = 0$.

DEFINITION 1.5.12 A topology $\mathcal{T}$ on a linear space E is called <u>locally convex</u> iff every neighborhood of the element 0 includes a convex neighborhood of 0.

A locally convex topology $\mathcal{T}$ on a linear space is <u>determined by a family of seminorms</u> $\{\rho_\alpha : \alpha \in I\}$, I being the index set.

Let E be a locally convex space, $x \in E$, $\{x_n\} \subset E$; $x_n \to x$ in E if and only if $\rho_\alpha(x_n - x) \to 0$ as $n \to \infty$, for every $\alpha \in I$.

A set $S \subset E$ is <u>bounded</u> if and only if the set of numbers $\{\rho_\alpha(x) : x \in S\}$ is bounded for every $\alpha \in I$.

DEFINITION 1.5.13 A complete metrizable locally convex space is called a <u>Frechet space</u>.

THEOREM 1.5.5 (Schauder and Tychonov) Let X be a locally convex topological linear space, C be a compact convex subset of X, and f: C → C be a continuous mapping with f(C) compact. Then f has a fixed point in C.

THEOREM 1.5.6 Let X be a locally convex topological space and let C be a compact convex nonempty subset of X. Suppose that f: C → X is a continuous mapping satisfying the following property: for each $y_0 \in C$ there exists a (real or complex) number z such that

$$|z| < 1 \quad \text{and} \quad zy_0 + (1 - z)f(y_0) \in C$$

Then f has a fixed point in C.

1.6 NOTES

The material in Section 1.1 is based on El'sgol'ts and Norkin's book [59] and Norkin's book [197]. See also Driver [57]. For various definitions of oscillation of solutions see Shevelo [234]. For various results on oscillation theory of ODE we refer to Barrett [10], Swanson [260], and Willett [287] for the linear case, Atkinson [7], Bulter [21], Macki and Wong [177], and Wong [289, 290] for the nonlinear case. See also [234]. The content of Section 1.4 can be found in [234] and Ladde and Zhang [163]. The material of Section 1.5 is taken from standard books in set theory and functional analysis. Remark (1.5.1) is based on Levitan [167]. Kartsatos [108] has several references and is a good source for nonlinear functional differential equations.

2

First Order Linear Equations

2.0 INTRODUCTION

Much of the work in the theory of oscillations centers around second or higher order ODEs, because of the fact that first order ODEs, in general, do not possess oscillatory solutions. The situation is quite different for <u>ODEs with deviating arguments</u> (ODEWDA). As we have indicated in Section 1.4, first order ODEWDA can have oscillatory solutions. One can easily see that the oscillations in this case are generated by deviating arguments. This interesting nature of ODEWDA has captured the attention of mathematical scientists. Oscillation problems for first order ODEWDA are interesting from a theoretical as well as the practical point of view. In fact, Bernoulli (1728) while studying the problem of sound vibrating in a tube with finite size, investigated the properties of solutions of the first order ODEWDA and was the first to work in this area. Myskis investigated several oscillation problems of first order ODEWDA and these are recorded in his book.

In this chapter, we attempt to present the state of the art in this rapidly growing area. Especially, we shall include some recent results, and indicate certain unsolved problems. We shall begin with first order linear ODEs with a deviating argument and then discuss the case with several deviating arguments presenting various available techniques.

2.1 STABLE TYPE EQUATIONS
WITH A SINGLE DELAY

We consider the oscillatory behavior of solutions of the following linear differential inequalities and equations with retarded argument

$$y'(t) + p(t)y(\tau(t)) \leq 0 \tag{2.1.1}$$

$$y'(t) + p(t)y(\tau(t)) \geq 0 \tag{2.1.2}$$

$$y'(t) + p(t)y(\tau(t)) = 0 \tag{2.1.3}$$

where p, $\tau \in C[R_+, R_+]$, $\tau(t) < t$, and $\lim_{t \to \infty} \tau(t) = +\infty$. Let us begin with the following result.

THEOREM 2.1.1 If

$$\lim_{t \to \infty} \int_{\tau(t)}^{t} p(s)\,ds > \frac{1}{e} \tag{2.1.4}$$

then

(i)   (2.1.1) has no eventually positive solutions;
(ii)   (2.1.2) has no eventually negative solutions;
(iii) all solutions of (2.1.3) are oscillatory.

Proof: Without loss of generality, we assume that $\tau(t)$ is nondecreasing, otherwise we set $\delta(t) = \max(\tau(s) = s \in [0,t])$. It is easy to prove that (2.1.4) is equivalent with $\lim_{t \to \infty} \int_{\delta(t)}^{t} p(s)\,ds > 1/e$. First, we prove the validity of statement (i). Assume that $y(t)$ is an eventually positive solution of (2.1.1) such that $y(\tau(t)) > 0$ for $t \geq t_1$. Because of (2.1.4), there exists a $t_2 \geq t_1$ such that

$$\int_{\tau(t)}^{t} p(s)\,ds \geq c > e^{-1} \quad \text{for} \quad t \geq t_2 \tag{2.1.5}$$

Since $y'(t) < 0$ for $t \geq t_1$, from (2.1.1) we get

$$y'(t) + p(t)y(t) \leq 0 \tag{2.1.6}$$

Dividing (2.1.6) by $y(t)$ and integrating from $\tau(t)$ to t, we obtain

$$\ln \frac{y(t)}{y(\tau(t))} + \int_{\tau(t)}^{t} p(s)\,ds \leq 0 \quad t \geq t_2$$

and hence

$$\ln \frac{y(\tau(t))}{y(t)} \geq \int_{\tau(t)}^{t} p(s)\,ds \geq c \quad t \geq t_2$$

Since $e^x \geq ex$, for $x \geq 0$, it follows that

$$\frac{y(\tau(t))}{y(t)} \geq ec \quad t \geq t_2$$

Repeating the above procedure, there exists a sequence $\{t_k\}$ such that

$$\frac{y(\tau(t))}{y(t)} \geq (ec)^k \quad t \geq t_k \tag{2.1.7}$$

From (2.1.5), there exists a t* such that

$$\int_{\tau(t)}^{t^*} p(s)\,ds \geq \frac{c}{2} \quad \text{and} \quad \int_{t^*}^{t} p(s)\,ds \geq \frac{c}{2} \quad \text{for } t \geq t_k$$

Integrating (2.1.1) from $\tau(t)$ to t* yields

$$y(t^*) - y(\tau(t)) + \int_{\tau(t)}^{t^*} p(s)y(\tau(s))\,ds \leq 0$$

This implies that

$$y(\tau(t)) \geq y(\tau(t^*))\frac{c}{2} \tag{2.1.8}$$

Similarly, we obtain

$$y(t) - y(t^*) + \int_{t^*}^{t} p(s)y(\tau(s))\,ds \leq 0$$

and consequently

$$y(t^*) \geq y(\tau(t))\frac{c}{2} \tag{2.1.9}$$

Combining (2.1.8) and (2.1.9), there results the inequality

$$y(t^*) \geq y(\tau(t^*))\left(\frac{c}{2}\right)^2 \tag{2.1.10}$$

From (2.1.7) and (2.1.10), it follows that

$$\left(\frac{2}{c}\right)^2 \geq \frac{y(\tau(t^*))}{y(t^*)} \geq (ec)^k, \quad \forall\, t \geq t_k \tag{2.1.11}$$

Now we choose k sufficiently large such that

$$(ec)^k > (2/c)^2 \tag{2.1.12}$$

which is possible because ec > 1. Therefore (2.1.11) is a contradiction.

A parallel argument holds for (2.1.2); therefore we obtain the conclusion (iii). The proof is complete.

We shall next discuss the special case with $p(t) \equiv p > 0$ and $\tau(t) \equiv t - \tau$, $\tau > 0$.

THEOREM 2.1.2 Assume that p and τ are positive numbers in (2.1.3). Further, assume that

$$p\tau e \leq 1 \tag{2.1.13}$$

Then (2.1.3) has a nonoscillatory solution.

Proof: Let us look at a solution of (2.1.3) of the form, $y(t) = \exp(\lambda t)$. It follows that

$$F(\lambda) \equiv \lambda + p \exp(-\lambda\tau) = 0$$

Observe that $F(0) = p > 0$ and

$$F\left(-\frac{1}{\tau}\right) = -\frac{1}{\tau} + pe = \frac{p\tau e - 1}{\tau} \leq 0$$

Hence there exists a negative real number $\lambda \in [-1/\tau, 0)$ such that $\exp(\lambda t)$ is a nonoscillatory solution of (2.1.3).

COROLLARY 2.1.1 If p and τ are positive numbers in (2.1.3), then

$$p\tau e > 1 \tag{2.1.14}$$

is necessary and sufficient for all solutions of (2.1.3) to oscillate.

EXAMPLE 2.1.1 The equation

$$y'(t) + \left(\frac{1}{e}\right) y(t - 1) = 0 \tag{2.1.15}$$

has a nonoscillatory solution $y(t) = \exp(-t)$, by Theorem 2.1.2, because $p\tau e = 1$.

EXAMPLE 2.1.2 We consider

$$y'(t) + \frac{1}{(e \ln 2)t}\, y\left(\frac{t}{2}\right) = 0 \tag{2.1.16}$$

where $p(t) = 1/(e \ln 2)t$. Obviously, we have

$$\int_{t/2}^{t} p(s)\, ds = e^{-1} \tag{2.1.17}$$

Hence (2.1.16) does not satisfy condition (2.1.4). In fact, equation (2.1.16) has nonoscillatory solution $y(t) = t^{\alpha}$ where $\alpha = -1/\ln 2$.

In view of Theorem 2.1.2 and above examples, condition (2.1.4) is the best possible condition for all solutions of (2.1.3) to be oscillatory.

COROLLARY 2.1.2 Consider the equation with the delay in a more complex form

$$y'(t) + p(t)y(\tau(t) - s(y(t))) = 0 \tag{2.1.18}$$

where $p(t)$ and $\tau(t)$ satisfy the conditions of Theorem 2.1.1. If $s(y)$ is a bounded nonnegative continuous function, then every solution of (2.1.18) is oscillatory under condition (2.1.4).

In fact, without loss of generality, assume that there is a positive solution $y(t)$ of (2.1.18). Then there exists a sufficiently large t_1 such that

$y(\tau(t) - s(y(t))) > 0$ for $t \geq t_1$. Hence $y'(t) \leq 0$ and $y(\tau(t)) \leq y(\tau(t) - s(y(t)))$. Therefore

$$y'(t) + p(t)y(\tau(t)) \leq 0$$

which contradicts Theorem 2.1.1. The proof is complete.

In case $\lim_{t \to \infty} \int_{\tau(t)}^{t} p(s)\, ds$ does not exist, we still have the following result.

THEOREM 2.1.3   If p, $\tau \in C[R_+, R_+]$, $\tau(t) < t$ and it is nondecreasing, $\lim_{t \to \infty} \tau(t) = +\infty$, and

$$\varlimsup_{t \to \infty} \int_{\tau(t)}^{t} p(s)\, ds > 1 \tag{2.1.19}$$

then every solution of (2.1.3) is oscillatory.

Proof: Without loss of generality, let $y(t) > 0$ be a nonoscillatory solution such that $y(\tau(t)) > 0$, $t \geq t_1$. Integrating (2.1.3) from $\tau(t)$ to t, we have

$$y(t) - y(\tau(t)) + \int_{\tau(t)}^{t} p(s)y(\tau(s))\, ds = 0$$

or equivalently

$$y(t) + y(\tau(t))\left[\int_{\tau(t)}^{t} p(s)\, ds - 1 \right] \leq 0 \tag{2.1.20}$$

From (2.1.20), $\int_{\tau(t)}^{t} p(s)\, ds \geq 1$, when t is sufficiently large; therefore (2.1.20) is a contradiction. The proof is complete.

EXAMPLE 2.1.3   Consider

$$y'(t) + \left(\left(\sqrt{2} + \frac{1}{e} \right)\left(\frac{2}{\pi} \right) + \cos t \right) y\left(t - \frac{2}{\pi} \right) = 0 \tag{2.1.21}$$

where $p(t) = (\sqrt{2} + 1/e)(2/\pi) + \cos t > 0$ for $t \in R_+$, and $\int_{t-\pi/2}^{t} p(s)\, ds = \int_{t-\pi/2}^{t} ((\sqrt{2} + 1/e)(2/\pi) + \cos s)\, ds = \sqrt{2} + 1/e + \sin t + \cos t$. Hence

$$\varliminf_{t \to \infty} \int_{t-\pi/2}^{t} p(s)\, ds = e^{-1}$$

which does not satisfy condition (2.1.4) of Theorem 2.1.1. However

$$\varlimsup_{t \to \infty} \int_{t-\pi/2}^{t} p(s)\, ds = 2\sqrt{2} + e^{-1} > 1$$

Consequently (2.1.21) satisfies condition (2.1.19) of Theorem 2.1.3. Therefore every solution of (2.1.21) is oscillatory.

This example shows us that if

$$\lim_{t \to \infty} \int_{\tau(t)}^{t} p(s)\, ds$$

does not exist, then conditions (2.1.4) and (2.1.19) intersect.

THEOREM 2.1.4 If p, $\tau \in C[R_+, R_+]$, $\tau(t) < t$ and $\lim_{t \to \infty} \tau(t) = +\infty$, and

$$\overline{\lim_{t \to \infty}} \int_{\tau(t)}^{t} p(s)\, ds < e^{-1} \tag{2.1.22}$$

then equation (2.1.3) has nonoscillatory solution.

Proof: We wish to find a solution of (2.1.3) that has the following form:

$$y(t) = \exp\left[\int_{t_0}^{t} \lambda(s)\, ds\right] \tag{2.1.23}$$

Then

$$\lambda(t) = -p(t) \exp\left[-\int_{\tau(t)}^{t} \lambda(s)\, ds\right] \tag{2.1.24}$$

Our objective is to show that there exists a real-valued continuous function $\lambda(t)$ such that $\lambda(t)$ satisfies (2.1.24).

We define an operator as follows:

$$(T\lambda)(t) = \begin{cases} -p(t) \exp\left[-\int_{\tau(t)}^{t} \lambda(s)\, ds\right] & t \geq t_0 \\[3ex] \phi(t) & t_0 - \tau \leq t \leq t_0\,, \ \inf_{t \geq t_0} \tau(t) = t_0 - \tau \end{cases} \tag{2.1.25}$$

where $\tau > 0$.

It is clear that T is nondecreasing and continuous operator defined on a space of continuous functions $C[t_0 - \tau, +\infty)$ into itself.

By condition (2.1.22), we can find $t_0 \in R_+$ such that

$$e \int_{\tau(t)}^{t} p(s)\, ds < 1 \quad \text{as } t \geq t_0 \tag{2.1.26}$$

Let

$$y_0(t) = -ep(t) \leq 0$$

and $\phi(t)$ in (2.1.25) satisfy

$$y_0(t) \leq \phi(t) \leq 0 \quad \text{on } [t_0 - \tau, t_0] \tag{2.1.27}$$

Obviously, $y_0 \in C[t_0 - \tau, +\infty)$, and from (2.1.25) to (2.1.27), we see that

$$(Ty_0)(t) = -p(t) \exp\left[-\int_{\tau(t)}^{t} y_0(s)\, ds\right]$$

$$\geq -p(t)e = y_0(t) \qquad t \geq t_0$$

Set $x_0(t) \equiv 0$ for $t \in [t_0 - \tau, +\infty)$, then

$$(Tx_0)(t) \leq x_0(t)$$

We note that $y_0 \leq x_0$. Hence $Ty_0 \leq Tx_0$, and

$$y_0 \leq Ty_0 \leq Tx_0 \leq x_0$$

Let $y_{n+1} = Ty_n$ be an increasing sequence satisfying

$$y_0 \leq y_n \leq y_{n+1} \leq x_0$$

Thus the sequence $\{y_n\}$ increases to a limit λ. By the Lebesgue convergence theorem, Ty_n converges to $T\lambda$. Therefore $T\lambda = \lambda$. Thus λ is a continuous function on $[t_0 - \tau, +\infty)$. Furthermore,

$$y_0(t) \leq \lambda(t) \leq x_0(t) \quad t \geq t_0 - \tau \tag{2.1.28}$$

This proves that (2.1.24) has a solution $\lambda(t)$ which is continuous on $[t_0 - \tau, +\infty)$ and

$$y(t) = \exp\left[\int_{t_0}^{t} \lambda(s)\, ds\right] \tag{2.1.29}$$

is a nonoscillatory solution of (2.1.3).

REMARK 2.1.1 When

$$\lim_{t \to \infty} \int_{\tau(t)}^{t} p(s)\, ds$$

does not exist, there is a gap between the condition (2.1.4) and (2.1.22) or conditions (2.1.19) and (2.1.22). How to fill this gap is an open problem.

REMARK 2.1.2 Myskis [176] first studied the oscillation of solution of (2.1.3). A sufficient condition was given as

$$\overline{\lim_{t\to\infty}} \, (t - \tau(t)) < +\infty \qquad \overline{\lim_{t\to\infty}} \, (t - \tau(t)) \, \underline{\lim_{t\to\infty}} \, p(t) > e^{-1} \tag{2.1.30}$$

Obviously, condition (2.1.4) is better than condition (2.1.30).

Now we shall investigate the asymptotic behavior of the solution of (2.1.3).

We consider

$$y'(t) + p(t)y(t - \tau) = 0 \qquad t \geq t_0 \tag{2.1.31}$$

where $\tau > 0$ is constant, $p \in C[R_+, R_+]$.

Let $y(t; t_0, \phi)$ denote the solution of (2.1.31) satisfying the initial condition $y(t) = \phi(t)$ for $t \in E_{t_0}$, ϕ is a continuous function on E_{t_0}, where $E_{t_0} = [t_0 - \tau, t_0]$. We need the following well-known result [57].

LEMMA 2.1.1 Let $p(t) \equiv p > 0$ and $0 \leq p\tau < \pi/2$. Then there exist positive constants M and ν such that

$$|y(t; t_0, \phi)| \leq M\|\phi\| \exp(-\nu(t - t_0)) \qquad t \geq t_0 \tag{2.1.32}$$

where

$$\|\phi\| = \sup_{t_0 - \tau \leq s \leq t_0} |\phi(s)|$$

Also, if $z(t; t_0, 0)$ denotes the solution of

$$z'(t) + pz(t - \tau) = h(t) \qquad t \geq t_0 \tag{2.1.33}$$

with zero initial function at t_0, then

$$|z(t; t_0, 0)| \leq \frac{M}{\nu} \exp(p + \nu)\tau \, \max_{t_0 \leq s \leq t} |h(s)| \tag{2.1.34}$$

LEMMA 2.1.2 Consider the retarded differential equation

$$y'(t) + y(t - \sigma(t)) = 0 \qquad t \geq t_0 \tag{2.1.35}$$

where $0 \leq \sigma(t) \leq t$ is continuous and $\lim_{t\to\infty} \sigma(t) = \tau < \infty$ exists. Assume that

$$\tau < \frac{\pi}{2} \tag{2.1.36}$$

Then every solution of (2.1.31) tends to zero as $t \to \infty$.

Proof: Let $y(t)$ be any solution of (2.1.31). Choose a $t_1 \geq t_0 + 2\tau + 2$ such that

$$\sigma(t) \leq \tau + 1 \qquad t \geq t_1$$

and

$$\frac{M}{\nu} \exp(1 + \nu)\tau \, |\tau - \sigma(t)| \; \leq \; \frac{1}{2} \qquad t \geq t_1 \qquad\qquad (2.1.37)$$

where the constants M and ν are as defined in Lemma 2.1.1 with $p = 1$.
Let $x(t)$ be the solution of

$$x'(t) + x(t - \tau) = 0 \qquad t \geq t_1$$

with initial function $x_{t_1} = y_{t_1}$ on $E_{t_1} = [t_1 - \tau, t_1]$. Because of condition
(2.1.36), by Lemma 2.1.1, we have that $x(t)$ tends to zero as $t \to \infty$. Set
$z(t) = y(t) - x(t)$; then $z(t)$ satisfies the following equation:

$$z'(t) + z(t - \tau) = y(t - \tau) - y(t - \sigma(t)) \qquad t \geq t_1$$

with zero initial function at t_1. Using (2.1.34), with $p = 1$ and
$h(s) = y(s - \tau) - y(s - \sigma(s))$, we find

$$|z(t)| \; \leq \; \frac{M}{\nu} \exp(1 + \nu)\tau \max_{t_1 \leq s \leq t} |y(s - \tau) - y(s - \sigma(s))| \qquad (2.1.38)$$

Applying the mean value theorem to equation (2.1.35), we obtain

$$|y(s - \tau) - y(s - \sigma(s))| = |\sigma(s) - \tau| \, |y'(\xi)|$$

$$= |\sigma(s) - \tau| \, |y(\xi - \sigma(\xi))|$$

where ξ is between $s - \tau$ and $s - \sigma(s)$. Then, setting

$$B_1 = \max_{t_0 \leq s \leq t_1} |y(s)|$$

we have

$$\max_{t_1 \leq s \leq t} |y(s - \tau) - y(s - \sigma(s))| \; \leq \; \max_{t_1 \leq s \leq t} |\sigma(s) - \tau| \max_{t_0 \leq s \leq t} |y(s)|$$

$$\leq \; \max_{t_1 \leq s \leq t} |\sigma(s) - \tau|[B_1 + \max_{t_1 \leq s \leq t} |y(s)|]$$

From (2.1.38), one gets

$$|y(t)| - |x(t)| \; \leq \; \frac{M}{\nu} \exp(1 + \nu)\tau \max_{t_1 \leq s \leq t} |\sigma(s) - \tau|[B_1 + \max_{t_1 \leq s \leq t} |y(s)|] \quad (2.1.39)$$

and, according to (2.1.37), it follows that

$$|y(t)| \; \leq \; |x(t)| + \frac{1}{2}[B_1 + \max_{t_1 \leq s \leq t} |y(s)|]$$

Hence, for every $T \geq t_1$ and $t_1 \leq t \leq T$, we obtain

$$|y(t)| \; \leq \; |x(t)| + \frac{1}{2}[B_1 + \max_{t_1 \leq s \leq T} |y(s)|]$$

Taking the maximum of both sides and rearranging terms, we find

$$\max_{t_1 \leq s \leq T} |y(t)| \leq 2 \max_{t_1 \leq s \leq T} |x(t)| + B_1$$

That is, $y(t)$ is a bounded function and so there exists a $B \geq B_1$ such that

$$|y(t)| \leq B \quad \text{for } t \geq t_0$$

Then, using (2.1.39), we have

$$|y(t)| \leq |x(t)| + 2B\left(\frac{M}{\nu}\right) \exp(1 + \nu)\tau \max_{t_1 \leq s \leq t} |\sigma(s) - \tau|$$

Because $\lim_{t \to \infty} \sigma(t) = \tau$, the above inequality implies that

$$\lim_{t \to \infty} y(t) = 0$$

This proves Lemma 2.1.2.

We can now prove the following result.

THEOREM 2.1.5 Assume that $p \in C[R_+, R_+]$, $p > 0$, $\tau > 0$ is constant, and

$$\int_{t_0}^{\infty} p(t)\, dt = +\infty \tag{2.1.40}$$

Suppose further that $\lim_{t \to \infty} \int_{t-\tau}^{t} p(s)\, ds$ exists, and

$$\lim_{t \to \infty} \int_{t-\tau}^{t} p(s)\, ds < \frac{\pi}{2} \tag{2.1.41}$$

Then every solution of (2.1.31) tends to zero as $t \to \infty$.

Proof: Set

$$u = \sigma(t) \equiv \int_{t_0}^{t} p(s)\, ds \qquad t \geq t_0$$

Because of (2.1.40), $\sigma^{-1}(t)$ exists, and $\lim_{t \to \infty} u(t) = \infty$. Furthermore,

$$\sigma(t - \tau) = \int_{t_0}^{t-\tau} p(s)\, ds = \int_{t_0}^{t} p(s)\, ds - \int_{t-\tau}^{t} p(s)\, ds = u(t) - \int_{\sigma^{-1}(u)-\tau}^{\sigma^{-1}(u)} p(s)\, ds$$

That is,

$$t - \tau = \sigma^{-1}\left(u - \int_{\sigma^{-1}(u)-\tau}^{\sigma^{-1}(u)} p(s)\, ds\right)$$

Then the transformation

$$z(u) = y(\sigma^{-1}(u)) \tag{2.1.42}$$

reduces (2.1.31) to

$$z'(u) + z\left(u - \int_{\sigma^{-1}(u)-\tau}^{\sigma^{-1}(u)} p(s)\ ds\right) = 0 \tag{2.1.43}$$

According to condition (2.1.41), the equation (2.1.43) satisfies the hypotheses of Lemma 2.1.2 and therefore

$$\lim_{u\to\infty} z(u) = 0$$

From (2.1.42), we have

$$\lim_{t\to\infty} y(t) = 0$$

The proof is complete.

EXAMPLE 2.1.4 Consider

$$y'(t) + y\left(t - \frac{\pi}{2}\right) = 0 \tag{2.1.44}$$

where $\int_{t-\pi/2}^{t} p(s)\ ds = \pi/2$. Note that condition (2.1.41) is not satisfied. In fact, (2.1.44) has the solution $y(t) = \sin t$ which does not tend to zero as $t \to \infty$. This example shows us that condition (2.1.41) is the best condition for every solution of equation (2.1.31) to tend to zero as $t \to \infty$.

EXAMPLE 2.1.5 Consider next

$$y'(t) + p(2 + \cos t)y(t - 2\pi) = 0 \tag{2.1.45}$$

where $0 < p < 1/8$ and $p(t) = p(2 + \cos t)$. Obviously, $\int_{0}^{\infty} p(t)\ dt = \infty$ and $\int_{t-2\pi}^{t} p(s)\ ds = 4p\pi < \pi/2$. Therefore every solution of (2.1.45) tends to zero as $t \to \infty$ by Theorem 2.1.5.

REMARK 2.1.3 When

$$\lim_{t\to\infty} \int_{t-\tau}^{t} p(s)\ ds$$

does not exist, condition (2.1.41) is replaced by condition

$$\overline{\lim_{t\to\infty}} \int_{t-\tau}^{t} p(s)\ ds < 1 \tag{2.1.46}$$

Then the conclusion of Theorem 2.1.5 remains valid [141].

EXAMPLE 2.1.6 Consider equation

$$y'(t) + p(2 + \cos t)y(t - \pi) = 0 \qquad t \geq 0$$

where $0 < p < 1/2(\pi + 1)$.

Observe that for $p(t) = p(2 + \cos t)$, we have $\int_0^\infty p(t)\, dt = \infty$ and $\int_{t-\pi}^t p(s)\, ds = \int_{t-\pi}^t p(2 + \cos s)\, ds = 2p(\pi + \sin t)$. Moreover,

$$\limsup_{t \to \infty} \int_{t-\pi}^t p(s)\, ds = 2p(\pi + 1) < 1$$

and, therefore, condition (2.1.46) is satisfied. Thus every solution of (2.1.47) tends to zero as $t \to \infty$, by Remark 2.1.3.

2.2 EQUATIONS WITH OSCILLATING COEFFICIENTS

Consider the retarded differential inequalities

$$y'(t) + p(t)y(t - \tau) \leq 0 \tag{2.2.1}$$

$$y'(t) + p(t)y(t - \tau) \geq 0 \tag{2.2.2}$$

and the equation

$$y'(t) + p(t)y(t - \tau) = 0 \tag{2.2.3}$$

The following result is true even when $p(t)$ is not assumed to be positive everywhere.

THEOREM 2.2.1 Assume that $p(t) > 0$, at least on a sequence of disjoint intervals $\{(\xi_n, t_n)\}_{n=1}^\infty$ with $t_n - \xi_n = 2\tau$. If

$$\limsup_{n \to \infty} \int_{t_n - \tau}^{t_n} p(s)\, ds \geq 1 \tag{2.2.4}$$

then

(i) (2.2.1) has no eventually positive solution;
(ii) (2.2.2) has no eventually negative solution;
(iii) (2.2.3) has oscillatory solutions only.

Proof: First we prove that (2.2.1) has no eventually positive solutions. For this purpose, suppose that $y(t)$ is a solution of (2.2.1) such that for t_0 sufficiently large $y(t) > 0$ for $t > t_0$. Then $y(t - \tau) > 0$ for $t > t_0 + \tau$, since

$\xi_n \to \infty$, there exists an N such that for $n \geq N$, $\xi_n > t_0 + \tau$ and, because $p(t) > 0$ on (ξ_n, t_n), $y'(t) < 0$ for $t \in (\xi_n, t_n)$. Hence $y(t - \tau) > y(t)$ for $t \in (\xi_n + \tau, t_n)$, $n \geq N$. Integrating (2.2.1) from $t_n - \tau$ to t_n, $n \geq N$, we obtain

$$y(t_n) - y(t_n - \tau) + \int_{t_n - \tau}^{t_n} p(s)y(s - \tau)\, ds \leq 0$$

and, because $y(t)$ is decreasing, it follows that

$$y(t_n) - y(t_n - \tau) + y(t_n - \tau) \int_{t_n - \tau}^{t_n} p(s)\, ds \leq 0$$

or

$$y(t_n) + y(t_n - \tau)\left[\int_{t_n - \tau}^{t_n} p(s)\, ds - 1\right] \leq 0$$

which, in view of (2.2.4), is a contradiction.

To prove that (2.2.2) has no eventually negative solutions it suffices to observe that if $y(t)$ is a solution of (2.2.2) then $-y(t)$ is a solution of (2.2.1).

From the foregoing discussion, it follows that (2.2.3) has neither eventually positive nor eventually negative solutions; therefore every solution of (2.2.3) oscillates. The proof of the theorem is complete.

In the next theorem we will assume that the coefficient is positive in intervals of arbitrarily large length but otherwise it may oscillate. The results of Section 2.1 can be viewed as a special case because the coefficient there is positive everywhere.

THEOREM 2.2.2 Assume that $p(t) > 0$ (at least) on a sequence of disjoint intervals $\{(\xi_n, t_n)\}_{n=1}^{\infty}$ with $t_n - \xi_n \geq 2\tau$ and $\lim_{n\to\infty}(t_n - \xi_n) = +\infty$. Further assume that

$$\liminf_{t\to\infty} \int_{t-\tau}^{t} p(s)\, ds > e^{-1} \quad \text{for} \quad t \in \bigcup_{n=1}^{\infty}(\xi_n + \tau, t_n) \tag{2.2.5}$$

Then the conclusion of Theorem 2.2.1 remains true.

Proof: Suppose that $y(t)$ is a solution of (2.2.1) such that for t_0 sufficiently large,

$$y(t) > 0 \qquad t > t_0$$

Then $y(t - \tau) > 0$ for $t > t_0 + \tau$. Since $\xi_n \to +\infty$ there exists an N such that for $n \geq N$, $\xi_n > t_0 + \tau$ and therefore for $t \in \bigcup_{n=N}^{\infty} (\xi_n, t_n)$, $y(t) > 0$, $y(t - \tau) > 0$ and $y'(t) < 0$.

We choose a number K such that

$$\liminf_{t \to \infty} \int_{t-\tau}^{t} p(s) \, ds > K > e^{-1} \quad \text{for } t \in \bigcup_{n=N}^{\infty} (\xi_n + \tau, t_n)$$

There exists N_1 such that

$$\int_{t-\tau}^{t} p(s) \, ds \geq K > e^{-1} \quad t \in \bigcup_{n=N_1}^{\infty} (\xi_n + \tau, t_n)$$

There also exist $\bar{\xi}_n \in (t_n - \tau, t_n)$ such that

$$\int_{t_n - \tau}^{\bar{\xi}_n} p(s) \, ds \geq \frac{K}{2} \qquad \int_{\bar{\xi}_n}^{t_n} p(s) \, ds \geq \frac{K}{2}$$

From (2.2.1), we have

$$y(\bar{\xi}_n) - y(t_n - \tau) + \int_{t_n - \tau}^{\bar{\xi}_n} p(s) y(s - \tau) \, ds \leq 0$$

which implies

$$-y(t_n - \tau) + y(\bar{\xi}_n - \tau) \int_{t_n - \tau}^{\bar{\xi}_n} p(s) \, ds \leq 0$$

$$-y(t_n - \tau) + y(\bar{\xi}_n - \tau) \frac{K}{2} \leq 0$$

Similarly, it follows that

$$y(t_n) - y(\bar{\xi}_n) + \int_{\bar{\xi}_n}^{t_n} p(s) y(s - \tau) \, ds \leq 0$$

and hence we obtain successively

$$-y(\bar{\xi}_n) + y(t_n - \tau) \int_{\bar{\xi}_n}^{t_n} p(s) \, ds \leq 0$$

and

$$-y(\bar{\xi}_n) + y(t_n - \tau)\frac{K}{2} \leq 0$$

Thus

$$y(\bar{\xi}_n) \geq y(t_n - \tau)\frac{K}{2} \geq \left(\frac{K}{2}\right)^2 y(\bar{\xi}_n - \tau)$$

or equivalently

$$\frac{y(\bar{\xi}_n - \tau)}{y(\bar{\xi}_n)} \leq \left(\frac{2}{K}\right)^2 \qquad \bar{\xi}_n \in (t_n - \tau, \ t_n) \tag{2.2.6}$$

for $n = N_1, \ N_1 + 1, \ \ldots$

On the other hand, from (2.2.1), we have

$$y'(t) + p(t)y(t) \leq 0 \qquad t \in \bigcup_{n=N_1}^{\infty} (\xi_n + \tau, \ t_n)$$

which implies

$$\frac{y'(t)}{y(t)} + p(t) \leq 0 \qquad t \in \bigcup_{n=N_1}^{\infty} (\xi_n + \tau, \ t_n)$$

Integrating, we get

$$\ln \frac{y(t)}{y(t - \tau)} + \int_{t-\tau}^{t} p(s) \ ds \leq 0 \qquad t \in \bigcup_{n=N_1}^{\infty} (\xi_n + 2\tau, \ t_n)$$

so that

$$\ln \frac{y(t - \tau)}{y(t)} \geq K \qquad t \in \bigcup_{n=N_1}^{\infty} (\xi_n + 2\tau, \ t_n)$$

or equivalently

$$\frac{y(t - \tau)}{y(t)} \geq e^K \geq eK \qquad t \in \bigcup_{n=N_1}^{\infty} (\xi_n + 2\tau, \ t_n)$$

Repeating the above procedure, we arrive at

$$\frac{y(t - \tau)}{y(t)} \geq (eK)^m \qquad t \in \bigcup_{n=N_1}^{\infty} (\xi_n + (m + 1)\tau, \ t_n)$$

taking m sufficiently large such that

$$(eK)^m > (2/K)^2$$

and $t_n - \tau < \xi_n + (m + 1)\tau < t_n$. Since $t_n - \xi_n \to +\infty$ as $n \to \infty$, this is possible. This is a contradiction because (2.2.6) holds.

The rest of the proof is similar to the proof of Theorem 2.2.1 and hence the proof is complete.

EXAMPLE 2.2.1 The equation

$$y'(t) + \sin t \, y\left(t - \frac{\pi}{2}\right) = 0 \qquad\qquad (2.2.7)$$

satisfies the conditions of Theorem 2.2.1, therefore all solutions of (2.2.7) are oscillatory.

EXAMPLE 2.2.2 The equation

$$y'(t) + p(t)y(t - 1) = 0 \qquad\qquad (2.2.8)$$

where

$$p(t) = \begin{cases} e^{-t} & \text{for } t \in \left[(2^n + k)\pi, \ (2^n + k)\pi + \frac{1}{2}\right] & k = 0, \ldots, 2^n - 1 \\[2ex] P > 2e^{-1} & \text{for } t \in \left[(2^n + k)\pi + \frac{1}{2}, \ (2^n + k + 1)\pi\right] & \text{and n odd} \\[2ex] \cos t & \text{for } t \in (2^n\pi, \ 2^{n+1}\pi) & \text{and n even} \end{cases}$$

satisfies the conditions of Theorem 2.2.2, therefore all solutions of (2.2.8) oscillate.

EXAMPLE 2.2.3 The equations

$$y'(t) + \sin t \, y(t - 2\pi) = 0$$

and

$$y'(t) + \frac{\sin t}{2 + \sin t} \, y\left(t - \frac{\pi}{2}\right) = 0$$

have the nonoscillatory solutions $y(t) = e^{\cos t}$ and $y(t) = 2 + \cos t$ respectively. We observe that the conditions of Theorems 2.2.1 and 2.2.2 are violated for these two equations.

2.3 UNSTABLE TYPE EQUATIONS WITH A SINGLE DELAY

An equation of the form (2.1.3) is called a stable type. Now we consider the following unstable type equation:

$$y'(t) = p(t)y(t - \tau) \qquad\qquad (2.3.1)$$

where $p(t) \geq 0$, $\tau > 0$. In Section 2.1, we have given conditions on $p(t)$ and τ to guarantee the oscillation of all solutions of (2.1.3). But there are no similar conditions for (2.3.1). We will explain the reason in the case of constant coefficients.

THEOREM 2.3.1 Assume that (2.3.1) with $p(t) \equiv p > 0$ and $\tau > 0$. Then equation (2.3.1) always has an unbounded nonoscillatory solution.

Proof: Set $y(t) = e^{\lambda t}$; we get the characteristic equation $F(\lambda) = \lambda - p \exp(-\lambda\tau) = 0$, $F'(\lambda) = 1 + p\tau \exp(-\lambda\tau) > 0$, i.e., $F(\lambda)$ is monotone increasing, and $F(0) = -p < 0$, $F(p) = p(1 - \exp(-p\tau)) > 0$, so $F(\lambda) = 0$ has a real root $\lambda_0 \in (0, p)$. That is, (2.3.1) has the nonoscillatory solution $y(t) = \exp(\lambda_0 t)$, $\lambda_0 > 0$.

Theorem 2.3.1 implies that the conditions to guarantee the oscillation of all solutions of (2.3.1) do not exist for the case with constant coefficient.
As before, it is clear, the characteristic equation of (2.3.1) has one and only one real positive root.

THEOREM 2.3.2 Consider equation (2.3.1) with constant coefficient $p > 0$. If

$$\frac{(1/2 + k)\pi}{\tau} = p(-1)^{k+1} \tag{2.3.2}$$

has an integer solution k, then (2.3.1) has a bounded oscillatory solution.

Proof: Let $\lambda = \alpha + i\beta$ with $\beta \neq 0$ and $y(t) = \exp(\lambda t)$. Then the characteristic equation becomes

$$\alpha = pe^{-\alpha t} \cos \beta\tau$$
$$\beta = -pe^{-\alpha t} \sin \beta\tau \tag{2.3.3}$$

We want to prove that (2.3.3) has real solution $(0, \beta)$.
From (2.3.3) with $\alpha = 0$, we have

$$\cos \beta\tau = 0$$

Therefore

$$\beta\tau = \left(\frac{1}{2} + k\right)\pi \qquad k = 0, \pm 1, \pm 2, \ldots$$

From the second equation of (2.3.3), we see that

$$\beta = -p \sin \beta\tau = p(-1)^{k+1}$$

Combining the above two equations, there results

$$\frac{(1/2 + k)\pi}{\tau} = p(-1)^{k+1}$$

If this has integer solution k, setting $\beta = p(-1)^{k+1}$, we see that $y = \cos \beta t$ is a solution of (2.3.1).

EXAMPLE 2.3.1 We consider

$$y'(t) = y\left(t - \frac{3}{2}\pi\right) \tag{2.3.4}$$

In this case, equation (2.3.2) has integer solution k = 1. Hence $\beta = 1$. In fact, y = cos t is a bounded oscillatory solution of (2.3.4).

REMARK 2.3.1 Driver [55] pointed out that if $p\tau < e$, in general, Eq. (2.3.1) has no oscillatory solution.

REMARK 2.3.2 Theorem 2.3.1 shows that there is no point in trying to obtain sufficient conditions to guarantee the oscillation of every solution. It is valuable to find sufficient conditions for existence of nonoscillatory or oscillatory solutions. Unfortunately, there are no useful results on this problem at the present time. Later, we shall give a sufficient condition for existence of a nonoscillatory solution of (2.3.1).

REMARK 2.3.3 Equations of the type

$$y'(t) = ay(t) + by(\lambda t) \tag{2.3.5}$$

where a, b, and λ are constants which have been the subject of considerable research interest. There are even some results on the asymptotic proper- ties of nonoscillatory solutions to (2.3.5).

Now we consider some properties of oscillatory solutions of (2.3.1). Let y(t) be a solution of (2.3.1) and define N(t) to be the number of zeros of y(t) in [t - τ, t], and $\nu(t)$ the number of zeros where the sign changes in [t - τ, t]. Obviously, if

$$\limsup_{t \to \infty} N(t) \geq 1 \tag{2.3.6}$$

then this solution y(t) is oscillatory, and if $N(t) \equiv 0$ for sufficiently large t, then y(t) is nonoscillatory and $\nu(t) \leq N(t)$.

LEMMA 2.3.1 In (2.3.1), let p(t) be a positive or negative constant sign function, and let t_0, t_1, t_2 be three successive zeros of y(t). Suppose y(t) fails to change sign at t_1: then $\nu(t_2) \leq \nu(t_0) - 2$.

Proof: Without loss of generality, assume $y(t) \geq 0$ on (t_0, t_2); then y' must reverse sign (from - to +) in any interval $(t_1 - \epsilon, t_1 + \epsilon)$, $\epsilon > 0$. Hence $y(t - \tau) = y'(t)/p(t)$. Therefore $\nu(t)$ must decrease by one at t_1. Moreover, y' must change sign from positive to negative before t_2, and hence $\nu(t)$ must decrease by one. Since y has constant sign, $\nu(t)$ cannot increase. This completes the proof.

LEMMA 2.3.2 In (2.3.1), if $p(t) > 0$, then $\nu(t)$ can increase at a (i.e., has a jump of +1 at a) only where $\nu(a)$ is odd; if $p(t) < 0$ then $\nu(t)$ can increase only where $\nu(a)$ is even.

Proof: If $\nu(a)$ has a jump of +1 at a, then $y(a - \tau) \neq 0$ (otherwise $\nu(a)$ could not increase). Moreover $y(t)$ must change sign at a, and $y(a^-)y'(a^-) < 0$, while $y(a^+)y'(a^+) > 0$. By (2.3.1), it follows that

$$p(a)y(a - \tau)y(a^-) < 0 \qquad p(a)y(a - \tau)y(a^+) > 0$$

Therefore, if $p(a) > 0$ then $y(a - \tau)$ and $y(a^-)$ must have the same sign, whence $\nu(a)$ must be even. Combining the preceding results, we get the following theorem.

THEOREM 2.3.3 If $p(t) > 0$, and $\nu(a) \leq 2k$, then $\nu(t) \leq 2k$ for all $t \geq a$. If $p(t) < 0$, and $\nu(a) \leq 2k + 1$, then $\nu(t) \leq 2k + 1$ for all $t \geq a$.

COROLLARY 2.3.1 If $p(t) > 0$ and $y(t)$ is oscillatory then $\nu(t) \geq 1$.

COROLLARY 2.3.2 If $p(t) < 0$, then either $\nu(t) \leq 1$ for all sufficiently large t ("slowly oscillatory" or if $\nu(t) \equiv 0$ "nonoscillatory") or $\nu(t) \geq 2$ for all t ("rapidly oscillatory").

What condition will guarantee that (2.3.1) ($p(t) \geq 0$) has oscillatory solution? There are no good results on this problem up to now.

The following comparison theorem is useful in proving existence of nonoscillatory solutions of (2.3.1).

THEOREM 2.3.4 Let $y_i(t)$ be solutions of

$$y_i'(t) = p_i(t)y_i(t - \tau) \qquad i = 1, 2$$

Suppose that $|p_1(t)| \leq p_2(t)$. If $|y_1(t)| \leq y_2(t)$ on $[a - \tau, a]$, then $|y_1(t)| \leq y_2(t)$ for $t \geq a$.

Proof: For any $t \in [a, a + \tau]$, we have

$$|y_1(t)| = \left| y_1(a) + \int_a^t p_1(t)y_1(t - \tau)\, dt \right|$$

$$\leq y_2(a) + \int_a^t p_2(t)y_2(t - \tau)\, dt = y_2(t)$$

The extension to all intervals $[a + n - 1, a + n]$ follows by induction.

THEOREM 2.3.5 If

$$p(t) \geq \frac{1}{\underbrace{t \ln t \, \ln\ln t \, \cdots \, \ln \, \cdots \, \ln t}_{n \text{ times}}} \qquad (2.3.7)$$

then (2.3.1) has an unbounded nonoscillatory solution.

Proof: In fact, equation

$$y'(t) = \frac{1}{t \ln t \, \ln\ln t \, \cdots \, \ln \, \cdots \, \ln t \, \underbrace{\ln \, \cdots \, \ln (t - \tau)}_{(n + 1) \text{ times}}} \, y(t - \tau) \qquad (2.3.8)$$

has nonoscillatory solution $y(t) = \ln \, \cdots \, \ln t$ $(n + 1$ times). Then we get the desired result from Theorem 2.3.4.

THEOREM 2.3.6 If $\int^{\infty} |p(t)| \, dt < \infty$ then every solution of (2.3.1) tends to a finite limit as $t \to \infty$.

Proof: From (2.3.1),

$$y(t) = y(t_0) + \int_{t_0}^{t} p(s) y(s - \tau) \, ds$$

Hence

$$|y(t)| \leq c + \int_{t_0}^{t} |p(s + \tau)| \, |y(s)| \, ds \qquad (2.3.9)$$

because we always assume that the initial function of any solution is continuous and so bounded on a closed interval. From (2.3.9), we have

$$|y(t)| \leq c \, \exp \int_{t_0}^{t} |p(s + \tau)| \, ds < \infty$$

i.e., every solution of (2.3.1) is bounded by some finite constants M.

If $t < t'$, we get the inequality

$$|y(t) - y(t')| \leq M \int_{t}^{t'} |p(s)| \, ds$$

Letting $t \to \infty$, we see that $|y(t) - y(t')|$ tends to zero if $\int^{\infty} |p(t)| \, dt < \infty$. The proof is complete.

REMARK 2.3.4 From Theorem 2.3.6, we see that the condition $\int^{\infty} |p(t)| \, dt = \infty$ is necessary for the existence of unbounded solutions of (2.3.1).

THEOREM 2.3.7 If $p(t) > 0$, $\int_{t_0}^{\infty} p(d)\ dt = \infty$, then every nonoscillatory solution of (2.3.1) tends to infinity as $t \to \infty$.

Proof: In fact, if $y(t) > 0$ and $y(t - \tau) > 0$ for $t \geq T$, then $y'(t) > 0$. From (2.3.1), one gets

$$y(t) \geq y(T) + y(T - \tau) \int_{T}^{t} p(s)\ ds$$

Letting $t \to \infty$, it follows that $y(t) \to \infty$.

If $y(t) < 0$ and $y(t - \tau) < 0$ for $t \geq T_1$ then $y'(t) < 0$. From (2.3.1), we see

$$y(t) - y(T_1) = \int_{T_1}^{t} p(s)y(s - \tau)\ ds$$

$$\leq y(T_1 - \tau) \int_{T_1}^{t} p(s)\ ds \qquad t \geq T_1$$

i.e., $y(t) \to -\infty$ as $t \to \infty$. The proof is complete.

REMARK 2.3.5 Theorem 2.3.7 may not be true for an oscillatory solution of (2.3.1).

2.4 UNSTABLE TYPE EQUATIONS WITH A SINGLE ADVANCED ARGUMENT

In this section we consider the oscillatory behavior of solutions of first order differential inequalities and equations with advanced argument.

Consider

$$y'(t)\ \operatorname{sgn} y(t) - p(t)|y(\tau(t))| \geq 0 \tag{2.4.1}$$

THEOREM 2.4.1 Assume that p, $\tau \in C[R_+, R_+]$, $\tau(t) > t$ and

$$\lim_{t \to \infty} \int_{t}^{\tau(t)} p(s)\ ds > e^{-1} \tag{2.4.2}$$

Then all solutions of (2.4.1) are oscillatory.

Proof: The method of proof is similar to that of Theorem 2.1.1. Assume that there exists an eventually positive solution $y(t)$ of (2.4.1). From (2.4.2), there exists a $t_2 \geq t_1$ such that

$$\int_t^{\tau(t)} p(s)\, ds \geq c > e^{-1} \qquad t \geq t_2$$

and $y(t) > 0$, $y'(t) > 0$ for $t \geq t_2$. Hence

$$y'(t) \geq p(t)y(\tau(t)) \geq p(t)y(t) \qquad t \geq t_2$$

Dividing by $y(t)$ and integrating from t to $\tau(t)$, we obtain

$$\ln \frac{y(\tau(t))}{y(t)} \geq \int_t^{\tau(t)} p(s)\, ds \qquad t \geq t_2$$

which is equivalent to

$$\frac{y(\tau(t))}{y(t)} \geq \exp\left(\int_t^{\tau(t)} p(s)\, ds\right) \geq \exp(c) \geq ec$$

for $t \geq t_2$. Repeating the above procedure, there exists a sequence t_k such that

$$\frac{y(\tau(t))}{y(t)} \geq (ec)^k \qquad t \geq t_k$$

This implies that

$$\lim_{t \to \infty} \frac{y(\tau(t))}{y(t)} = +\infty$$

On the other hand, using the argument in the proof of Theorem 2.1.1, we can get

$$\frac{y(\tau(t))}{y(t)} \leq \left(\frac{2}{c}\right)^2$$

for large t; this leads to a contradiction.

The proof of the remaining part can be constructed by following the proof of Theorem 2.1.1.

We can obtain the following results by utilizing the ideas of Section 2.1. We shall merely state these results and omit the proofs.

THEOREM 2.4.2 If

$$\overline{\lim_{t \to \infty}} \int_t^{\tau(t)} p(s)\, ds < e^{-1} \tag{2.4.3}$$

then

$$y'(t) = p(t)y(\tau(t)) \tag{2.4.4}$$

has a nonoscillatory solution, where $0 \leq p(t)$, $\tau(t) > t$ are continuous.

THEOREM 2.4.3 If

$$\overline{\lim_{t \to \infty}} \int_t^{\tau(t)} p(s) \, ds > 1 \tag{2.4.5}$$

then every solution of (2.4.4) is oscillatory.

THEOREM 2.4.4 If $p(t) \equiv p > 0$, $\tau(t) = t + \tau$, $\tau > 0$, the condition

$$p\tau e > 1 \tag{2.4.6}$$

is necessary and sufficient for all solutions of (2.4.4) to oscillate.

In the following result, we establish the asymptotic behavior of solutions of (2.4.4) with $\tau(t) = t + \tau$, $\tau > 0$.

THEOREM 2.4.5 Assume that $p(t) > 0$ and

$$\overline{\lim_{t \to \infty}} \int_t^{t+\tau} p(s) \, ds < 1 \tag{2.4.7}$$

Then the amplitude of every oscillatory solution of (2.4.4) tends to ∞ as $t \to \infty$.

Proof: Let $y(t)$ be an oscillatory solution of (2.4.4). Then there exists a sequence t_n, $n = 1, 2, \ldots$ of zeros of $y(t)$ with the property that $t_{n+1} - t_n \geq \tau$ and $y(t) \neq 0$ on (t_n, t_{n+1}) for $n = 1, 2, \ldots$

Setting $S_n = \max_{t_n \leq t \leq t_{n+1}} |y(t)|$, $n = 1, 2, \ldots$, we see that $S_n = |y(\xi_n)|$ for some $\xi_n \in (t_n, t_{n+1})$ and $y'(\xi_n) = 0$. Hence $y(\xi_n + \tau) = 0$. Let $\tau_n = \min \{t_{n+1}, \xi_n + \tau\}$, $n = 1, , \ldots$

Integrating (2.4.4) from ξ_n to τ_n, we get

$$-y(\xi_n) = \int_{\xi_n}^{\tau_n} p(s) y(s + \tau) \, ds$$

Hence

$$|y(\xi_n)| \leq \int_{\xi_n}^{\tau_n} p(s) |y(s + \tau)| \, ds \leq \left(\max_{[t_n, t_{n+2}]} |y(t)| \right) \int_{\xi_n}^{\xi_n + \tau} p(s) \, ds$$

which yields

$$S_n \leq \max \{S_n, S_{n+1}\} \int_{\xi_n}^{\xi_n + \tau} p(s) \, ds \tag{2.4.8}$$

From (2.4.7), we have

$$\int_{\xi_n}^{\xi_n + \tau} p(s) \, ds \le \mu \; < 1$$

for sufficiently large n, say $n \ge N$. From (2.4.8), $S_n > S_{n+1}$ is impossible. Therefore

$$S_n \le S_{n+1} \mu$$

This implies that

$$S_{n+1} \ge \frac{1}{\mu} S_n \ge \left(\frac{1}{\mu}\right)^2 S_{n-1} \ge \cdots \ge \left(\frac{1}{\mu}\right)^{n-N+1} S_N \qquad n \ge N$$

Letting $n \to \infty$, we get $\lim_{n \to \infty} S_n = \infty$. The proof is complete.

Next, we present an example that will illustrate the sharpness of conditions of Theorem 2.4.1.

EXAMPLE 2.4.1 We consider

$$y'(t) - \frac{2}{e(\ln 2)t} y(2t) = 0 \qquad t \ge t_0 > 0 \tag{2.4.9}$$

where $p(t) = \dfrac{2}{e(\ln 2)t} > 0$, $\tau(t) = 2t$, and therefore

$$\lim_{t \to \infty} \int_t^{\tau(t)} p(s) \, ds = \lim_{t \to \infty} \int_t^{2t} \frac{2}{e(\ln 2)} \frac{ds}{s} = \frac{2}{e} > \frac{1}{e}$$

So all solutions of (2.4.9) are oscillatory.

But consider the equation

$$y'(t) - \frac{1}{e(\ln 2)t} y(2t) = 0 \tag{2.4.10}$$

where $p(t) = [e(\ln 2)t]^{-1}$ and hence $\lim_{t \to \infty} \int_t^{2t} p(s) \, ds = 1/e$. Consequently, (2.4.10) does not satisfy the conditions of Theorem 2.4.1, and therefore (2.4.10) has the nonoscillatory solution $y(t) = t^\alpha$, $\alpha = 1/\ln 2$.

REMARK 2.4.1 Condition (2.4.7) guarantees that the amplitude of every oscillatory solution tends to infinity. But it is possible that the equation has a bounded nonoscillatory solution even though condition (2.4.7) holds.

EXAMPLE 2.4.2 Equation

$$y'(t) = \frac{Ny(t + 1)}{e^{Nt} - e^{-N}} \tag{2.4.11}$$

satisfies condition (2.4.7), but it has the bounded nonoscillatory solution

$$y(t) = A(1 - e^{-Nt}) \tag{2.4.12}$$

where N is a positive integer and A is any constant.

We shall now try to extend the above results to the case of a more complicated advanced argument.

Consider

$$y'(t) = p(t)y(\Delta(t, y(t))) \tag{2.4.13}$$

where $p \in C[R_+, R_+]$, $\Delta \in C[R_+ \times R, R]$, Δ is nondecreasing in t for fixed v and $\Delta(t, v) > t$, and

$$\Delta(t, v_1) \leq \Delta(t, v_2) \quad \text{for} \quad |v_2| \geq |v_1|, \quad v_1 v_2 \geq 0$$

COROLLARY 2.4.1 In addition to the above conditions, if

$$\lim_{t \to \infty} \int_t^{\Delta(t, \eta)} p(s) \, ds > e^{-1} \quad \text{for any } \eta \tag{2.4.14}$$

then all solutions of (2.4.13) oscillate.

Proof: Without loss of generality, assume that there exists a positive solution $y(t) > 0$ for $t \geq t_1 \geq t_0$, then $y'(t) \geq 0$ and hence $y(t) \geq y(t_1) = \eta$, $\Delta(t, y(t)) \geq \Delta(t, \eta)$. Thus

$$y(t) \geq p(t)y(\Delta(t, \eta))$$

which contradicts Theorem 2.4.1.

EXAMPLE 2.4.3 Consider

$$y'(t) = \sqrt{t}\, y(t + y^2(t)) \tag{2.4.15}$$

where $\Delta(t, v) = t + v^2$, $p(t) = \sqrt{t}$, (2.4.15) satisfies the conditions of Corollary 2.4.1. Therefore all solutions oscillate.

Consider

$$y'(t) = p(t)y(g(t) + ks(y(t))) \tag{2.4.16}$$

where $p(t) \geq 0$, $g(t) > t$, $k \geq 0$, $s(y(t)) \geq 0$ are continuous.

COROLLARY 2.4.2 If

$$\lim_{t \to \infty} \int_{t}^{g(t)} p(s)\, ds > e^{-1} \tag{2.4.17}$$

then all solutions of (2.4.16) are oscillatory.

Proof: Otherwise, there is a positive solution $y(t)$, for $t \geq t_0$. Then $y'(t) \geq 0$, so

$$y(g(t) + ks(g(t))) \geq y(g(t))$$

Hence

$$y'(t) \geq p(t)y(g(t))$$

This contradicts Theorem 2.4.1.

2.5 STABLE EQUATIONS WITH A SINGLE ADVANCED ARGUMENT

Consider the stable type first order advanced differential equation

$$y'(t) + p(t)y(t + \tau(t)) = 0 \tag{2.5.1}$$

where $p(t) > 0$ and $\tau(t) > t$ are continuous.
    First, let us present an example.

EXAMPLE 2.5.1 The equation

$$y'(t) + \frac{3\pi}{2}y(t + 1) = 0 \tag{2.5.2}$$

has the oscillatory solution

$$y(t) = \cos\frac{3\pi t}{2} + \sin\frac{3\pi t}{2}$$

and the bounded nonoscillatory solution

$$y(t) = Ae^{st}$$

where A is any constant and s is a root of the equation $s + (3/2)\pi e^{s} = 0$ $(s = -1.2931)$.

THEOREM 2.5.1 Assume that $p(t) > 0$, and

$$\lim_{t \to \infty} \int_{t}^{t+\tau(t)} p(s)\, ds \tag{2.5.3}$$

exists; then (2.5.1) has a bounded nonoscillatory solution.

Proof: Set

$$y(t) = \exp \int^{t} \lambda(s)\, ds \qquad (2.5.4)$$

Then from (2.5.1), we have

$$\lambda(t) + p(t) \exp\left(\int_{t}^{t+\tau(t)} \lambda(s)\, ds\right) = 0$$

Define a map of the form

$$(T\lambda)(t) = -p(t) \exp\left(\int_{t}^{t+\tau(t)} \lambda(s)\, ds\right)$$

Let C be the locally convex space of all continuous functions defined on R_+ into R, and S be subset of C given by

$$S = \left\{ \lambda \in C : -p(t) \leq \lambda(t) \leq 0 \ \text{ and } \ \lim_{t\to\infty} \int_{t}^{t+\tau(t)} \lambda(s)\, ds \ \text{ exists} \right\}.$$

(i) We note that $TS \subset S$.

In fact, $-p(t) \leq (T\lambda)(t) \leq 0$, and

$$\int_{t}^{t+\tau(t)} (T\lambda)(s)\, ds = -\int_{t}^{t+\tau(t)} p(s) \exp \int_{s}^{s+\tau(s)} \lambda(\sigma)\, d\sigma\, ds$$

This, in view of

$$0 \leq p(s) \exp \int_{s}^{s+\tau(s)} \lambda(\sigma)\, d\sigma \leq p(s)$$

implies

$$\lim_{t\to\infty} \int_{t}^{t+\tau(t)} (T\lambda)(s)\, ds$$

exists.

(ii) T is continuous.

(iii) $\overline{TS}$ is relatively compact in S.

In fact,

$$|T\lambda(t_2) - T\lambda(t_1)| = \left| p(t_1) \exp \int_{t_1}^{t_1 + \tau(t_1)} \lambda(s)\, ds - p(t_2) \exp \int_{t_2}^{t_2 + \tau(t_2)} \lambda(s)\, ds \right|$$

$$\leq p(t_1) \exp \int_{t_1}^{t_1 + \tau(t_1)} \lambda(s)\, ds \left| 1 - \frac{p(t_2)}{p(t_1)} \exp \left(\int_{t_1 + \tau(t_1)}^{t_2 + \tau(t_2)} \lambda(s)\, ds + \int_{t_1}^{t_2} \lambda(s)\, ds \right) \right|$$

for any finite interval $[T, T_1]$, $T_1 > T$.

For any given $\epsilon > 0$ there exist δ such that $\left| \int_{t_1}^{t_2} \lambda(s)\, ds \right| < \epsilon$, $t_1, t_2 \in$ $[T_1, T]$ and $|t_2 - t_1| < \delta$ for any $\lambda \in S$. Therefore for any given $\eta > 0$, there exists a $\delta > 0$ such that

$$|T\lambda(t_2) - T\lambda(t_1)| < \eta$$

for any $\lambda \in S$, every $t_1, t_2 \in [T, T_1]$, whenever $|t_1 - t_2| < \delta$. By the Ascoli-Arzela theorem $\overline{TS}$ is relatively compact in S. Finally, by the application of the Schauder-Tychonov fixed point theorem, there exists a fixed point $\lambda(t)$ for T. Therefore (2.5.1) has a nonoscillatory solution $y(t) = \exp \int_T^t \lambda(s)\, ds$, which is bounded.

2.6 EQUATIONS WITH SEVERAL DEVIATING ARGUMENTS AND CONSTANT COEFFICIENTS

In this section, we develop several results concerning oscillatory and nonoscillatory behavior of solutions of first order differential equations with several deviating arguments and constant coefficients.

Consider the first order differential equation with several delays

$$y'(t) + \sum_{i=1}^{n} p_i y(t - \tau_i) = 0 \tag{2.6.1}$$

where $p_i > 0$, $\tau_i > 0$ are constants.

Before we present sufficient conditions for oscillation of solutions of (2.6.1), we need to prove the following lemma.

LEMMA 2.6.1 If a continuous positive function y defined on $[-\tau, +\infty)$ is a solution of the retarded differential inequality

$$y'(t) + py(t - \tau) \leq 0 \qquad t \geq 0 \tag{2.6.2}$$

where p and τ are positive numbers, then

$$\left(\frac{p\tau}{2} \right)^2 y(t - \tau) \leq y(t) \qquad t \geq \frac{3\tau}{2} \tag{2.6.3}$$

Proof: For a given $s \geq \tau$, we integrate both sides of (2.6.2) from s to

$s + \tau/2$. Using the fact that y is decreasing on $[0,\infty)$, we obtain for $s > \tau$,

$$-y(s) + \left(\frac{p\tau}{2}\right) y\left(s - \frac{\tau}{2}\right) \leq y\left(s + \frac{\tau}{2}\right) - y(s) + \left(\frac{p\tau}{2}\right) y\left(s - \frac{\tau}{2}\right) \leq 0$$

For a given $t \geq 3\tau/2$, $s = t$, and get

$$y\left(t - \frac{\tau}{2}\right) \geq \left(\frac{p\tau}{2}\right) y(t - \tau)$$

and

$$y(t) \geq \left(\frac{p\tau}{2}\right) y\left(t - \frac{\tau}{2}\right) \qquad t \geq \frac{3\tau}{2}$$

The desired inequality (2.6.3) follows by combining these two inequalities.

THEOREM 2.6.1 All solutions of (2.6.1) oscillate if and only if

$$-\lambda + \sum_{i=1}^{n} p_i \exp(\lambda \tau_i) > 0 \qquad \text{for all } \lambda > 0 \tag{2.6.4}$$

Proof: Assume first that (2.6.4) does not hold. We may then choose $\lambda_0 > 0$ such that

$$-\lambda_0 + \sum_{i=1}^{n} p_i \exp(\lambda_0 \tau_i) = 0$$

But then y, defined on $[-\tau_n, \infty)$ by $y(t) = e^{-\lambda_0 t}$, is a nonoscillatory solution of (2.6.1).

Assume conversely, that not all solutions of (2.6.1) oscillate. This would imply that there exists at least one nonoscillatory solution of (2.6.1). Without loss of generality (WLOG) we assume that $y(t)$ is an eventually positive solution. By (2.6.1) we note that this solution is decreasing. Since every left-translate of a solution of (2.6.1) is also a solution, we may choose a decreasing positive solution $y(t)$ of (2.6.1).

When all τ_i are zero, obviously, the relation (2.6.4) does not hold. We shall therefore assume that $\tau_n = \max(\tau_1, \ldots, \tau_n) > 0$. We define the set

$$\Lambda = \{\lambda > 0 : y'(t) + \lambda y(t) < 0 \text{ eventually}\}$$

From (2.6.1), we have

$$y'(t) + p_n y(t - \tau_n) \leq 0 \qquad t \geq 0 \tag{2.6.5}$$

On the other hand, since y is decreasing, it follows from (2.6.5) that

$$y'(t) + p_n y(t) < 0 \qquad t \geq 0$$

so that $p_n \in \Lambda$. Also, from Lemma 2.6.1, we get

$$y(t - \tau_n) \le \left(\frac{2}{p_n \tau_n}\right)^2 y(t) \qquad t \ge \frac{3\tau_n}{2}$$

Since y is decreasing, this implies that

$$0 = y'(t) + \sum_{i=1}^{n} p_i y(t - \tau_i) \le y'(t) + \sum_{i=1}^{n} p_i y(t - \tau_n)$$

$$\le y'(t) + \left(\frac{2}{p_n \tau_n}\right)^2 \left(\sum_{i=1}^{n} p_i\right) y(t) \qquad t \ge \frac{3\tau_n}{2}$$

Therefore $(2/p_n \tau_n)^2 (\Sigma_{1 \le i \le n} p_i)$ is an upper bound of Λ. Since Λ is nonempty and bounded, we may set $\lambda_0 = \sup \Lambda$.

Let $\lambda \in \Lambda$ be given, and define z on $[-\tau_n, +\infty)$ by $z(t) = y(t)e^{\lambda t}$. Then there is a suitable $T \in (0, \infty)$ such that

$$z'(t) = (y'(t) + \lambda y(t))e^{\lambda t} < 0 \qquad t \ge T$$

and therefore z is decreasing on $[T, \infty)$. It follows that

$$0 = y'(t) + \sum_{i=1}^{n} p_i y(t - \tau_i) = y'(t) + \sum_{i=1}^{n} p_i z(t - \tau_i)e^{-\lambda(t-\tau_i)}$$

$$> y'(t) + \sum_{i=1}^{n} p_i z(t)e^{-\lambda(t-\tau_i)} = y'(t) + \sum_{i=1}^{n} p_i e^{\lambda \tau_i} y(t)$$

for $t \ge T + \tau_n$. This shows that $\Sigma_{1 \le i \le n} p_i e^{\lambda \tau_i} \in \Lambda$ and $\Sigma_{1 \le i \le n} p_i e^{\lambda \tau_i} \le \lambda_0$. Since $\lambda \in \Lambda$ is arbitrary, we conclude that

$$\sum_{i=1}^{n} p_i e^{\lambda_0 \tau_i} \le \lambda_0$$

and therefore (2.6.4) does not hold. The proof is complete.

Theorem 2.6.1 is of theoretical interest only. It is not attractive and the assumptions are not easy to verify. Hence it is not valuable in applications. Therefore, in the following we want to find conditions expressed in terms of p_i and τ_i. The following results provide such information.

THEOREM 2.6.2 If

$$F(\lambda_0) = \lambda_0 + \sum_{i=1}^{n} p_i \exp(-\lambda_0 \tau_i) > 0 \qquad (2.6.6)$$

where λ_0 satisfies the equation

$$\sum_{i=1}^{n} p_i \tau_i \exp(-\lambda_0 \tau_i) = 1 \tag{2.6.7}$$

then all solutions of (2.6.1) oscillate.

Proof: The characteristic equation of (2.6.1) is

$$F(\lambda) = \lambda + \sum_{i=1}^{n} p_i \exp(-\lambda \tau_i) = 0 \tag{2.6.8}$$

We have

$$F'(\lambda) = 1 - \sum_{i=1}^{n} p_i \tau_i \exp(-\lambda \tau_i) = 0 \tag{2.6.9}$$

and

$$F''(\lambda) = \sum_{i=1}^{n} p_i \tau_i^2 \exp(-\lambda \tau_i) > 0$$

Therefore λ_0 in (2.6.6) is a minimum point of the function $F(\lambda)$. Consequently condition (2.6.6) implies that the characteristic equation has no real root. The proof is complete.

REMARK 2.6.1 Theorem 2.6.2 is applicable, because equation (2.6.7) can be solved at least using numerical methods or graphic methods. After getting λ_0 then one can check condition (2.6.6) easily. One can observe that condition (2.6.6) is the best possible sufficient condition for the oscillation of (2.6.1). In fact, if the left-hand side of (2.6.6) is equal to zero, then $\exp(\lambda_0 t)$ is a nonoscillatory solution of (2.6.1).

In the following, we present very general sufficient conditions which have a wide range of applications for oscillations of (2.6.1).

THEOREM 2.6.3 If there exist $N_i > 0$, $\Sigma_{1 \leq i \leq n} N_i = 1$ such that

$$\sum_{i=1}^{n} \frac{N_i}{\tau_i}\left(1 - \ln \frac{N_i}{p_i \tau_i}\right) > 0 \tag{2.6.10}$$

then all solutions of (2.6.1) oscillate.

Proof: We write

$$F(\lambda) = \lambda + \sum_{i=1}^{n} p_i \exp(-\lambda \tau_i) = \sum_{i=1}^{n} (N_i \lambda + p_i \exp(-\lambda \tau_i))$$

$$= \sum_{i=1}^{n} f_i(\lambda)$$

where

$$f_i(\lambda) = N_i \lambda + p_i \exp(-\lambda \tau_i)$$

Obviously,

$$\min F(\lambda) \geq \sum_{i=1}^{n} \min f_i(\lambda) \tag{2.6.11}$$

We have

$$f_i'(\lambda) = N_i - p_i \tau_i \exp(-\lambda \tau_i)$$

Since the extreme point λ_i of $f_i(\lambda)$ is

$$\lambda_i = -\frac{1}{\tau_i} \ln \frac{N_i}{p_i \tau_i}$$

so

$$\min f_i(\lambda) = \frac{N_i}{\tau_i}\left(1 - \ln \frac{N_i}{p_i \tau_i}\right)$$

Hence, (2.5.10) implies

$$\min F(\lambda) > 0$$

i.e., characteristic equation of (2.6.1) has no real root. The proof is complete.

REMARK 2.6.2 We observe that for $\lambda_1 = \lambda_2 = \cdots = \lambda_n$ we get the best possible sufficient condition for oscillation for (2.6.1).

From Theorem 2.6.3, we can obtain many sufficient conditions that guarantee oscillation of all solutions of (2.6.1) depending on the choice of N_i. The following results are obtained by using different N_i.

THEOREM 2.6.4 Each of the following is a sufficient condition for all the solutions of (2.6.1) to be oscillatory.

(i) $$\sum_{i=1}^{n} p_i \tau_i > e^{-1}$$

(ii) $\quad \left(\prod_{i=1}^{n} p_i\right)^{1/n} \left(\sum_{i=1}^{n} \tau_i\right) > e^{-1}$

(iii) There exists some j such that

$$\sum_{i \neq j} p_i \tau_i + p_j \tau_j e > \exp\left(-\frac{\sum_{i \neq j} p_i}{\sum_{i \neq j} p_i + p_j e}\right)$$

(iv) There exists some j such that

$$\ln\left(\sum_{i \neq j} (p_i \tau_i) + p_j \tau_{max}\right) > \frac{p_j(\tau_{max}/\tau_j) \ln (\tau_{max}/\tau_j)}{\sum_{i \neq j} p_i + p_j(\tau_{max}/\tau_j)} - 1$$

(v) There exists some j such that

$$\ln\left(\sum_{i\ j} (p_i \tau_i) + p_j \tau_{min}\right) > \frac{p_j(\tau_{min}/\tau_j) \ln (\tau_{min}/\tau_j)}{\sum_{i \neq j} p_i + p_j(\tau_{min}/\tau_j)} - 1$$

(vi)

$$\ln\left(\tau_{max} \sum_{j=1}^{n} p_j \exp\left(\frac{\tau_j}{\tau_{max}}\right)\right) > \frac{\sum_{j=1}^{n} \frac{\tau_{max}}{\tau_j} p_j e^{\tau_j/\tau_{max}} \ln\left(\frac{\tau_{max}}{\tau_j}\right) e^{\tau_j/\tau_{max}}}{\sum_{j=1}^{n} \frac{\tau_{max}}{\tau_j} p_j e^{\tau_j/\tau_{max}}} - 1$$

Proof: The proof of the theorem follows by an application of Theorem 2.6.3. For this purpose, it is enough to verify the following choices of N_i satisfy the conditions of Theorem 2.6.3.

(i) $\quad N_i = \dfrac{p_i \tau_i}{\sum_{i=1}^{n} p_i \tau_i}, \quad i = 1, 2, \ldots, n$

(ii) $\quad N_i = \dfrac{\tau_i}{\sum_{i=1}^{n} \tau_i}, \quad i = 1, 2, \ldots, n$

(iii) $\quad N_i = \dfrac{p_i \tau_i}{\sum_{k \neq j} p_k \tau_k + p_j \tau_j e}, \quad i \neq j, \text{ and } N_j = \dfrac{p_j \tau_j e}{\sum_{k \neq j} p_k \tau_k + p_j \tau_j e}$

(iv) $\quad N_i = \dfrac{p_i \tau_i}{\sum\limits_{k \neq j} p_k \tau_k + p_j \tau_{max}}, \quad i \neq j, \text{ and } \quad N_j = \dfrac{p_j \tau_{max}}{\sum\limits_{k \neq j} p_k \tau_k + p_j \tau_{max}}$

(v) $\quad N_i = \dfrac{p_i \tau_i}{\sum\limits_{k \neq j} p_k \tau_k + p_j \tau_{min}}, \quad i \neq j, \text{ and } \quad N_j = \dfrac{p_j \tau_{min}}{\sum\limits_{k \neq j} p_k \tau_k + p_j \tau_{min}}$

(vi) $\quad N_j = \dfrac{\tau_{max} \, p_j \, \exp(\tau_j / \tau_{max})}{\tau_{max} \sum\limits_{i=1}^{n} p_i \, \exp(\tau_i / \tau_{max})}, \quad j = 1, 2, \ldots, n$

From these definitions of N_i, it is obvious that $\Sigma_{1 \leq i \leq n} \, N_i = 1$. Moreover, in view of conditions (i)-(vi), the corresponding N_i in (i)-(vi) satisfy condition (2.6.10).

REMARK 2.6.3 We note that (iv) $\to$ (i), whenever $\tau_{max} = \tau_j$. Condition (v) is better than condition

$$\tau_{min} \left(\sum_{i=1}^{n} p_i \right) > e^{-1} \tag{2.6.12}$$

THEOREM 2.6.5 If

$$\tau_{max} \sum_{j=1}^{n} p_j \, \exp(\tau_j / \tau_{max}) < 1 \tag{2.6.13}$$

then (2.6.1) has a nonoscillatory solution.

Proof: We consider the characteristic equation of (2.6.1)

$$F(\lambda) = \lambda + \sum_{i=1}^{n} p_i \, \exp(-\lambda \tau_i) = 0$$

Obviously,

$$F(0) = \sum_{i=1}^{n} p_i > 0$$

and

$$f\left(-\frac{1}{\tau_{max}} \right) = -\frac{1}{\tau_{max}} + \sum_{i=1}^{n} p_i \, \exp \frac{\tau_i}{\tau_{max}} < 0$$

Hence $F(\lambda) = 0$ has a real root $\lambda_0 \in (-1/\tau_{max}, 0)$, i.e., (2.6.1) has a nonoscillatory solution $y(t) = \exp(\lambda_0 t)$.

We compare condition (vi) of Theorem 2.6.4 and (2.6.13) to see that there exists a gap between condition (vi) and (2.6.13).

EXAMPLE 2.6.1 The equation

$$y'(t) + \frac{1}{2} y\left(t - \frac{1}{e}\right) + y\left(t - \frac{1}{2e}\right) = 0 \tag{2.6.14}$$

with $p_1 = 1/2$, $p_2 = 1$, $\tau_1 = 1/e$, $\tau_2 = 1/2e$, satisfies

$$\sum_{i=1}^{n} p_i \tau_i = \frac{1}{e}$$

It does not satisfy condition (i) of Theorem 2.6.4, but does satisfy condition (ii) of Theorem 2.6.4.

EXAMPLE 2.6.2 The equation

$$y'(t) + e^{-a(\pi/2)} y\left(t - \frac{\pi}{2}\right) + ae^{-a2\pi} y(t - 2\pi) = 0 \tag{2.6.15}$$

has the oscillatory solutions $y_1(t) = e^{at} \sin t$ and $y_2(t) = e^{-at} \cos t$, when $a = 111/120$. It satisfies the condition (i) of Theorem 2.6.4, and does not satisfy condition (ii) of Theorem 2.6.4.

EXAMPLE 2.6.3 Consider the equation

$$y'(t) + \frac{1}{10e} (y(t - 1) + y(t - 9)) = 0 \tag{2.6.16}$$

It does not satisfy conditions (i) and (ii) of Theorem 2.6.4, but does satisfy condition (iii) of Theorem 2.6.4. In fact, set $p_1 = p_2 = 1/10e$, $\tau_1 = 1$, $\tau_2 = 9$,

$$\ln(p_1 \tau_1 + p_2 \tau_2 e) = \ln\left(\frac{1}{10e} + \frac{9}{10}\right)$$

and

$$-\frac{p_1}{p_1 + p_2 e} = -\frac{1}{1 + e}$$

but

$$\ln\left(\frac{1}{10e} + \frac{9}{10}\right) > -\frac{1}{1 + e}$$

Therefore (2.6.16) satisfies condition (iii) of Theorem 2.6.4, hence all solutions of (2.6.16) oscillate.

We consider first order advanced type equation

$$y'(t) = \sum_{i=1}^{n} p_i y(t + \tau_i) \tag{2.6.17}$$

where $p_i > 0$, $\tau_i > 0$, $i \in I_n = \{1, 2, \ldots, n\}$.

THEOREM 2.6.6   If

$$F(\lambda_0) = \lambda_0 - \sum_{i=1}^{n} p_i \exp(\lambda_0 \tau_i) < 0 \tag{2.6.18}$$

where λ_0 satisfies the equation

$$1 = \sum_{i=1}^{n} p_i \tau_i \exp(\lambda_0 \tau_i) \tag{2.6.19}$$

then all solutions of (2.6.17) oscillate.

Proof: The characteristic equation of (2.6.17) is

$$F(\lambda) = \lambda - \sum_{i=1}^{n} p_i \exp(\lambda \tau_i) = 0 \tag{2.6.20}$$

and

$$F'(\lambda) = 1 - \sum_{i=1}^{n} p_i \tau_i \exp(\lambda \tau_i)$$

$$F''(\lambda) = - \sum_{i=1}^{n} p_i \tau_i^2 \exp(\lambda \tau_i)$$

so $F(\lambda)$ is concave and has maximum value. The relation (2.6.19) shows that λ_0 is a maximum point of $F(\lambda)$ and (2.6.18) shows that the maximum value of $F(\lambda)$ is less than zero, that is, the characteristic equation (2.6.20) has no real root. Hence all solutions of (2.6.17) oscillate. The proof is complete.

THEOREM 2.6.7   If there exist $N_i > 0$, $\sum_{1 \leq i \leq n} N_i = 1$ such that (2.6.10) holds, then all solutions of (2.6.17) oscillate.

The proof is similar to the proof of Theorem 2.6.3 and hence we omit the details. Also, we note that results similar to Theorems 2.6.4 and 2.6.5 are valid for equation (2.6.17).

2.7 EQUATIONS WITH SEVERAL DEVIATING ARGUMENTS AND VARIABLE COEFFICIENTS

We consider the equation with several delays

$$y'(t) + \sum_{i=1}^{n} p_i(t)y(t - \tau_i(t)) = 0 \qquad (2.7.1)$$

where $p_i(t) \geq 0$ and $\tau_i(t) > 0$ are continuous $i \in I_n$.

THEOREM 2.7.1 If for some $i \in I_n$, either

$$\lim_{t \to \infty} \int_{t-\tau_i(t)}^{t} p_i(s)\, ds > e^{-1} \qquad (2.7.2)$$

or

$$\lim_{t \to \infty} \int_{t-\tau_{min}(t)}^{t} \sum_{i=1}^{n} p_i(s)\, ds > e^{-1} \qquad (2.7.3)$$

then all solutions of (2.7.1) oscillate, where

$$\tau_{min}(t) = \min(\tau_1(t), \cdots, \tau_n(t))$$

Proof: Without loss of generality, assume that there exists a positive
nonoscillatory solution $y(t) > 0$, for $t \geq t_0$. This implies that there exists
a t_1 such that $y(t - \tau_i(t)) > 0$ for $t \geq t_1$, $i \in I_n$. From (2.7.1), we have

$$y'(t) + p_i(t)y(t - \tau_i(t)) \leq 0 \qquad (2.7.4)$$

and

$$y'(t) + y(t - \tau_{min}(t)) \sum_{i=1}^{n} p_i(t) \leq 0 \qquad (2.7.5)$$

Comparing (2.7.4) and (2.7.5), we obtain a contradiction to Theorem 2.1.1.
The proof is complete.

Now we try to find some other sufficient conditions for oscillation of
(2.7.1).

Define

$$\lambda_i^* = \lim_{t \to \infty} \int_{t-\tau_i(t)}^{t} \lambda(s)\, ds \qquad (2.7.6)$$

and

$$P^*_{ij} = \lim_{t \to \infty} \int_{t-\tau_j(t)}^{t} p_i(s) \, ds \qquad\qquad (2.7.7)$$

LEMMA 2.7.1 Assume that $\tau_i(t) = \tau_i > 0$ in (2.7.1), $i \in I_n$, it is increasing and there exists $\tau_i^* < \tau_i$, for some $i \in I_n$, such that

$$\lim_{t \to \infty} \int_{t}^{t+\tau_i^*} p_i(s) \, ds > 0 \qquad\qquad (2.7.8)$$

Then there exists $N > 0$ such that

$$\frac{y(t)}{y(t - \tau)} \geq N \quad \text{for any } \tau \geq 0 \qquad\qquad (2.7.9)$$

where $y(t)$ is any positive solution of (2.7.1).

Proof: Integrating (2.7.1) from t to ∞, we have

$$y(t) \geq \int_{t}^{\infty} \sum_{i=1}^{n} p_i(s)y(s - \tau_i) \, ds$$

$$\geq \int_{t}^{t+\tau_i^*} p_i(s)y(s - \tau_i) \, ds$$

$$\geq y(t + \tau_i^* - \tau_i) \int_{t}^{t+\tau_i^*} p_i(s) \, ds$$

that is,

$$\frac{y(t)}{y(t + \tau_i^* - \tau_i)} \geq \int_{t}^{t+\tau_i^*} p_i(s) \, ds > d > 0$$

For any $\tau \geq 0$, there exists a k such that $k(\tau_i - \tau_i^*) > \tau$. Therefore

$$\frac{y(t)}{y(t - \tau)} \geq \frac{y(t)}{y(t - k(\tau_i - \tau_i^*))}$$

$$= \prod_{j=1}^{k} \frac{y(t - (j - 1)(\tau_i - \tau_i^*))}{y(t - j(\tau_i - \tau_i^*))}$$

$$\geq d^k > 0$$

completing the proof.

THEOREM 2.7.2 In addition to the conditions of Lemma 2.7.1, suppose that any one of the following conditions holds:

1. $\quad P_{ij}^{*} = \lim\limits_{t\to\infty} \int_{t-\tau_j}^{t} p_i(s)\, ds > e^{-1}$ for some $i,\, j \in I_n$ $\qquad(2.7.10)$

2. $\quad \left[\prod\limits_{i=1}^{n} \sum\limits_{j=1}^{n} P_{ij}^{*}\right]^{1/n} > e^{-1}$ $\qquad(2.7.11)$

3. $\quad \sum\limits_{i=1}^{n} P_{ij}^{*} + 2 \sum\limits_{i<j}^{n} \sqrt{P_{ij}^{*} P_{ji}^{*}} > \dfrac{n}{e}$ $\qquad(2.7.12)$

Then every solution of (2.7.1) oscillates.

Proof: Assume that (2.7.1) has nonoscillatory solution $y(t)$; without loss of generality, let $y(t)$ be a positive solution of the form

$$y(t) = \exp\left(-\int_{t_0}^{t} \lambda(s)\, ds\right) \qquad(2.7.13)$$

Putting (2.7.13) into equation (2.7.1), we have

$$\lambda(t) - \sum\limits_{i=1}^{n} p_i(t) \exp \int_{t-\tau_i}^{t} \lambda(s)\, ds = 0 \qquad(2.7.14)$$

We see that oscillation of all solutions of (2.7.1) is equivalent to having no solution $\lambda(t)$ for the equation (2.7.14). By Lemma 2.7.1, λ_i^{*}, $i \in I_n$ is bounded. From (2.7.14), we obtain for $j \in I_n$

$$\lambda_j^{*} = \lim\limits_{t\to\infty} \int_{t-\tau_j}^{t} \lambda(s)\, ds = \lim\limits_{t\to\infty} \int_{t-\tau_j}^{t} \sum\limits_{i=1}^{n} p_i(s) \exp \int_{s-\tau_i}^{s} \lambda(\sigma)\, d\sigma\, ds$$

$$\geq \sum\limits_{i=1}^{n} P_{ij}^{*} \exp \lambda_i^{*} \qquad(2.7.15)$$

It is well known that for $x \geq 0$, $\max x \exp(-x) = e^{-1}$. From (2.7.15), we have

$$P_{ij}^{*} \leq \lambda_i^{*} \exp(-\lambda_i^{*}) \leq e^{-1} \qquad(2.7.16)$$

By summing inequalities of (2.7.15), we get

$$\sum_{j=1}^{n} \lambda_j^* \geq \sum_{j=1}^{n} \left(\sum_{i=1}^{n} P_{ij}^* \exp \lambda_i^* \right)$$

$$= \sum_{i=1}^{n} \left(\sum_{j=1}^{n} P_{ij}^* \right) \exp \lambda_i^*$$

$$\geq n \left[\prod_{i=1}^{n} \sum_{j=1}^{n} P_{ij}^* \right]^{1/n} \exp \frac{\Sigma_{i=1}^{n} \lambda_i^*}{n}$$

which implies

$$\left[\prod_{i=1}^{n} \sum_{j=1}^{n} P_{ij}^* \right]^{1/n} \leq \frac{\Sigma_{j=1}^{n} \lambda_j^*}{n} \exp \left(- \frac{\Sigma_{j=1}^{n} \lambda_j^*}{n} \right) \leq e^{-1} \qquad (2.7.17)$$

From (2.7.15),

$$\lambda_j^* \exp(-\lambda_j^*) \geq \sum_{i=1}^{n} P_{ij}^* \exp(\lambda_i^* - \lambda_j^*)$$

Summing up these inequalities, we see that

$$\frac{n}{e} \geq \sum_{j=1}^{n} \lambda_j^* \exp(-\lambda_j^*) \geq \sum_{j=1}^{n} \sum_{i=1}^{n} P_{ij}^* \exp(\lambda_i^* - \lambda_j^*)$$

$$= \sum_{i=1}^{n} P_{ii}^* + \sum_{i \neq j} P_{ij}^* \exp(\lambda_i^* - \lambda_j^*)$$

$$= \sum_{i=1}^{n} P_{ii} + \frac{1}{2} \sum_{i \neq j} [P_{ij}^* \exp(\lambda_i^* - \lambda_j^*) + P_{ji}^* \exp(\lambda_j^* - \lambda_i^*)]$$

$$\geq \sum_{i=1}^{n} P_{ii}^* + \sum_{i \neq j} \sqrt{P_{ij}^* P_{ji}^*} = \sum_{i=1}^{n} P_{ii}^* + 2 \sum_{\substack{i<j \\ i,j=1}}^{n} \sqrt{P_{ij}^* P_{ji}^*} \qquad (2.7.18)$$

The relations (2.7.16), (2.7.17), and (2.7.18) contradict the conditions (2.7.10), (2.7.11), and (2.7.12), respectively. The proof is complete.

REMARK 2.7.1 Similar to what we did in Section 2.1, we can consider

$$y'(t) + \sum_{i=1}^{n} p_i(t)y(t - \tau_i) \leq 0 \qquad (2.7.19)$$

$$y'(t) + \sum_{i=1}^{n} p_i(t)y(t - \tau_i) \geq 0 \tag{2.7.20}$$

Then Theorems 2.7.1 and 2.7.2 give conditions to guarantee that (2.7.19) has no eventually positive solution and (2.7.20) has no eventually negative solution.

REMARK 2.7.2 Consider the mixed type equation

$$y'(t) + \sum_{i=1}^{n} p_i(t)y(t - \tau_i) + \sum_{j=1}^{m} q_j(t)y(t + \sigma_j) = 0 \tag{2.7.21}$$

where $p_i(t)$ and $q_j(t)$ are continuous and τ_i, σ_j are constants, $i \in I_n$, $j \in I_m$. Then Theorems 2.7.1 and 2.7.2 give conditions which guarantee that all solutions of (2.7.21) oscillate. This assertion follows from Remark 2.7.1.

THEOREM 2.7.3 If $p_i(t) > 0$, $\tau_i(t) > 0$ are continuous, $\tau_i(t)$ has a uniform upper bound τ_0, and

$$\lim_{t \to \infty} \sum_{i=1}^{n} p_i(t)\tau_i(t) > e^{-1} \tag{2.7.22}$$

then all solutions of (2.7.1) are oscillatory.

Obviously, Theorem 2.7.3 is a natural generalization of Theorem 2.6.4.

Proof: Without loss of generality, assume that (2.7.1) has a positive solution $y(t) > 0$ on $[-\tau_0, +\infty)$. Since condition (2.7.22) holds, we may also assume that there exists a $\mu > e^{-1}$ such that $\sum_{i=1}^{n} p_i(t)\tau_i(t) > \mu$ for $t \geq 0$. Then

$$x(t) = -\ln y(t)$$

exists and is increasing on $[-\tau_0, \infty)$ and hence

$$x'(t) = \sum_{i=1}^{n} p_i(t) \exp[x(t) - x(t - \tau_i(t))] \tag{2.7.23}$$

We wish to construct a "delay" $\tau(t)$ such that $\tau(t) \leq \tau_0$. For this $\tau(t)$, a solution $x_0(t)$ exists for the problem

$$x_0'(t) = p(t) \exp[x_0(t) - x_0(t - \tau(t))]$$

where $p(t) \equiv \mu/\tau(t)$, and $x_0(t)$ grows as slowly as possible. In particular, we will have $x_0(t) \leq x(t)$ when both are defined. We will reach a contradiction when we show that $x_0(t) \to \infty$ for $t \to t^* < \infty$. Specifically let $x_0(t)$ satisfy

$$x'(t) = \inf_{0 < \tau \le \tau_0} \frac{\mu}{\tau} \exp[x_0(t) - x_0(t - \tau)] \qquad t > 0 \qquad (2.7.24)$$

where on $[-\tau_0, 0]$, $x_0(t)$ is constant and less than $x(t)$. Then letting

$$f(t, \tau) = \frac{\mu}{\tau} \exp[x_0(t) - x_0(t - \tau)]$$

and observing that for all t for which $x_0(t)$ is defined, $f(t, \tau) \to \infty$ as $\tau \to 0$, we see that there is in fact a function $\tau(t)$ on $(0, \infty)$ for which

$$x'(t) = f(t, \tau(t)) = \frac{\mu}{\tau(t)} \exp[x_0(t) - x_0(t - \tau(t))] \qquad t > 0 \qquad (2.7.25)$$

and $0 < \tau(t) \le \tau_0$.

Since $x(t)$ is increasing and $x_0'(t) = 0$ on $[-\tau_0, 0]$, it follows that $x'(t) > x_0'(t)$ on this interval. Assume that for some $t \ge 0$, $x(t)$ and $x_0(t)$ are defined and $x'(t) \le x_0'(t)$. Let t_0 be the infimum of all such t. Then $x'(t_0) \le x_0'(t_0)$ while $x'(t) > x_0'(t)$ on $[-\tau_0, t_0)$. But

$$x'(t_0) = \sum_{i=1}^{n} p_i(t) \exp[x(t_0) - x(t_0 - \tau_i(t))]$$

$$> \sum_{i=1}^{n} p_i(t) \exp[x_0(t_0) - x_0(t_0 - \tau_i(t))]$$

$$\ge \sum_{i=1}^{n} p_i(t) \tau_i(t) \inf_{0 < \tau \le \tau_0} \frac{1}{\tau} \exp[x_0(t_0) - x_0(t_0 - \tau)]$$

$$= \frac{\sum_{i=1}^{n} p_i(t) \tau_i(t)}{\mu} x_0'(t_0) \ge x_0'(t_0)$$

This is a contradiction. So $x'(t) > x_0'(t)$ whenever the two are defined.

The continuous function $x_0(t)$ satisfies $x_0'(t) = 0$ on $[-\tau_0, 0]$ while $x_0'(t) > 0$ on $(0, \infty)$. Suppose that $x_0'(t)$ is not an increasing function on $(0, \infty)$. Let t_0 be the supremum of all t for which $x_0'(t)$ in increasing on $(0, t]$. Since $x_0(t)$ and f are continuous, $x_0'(t)$ is continuous on $(0, \infty)$. Therefore, there exists a $t_1 > t_0$ such that $x_0'(t) < 2x_0'(t_0)$ for $t \in [t_0, t_1]$. Now if $\tau < \tau_1$, where $\tau_1 = \mu/(2x_0'(t_0))$, then

$$f(t, \tau) > \frac{\mu}{\tau} > \frac{\mu}{\tau_1} = 2x_0'(t_0)$$

In fact for all $t \in [t_0, t_1]$, $x_0'(t) = \inf_{\tau \in [\tau_1, \tau_0]} f(t, \tau)$. We may choose $t_2 \in (t_0, t_1)$ close enough to t_0 so that $t_2 - \tau_1 < t_0$ and $x_0'(t) > x_0'(t_2 - \tau_1)$ for $t \in [t_0, t_2]$, because $x_0'(t)$ is increasing on $[0, t_0]$ and continuous on $[0, \infty)$.

Since $x_0'(t)$ is continuous, $\partial f / \partial t$ exists and is continuous, if $\tau \in [\tau_1, \tau_0]$ and $t \in [t_0, t_2]$, then

$$\left. \frac{\partial f}{\partial t} \right|_{(t, \tau)} = \frac{\mu}{\tau} (x_0'(t) - x_0'(t - \tau)) \exp[x_0(t) - x_0(t - \tau)]$$

$$\geq \frac{\mu}{\tau} (x_0'(t) - x_0'(t - \tau)) \exp[x_0(t) - x(t - \tau)] > 0$$

because $t - \tau < t_2 - \tau_1 < t_1$. Thus $x_0'(t)$ is increasing on $[t_0, t_2]$, and therefore on $[0, t_2]$, contradicting the definition of t_0. We conclude that $x_0'(t)$ is increasing on $[0, \infty)$.

Recall that we defined $\tau(t) \leq \tau_0$ on $[0, \infty)$ so that

$$f(t, \tau(t)) = \inf_{0 < \tau \leq \tau_0} f(t, \tau)$$

Since $x_0(t)$ is continuous, $\partial f / \partial t$ exists and is continuous, if $\tau(t) \neq \tau_0$,

$$0 = \left. \frac{\partial f}{\partial \tau} \right|_{(t, \tau(t))} = \frac{\mu}{(\tau(t))^2} (\tau(t) x_0'(t - z(t)) - 1) \exp(x_0(t) - x_0(t - \tau(t)))$$

and hence

$$\tau(t) = \frac{1}{x_0'(t - \tau(t))}$$

If $\tau(t) = \tau_0$, $(\partial f / \partial \tau)_{(t, \tau_0)} \leq 0$, then $\tau_0 \leq 1/x_0'(t - \tau_0)$. From (2.7.25)

$$x_0'(\tau_0) = \frac{\mu}{\tau(\tau_0)} \exp(x_0(\tau_0) - x_0(\tau_0 - \tau(\tau_0)))$$

$$\geq \frac{\mu}{\tau(\tau_0)} \exp[\tau(\tau_0) x_0'(\tau_0 - \tau(\tau_0))]$$

$$\geq \mu e \, x_0'(\tau_0 - \tau(\tau_0)) \geq \mu e \, x_0'(0)$$

where $x_0'(0)$ is the right-hand derivative of x at 0. Similarly

$$x_0'(2\tau_0) \geq \mu e \, x_0'(\tau_0) \geq (\mu e)^2 x_0'(0)$$

and, in general, if k is an integer

$$x_0'(k\tau_0) \geq (\mu e)^k x_0'(0)$$

Since $\mu e > 1$, there exists a t_0 for which $x_0'(t_0) > 1/\tau_0$.

Let $a = x_0'(t_0)$, $b = \mu e$, $t_1 = t_0 + \tau_0$ and

$$t_{n+1} = t_n + \frac{1}{ab^{n-1}} \quad \text{for any } n \geq 1$$

Then if $t \geq t_1$

$$x_0'(t) = \frac{\mu}{\tau(t)} \, \exp[x_0(t) - x_0(t - \tau(t))]$$

$$\geq \frac{\mu}{\tau(t)} \, \exp[\tau(t)x_0'(t - \tau(t))]$$

$$\geq \mu e \, x_0'(t - \tau(t)) \geq b x_0'(t_0) = ab$$

and since $x_0'(t - \tau(t)) \geq a$, $\tau(t) \leq 1/a$. Similarly, if $t \geq t_2$, then $t - \tau(t) \geq t_2 - 1/a = t_1$, so $x_0'(t) \geq b x_0'(t_1) \geq ab^2$ and $\tau(t) \leq 1/x_0'(t_1) \leq 1/ab$. Reasoning inductively, we conclude that for all n, $x_0'(t_n) \geq ab^n$. But as $n \to \infty$,

$$t_n \to t_1 + \frac{1/a}{1 - 1/b}$$

while $x_0'(t_n) \geq ab^n \to \infty$. Equation (2.7.24) implies that $x_0(t) \to \infty$ when $x_0'(t)$ does. Then $x_0(t) \to \infty$ as t approaches some $t_* < \infty$, and we are finished.

REMARK 2.7.3 Similar to Theorem 2.1.3, it is easy to see that the condition

$$\overline{\lim_{t \to \infty}} \int_{t-\tau_{min}(t)}^{t} \sum_{i=1}^{n} p_i(s) \, ds > 1 \tag{2.7.26}$$

is sufficient for oscillation of all solutions to (2.7.1).

The following gives sufficient conditions for existence of nonoscillatory solutions of (2.7.1).

THEOREM 2.7.4   If

$$\overline{\lim_{t \to \infty}} \int_{t-\tau_{max}(t)}^{t} \sum_{i=1}^{n} p_i(s) \, ds < \frac{1}{e} \tag{2.7.27}$$

then (2.7.1) has nonoscillatory solutions, where $\tau_{max}(t) = \max_{1 \leq i \leq n} (\tau_1(t), \ldots, \tau_n(t))$.

Proof: Let us assume that a solution of (2.7.1) has the following form

$$y(t) = \exp\left[\int_0^t \lambda(s) \, ds\right] \tag{2.7.28}$$

Therefore, $\lambda(t)$ satisfies the equation

$$\lambda(t) = -\sum_{i=1}^{n} p_i(t) \exp\left[-\int_{t-\tau_i(t)}^{t} \lambda(s) \, ds\right] \tag{2.7.29}$$

Our objective is to show that there exist a real-valued continuous function $\lambda(t)$ such that it satisfies equation (2.7.29). We define

$$(T\lambda)(t) = \begin{cases} -\displaystyle\sum_{i=1}^{n} p_i(t) \exp\left[-\int_{t-\tau_i(t)}^{t} \lambda(s)\, ds\right] & \text{for } t \geq t_0 \\[1em] \phi(t) & \text{for } t_0 - \tau \leq t \leq t_0 \end{cases} \tag{2.7.30}$$

where $\tau(t) = \min_{1\leq i\leq n,\, t\geq t_0} \tau_i(t)$.

T is a nondecreasing and continuous operator defined on a space of continuous functions $C[t_0 - \tau, +\infty)$ into itself.

From (2.7.27), we can find t_0 such that

$$\int_{t-\tau_{max}(t)}^{t} \sum_{i=1}^{n} p_i(s)\, ds < e^{-1} \quad \text{for } t \geq t_0 \tag{2.7.31}$$

Set

$$y_0(t) = -e \sum_{i=1}^{n} p_i(t) \quad \text{and} \quad x_0(t) \equiv 0, \quad t \geq t_0 - \tau$$

and

$$y_0(t) \leq \phi(t) \leq x_0(t) \quad t \in [t_0 - \tau, t_0]$$

By noting the proof of Theorem 2.1.4, it is easy to observe that

$$y_0 \leq Ty_0 \leq Tx_0 \leq x_0 \tag{2.7.32}$$

The sequence $\{y_n\}$, defined by $Ty_n = y_{n+1}$, is an increasing sequence and satisfies

$$y_0 \leq y_n \leq y_{n+1} \leq x_0 \tag{2.7.33}$$

By following the argument in the proof of Theorem 2.1.4, one can conclude that $T\lambda - \lambda$. Obviously,

$$y_0(t) \leq \lambda(t \leq x_0(t) \quad t \geq t_0 - \tau \tag{2.7.34}$$

This proves that (2.7.29) has a solution $\lambda(t)$ which is a continuous function on $[t_0 - \tau, +\infty)$ and satisfies (2.7.34). Thus (2.7.28) is a nonoscillatory solution of (2.7.1). The proof is complete.

REMARK 2.7.4 Theorem 2.7.4 shows that under condition (2.7.27) there exists a positive solution corresponding to some initial point t_0 (sufficiently large number). If condition (2.7.27) is replaced by condition

$$\int_{t-\tau_{max}(t)}^{t} \sum_{i=1}^{n} p_i(s) \, ds \leq e^{-1} \qquad t \in H_{t_0} \tag{2.7.35}$$

where $H_{t_0} = \{t \geq t_0, \ t - \tau_{max}(t) \geq t_0\}$, then equation (2.7.1) has a positive solution corresponding to an initial point t_0.

We next consider the advanced type equation

$$y'(t) = \sum_{i=1}^{n} p_i(t) y(t + \tau_i(t)) \tag{2.7.36}$$

where $p_i(t) \geq 0$, $\tau_i(t) > 0$ are continuous, $i \in I_n$.

A parallel argument can prove that the foregoing results with respect to the equation with delay (2.7.1) are true for the equation of advanced type (2.7.36). We merely state a theorem, omitting its proof.

THEOREM 2.7.5 Theorems 2.7.1 to 2.7.4 remain valid for (2.7.36).

Consider the more general type equation

$$y'(t) + p(t)y(t) + \sum_{i=1}^{n} p_i(t) y(t - \tau_i(t)) = 0 \tag{2.7.37}$$

where $p(t)$, $p_i(t) \geq 0$, $\tau_i(t) > 0$ are continuous, $i \in I_n$. Let

$$y(t) = \exp\left[-\int_{t_0}^{t} p(u) \, du\right] z(t) \qquad t \geq t_0 \tag{2.7.38}$$

then (2.7.37) becomes

$$z'(t) + \sum_{i=1}^{n} p_i(t) \exp\left[\int_{t-\tau_i(t)}^{t} p(u) \, du\right] z(t - \tau_i(t)) = 0 \tag{2.7.39}$$

for $t \geq t_0$. Setting

$$q_i(t) = p_i(t) \exp \int_{t-\tau_i(t)}^{t} p(u) \, du \qquad i \in I_n \tag{2.7.40}$$

we find that (2.7.39) becomes

$$z'(t) + \sum_{i=1}^{n} q_i(t) z(t - \tau_i(t)) = 0 \tag{2.7.41}$$

We see that the transformation (2.7.38) preserves oscillation. Therefore, we can apply the above results with respect to (2.7.1) to equation (2.7.37). For example, we have the following theorem.

THEOREM 2.7.6 If any one of the following conditions holds:

$$(1) \quad \lim_{t\to\infty} \int_{t-\tau_i(t)}^{t} q_i(s)\, ds > e^{-1} \quad \text{for some } i \in I_n \tag{2.7.42}$$

$$(2) \quad \lim_{t\to\infty} \int_{t-\tau_{min}(t)}^{t} \sum_{i=1}^{n} q_i(s)\, ds > e^{-1} \tag{2.7.43}$$

$$(3) \quad \lim_{t\to\infty} \left(\sum_{i=1}^{n} q_i(t)\tau_i(t) \right) > e^{-1} \tag{2.7.44}$$

$$(4) \quad \left[\prod_{i=1}^{n} \sum_{j=1}^{n} \lim_{t\to\infty} \int_{t-\tau_j(t)}^{t} q_i(s)\, ds \right]^{1/n} > e^{-1}$$

and $q_i(t)$ satisfies the condition of Lemma 2.7.1,

$$(5) \quad \overline{\lim_{t\to\infty}} \int_{t-\tau_{max}(t)}^{t} \sum_{i=1}^{n} q_i(s)\, ds > 1 \tag{2.7.45}$$

then all solutions of (2.7.37) oscillate, where $q_i(t)$ is defined by (2.7.40).

REMARK 2.7.5 We can use a similar transformation for advanced type equations of the form

$$y'(t) = p(t)y(t) + \sum_{i=1}^{n} p_i(t)y(t + \tau_i(t)) \tag{2.7.46}$$

2.8 EQUATIONS WITH FORCING TERMS

Consider the equation with a forcing term of the form

$$y'(t) + \sum_{i=1}^{n} p_i(t)y(t - \tau_i(t)) = q(t) \tag{2.8.1}$$

and prove the following result.

THEOREM 2.8.1 Assume that

(i) $q(t)$, $p_i(t) \geq 0$, $\tau_i(t) > 0$ are continuous, $i \in I_n$;

(ii) There exists a function $Q(t)$ and two constants q_1, q_2 and sequences $\{t'_m\}$, $\{t''_m\}$ such that $Q'(t) = q(t)$, $Q(t'_m) = q_1$, $Q(t''_m) = q_2$, $\lim_{m \to \infty} t'_m = \infty$, $\lim_{m \to \infty} t''_m = \infty$, and $q_1 \leq Q(t) \leq q_2$ for $t \geq 0$

(iii) $p_i(t)$, $i \in I_n$ satisfy any one of the conditions of Theorems 2.7.1, 2.7.2, 2.7.3, and relation (2.7.26). Then every solution of (2.8.1) oscillates.

Proof: Let $y(t)$ be a nonoscillatory solution such that $y(t) > 0$, $y(t - \tau_i(t)) > 0$ for $t \geq t_1$, and let

$$x(t) \equiv y(t) - Q(t)$$

Then $x'(t) = y'(t) - Q'(t) = -\Sigma p_i(t)y(t - \tau_i(t)) \leq 0$ for $t \geq t_1$. Suppose $x(t) + q_1 \leq 0$ for $t \geq t_2 \geq t_1$. Since $x(t) + Q(t) \equiv y(t) > 0$, especially, $x(t'_m) + Q(t'_m) = y(t'_m)$, $t'_m > t_2$, this is a contradiction. So $x(t) + q_1 > 0$ for all $t \geq t_2$.

Let $z(t) \equiv x(t) + q_1$. Then

$$z'(t) = x'(t) = y'(t) - Q'(t) = -\sum_{i=1}^{n} p_i(t)y(t - \tau_i(t))$$

$$= -\sum_{i=1}^{n} p_i(t)[x(t - \tau_i(t)) + Q(t - \tau_i(t))]$$

$$\leq -\sum_{i=1}^{n} p_i(t)z(t - \tau_i(t))$$

That is,

$$z'(t) + \sum_{i=1}^{n} p_i(t)z(t - \tau_i(t)) \leq 0$$

has an eventual positive solution. But it is impossible according to condition (iii). The proof is therefore complete.

EXAMPLE 2.8.1 The equation

$$y'(t) + y(t - \pi) = \sin t \tag{2.8.2}$$

satisfies all conditions of Theorem 2.8.1. Therefore, every solution of (2.8.2) oscillates.

2.9 EQUATIONS WITH DISTRIBUTED TYPE DEVIATING ARGUMENTS

We consider a very general linear differential equation with distributed type retarded arguments

$$y'(t) + p(t)y(t) + \int_{\alpha(t)}^{\beta(t)} y(t+s)\, d\eta(t,\, s) = 0 \tag{2.9.1}$$

where (i) p is locally integrable; (ii) $\alpha, \beta \in C[R_+, R]$ and $0 > \beta(t) > \alpha(t)$ on R_+. (iii) $\eta(t,s)$ is a measurable in (t,s) and a function of bounded variation in $s \in [\alpha(t),\, \beta(t)]$ for fixed t and satisfies mean continuity in t, i.e.,

$$\lim_{t \to t_0} \int_0^{\min(-\alpha(t),\, -\alpha(t_0))} |\,\eta(t,s) - \eta(t_0,s)\,|\, ds = 0 \quad \text{for any } t_0 \in R_+$$

Moreover

$$\operatorname*{Var}_{\alpha(t)}^{\beta(t)} (\eta(t,s)) \leq m(t) \quad \text{for } t \in R_+$$

where m is locally integrable on R_+. The integral in (2.9.1) is a Stieltjes integral.

DEFINITION 2.9.1 An absolutely continuous function y on $[t_0, \infty)$ is called a solution of (2.9.1), if it satisfies (2.9.1) almost everywhere together with the initial conditions

$$y(a) = \Phi(a)$$

and

$$y(t - \tau) \equiv \Phi(t - \tau) \qquad t - \tau < a$$

where $\Phi(t)$ is the given continuous initial function on $[\inf_{t \geq a} t + \alpha(t),\, a]$.

These conditions with respect to (2.9.1) guarantee global existence and uniqueness of solutions of (2.9.1) to the right of a [176].

Let y(t) be a solution of (2.9.1). Using the transformation

$$y(t) = \exp\left(-\int_{t_0}^{t} p(s)\, ds\right) z(t) \tag{2.9.2}$$

we see that z(t) satisfies the equation

$$z'(t) + \int_{\alpha(t)}^{\beta(t)} z(t+s) \exp\left(\int_{t+s}^{t} p(\sigma)\, d\sigma\right) d\eta(t,s) = 0 \tag{2.9.3}$$

Obviously, the oscillatory properties of $y(t)$ and $z(t)$ are the same.

THEOREM 2.9.1 Assume that (2.9.1) satisfies above mentioned conditions (i), (ii), and (iii). In addition, let $\eta(t, s)$ be nondecreasing in s for fixed t and nonnegative in t for fixed s such that $t + \alpha(t) \to \infty$ as $t \to \infty$. If

$$\lim_{t \to \infty} \int_{t+\beta(t)}^{t} \left[\int_{\alpha(u)}^{\beta(u)} \exp\left(\int_{u+s}^{u} p(\sigma)\, d\sigma \right) d\eta(u, s) \right] du > e^{-1} \tag{2.9.4}$$

then every solution of (2.9.1) oscillates.

Proof: Because of (2.9.2), the oscillatory behavior of (2.9.1) is equivalent to the oscillatory behavior of (2.9.3). Therefore it is enough to verify the oscillatory property of (2.9.3). Without loss of generality, let $z(t) > 0$, $t \geq t_0$ be a nonoscillatory solution of (2.9.3). This implies $z'(t) \leq 0$, in view of the nature of $\eta(t, s)$ in (2.9.1). By dividing equation (2.9.3) by $z(t)$ and integrating from $t + \beta(t)$ to t, we have

$$\ln \frac{z(t)}{z(t + \beta(t))} + \int_{t+\beta(t)}^{t} \frac{1}{z(u)} \left[\int_{\alpha(u)}^{\beta(u)} z(u+s) \exp\left(\int_{u+s}^{u} p(\sigma)\, d\sigma \right) d\eta(u, s) \right] du = 0$$

Let $w(t) = z(t + \beta(t))/z(t) \geq 1$ and $\lim_{t \to \infty} w(t) = \ell$. From the above equation, we have

$$\ln w(t) \geq \int_{t+\beta(t)}^{t} w(u) \left[\int_{\alpha(u)}^{\beta(u)} \exp\left(\int_{u+s}^{u} p(\sigma)\, d\sigma \right) d\eta(u, s) \right] du$$

$$= w(\xi) \int_{t+\beta(t)}^{t} \left[\int_{\alpha(u)}^{\beta(u)} \exp\left(\int_{u+s}^{u} p(\sigma)\, d\sigma \right) d\eta(u, s) \right] du \tag{2.9.5}$$

where $\xi \in [t + \beta(t), t]$.

We discuss two possible cases:

(i) ℓ is finite. Then there exists a sequence $\{t_n\}$, $t_n \to \infty$ as $n \to \infty$ such that $\lim_{n \to \infty} w(t_n) = \ell$. From (2.9.5), it follows that

$$\ln w(t_n) \geq w(\xi_n) \int_{t_n+\beta(t_n)}^{t_n} \left[\int_{\alpha(u)}^{\beta(u)} \exp\left(\int_{u+s}^{u} p(\sigma)\, d\sigma \right) d\eta(u, s) \right] du \tag{2.9.6}$$

where $t_n + \beta(t_n) \leq \xi_n \leq t_n$. Taking the lower limit inferior on both sides of (2.9.6) we obtain

$$\ln \ell \geq \ell \lim_{t \to \infty} \int_{t+\beta}^{t} \left[\int_{\alpha(u)}^{\beta(u)} \exp\left(\int_{u+s}^{u} p(\sigma)\, d\sigma \right) d\eta(u, s) \right] du$$

But $\max_{\ell \geq 1} (\ln \ell/\ell) = e^{-1}$, and hence the above inequality implies that

$$\frac{1}{e} \geq \lim_{t \to \infty} \int_{t+\beta(t)}^{t} \left[\int_{\alpha(u)}^{\beta(u)} \exp\left(\int_{u+s}^{u} p(\sigma)\, d\sigma \right) d\eta(u,s) \right] du$$

This is a contradiction to condition (2.9.4).

(ii) ℓ is infinite. Then there exists a sequence $\{t_n\}$ such that $\lim_{n \to \infty} t_n = \infty$ and $\lim_{n \to \infty} w(t_n) = \underline{\lim}_{t \to \infty} w(t) = \infty$. From (2.9.6), we see that

$$\ln w(t_n) \geq w(\xi_n) \int_{t_n+\beta(t_n)}^{t_n} \left[\int_{\alpha(u)}^{\beta(u)} \exp\left(\int_{u+s}^{u} p(\sigma)\, d\sigma \right) d\eta(u,s) \right] du \quad (2.9.7)$$

where $t_n + \beta(t_n) \leq \xi_n \leq t_n$. Obviously, $\lim_{n \to \infty} \ln w(t_n)/w(\xi_n) \geq 0$. However, $\underline{\lim}_{n \to \infty} \ln w(t_n)/w(\xi_n) \leq \overline{\lim}_{n \to \infty} \ln w(t_n)/\underline{\lim}_{n \to \infty} w(\xi_n)$

$\leq \overline{\lim}_{n \to \infty} \ln w(t_n)/\underline{\lim}_{n \to \infty} w(t_n) = 0$, because $\lim_{n \to \infty} w(t_n) = \overline{\lim}_{n \to \infty} w(t_n) = \underline{\lim}_{n \to \infty} w(t_n)$. From (2.9.7), it follows that

$$\lim_{n \to \infty} \int_{t+\beta(t)}^{t} \left[\int_{\alpha(u)}^{\beta(u)} \exp\left(\int_{u+s}^{u} p(\sigma)\, d\sigma \right) d\eta(u,s) \right] du = 0$$

which is a contradiction. The proof is complete.

The following result provides sufficient condition for the existence of nonoscillatory solution of (2.9.1).

THEOREM 2.9.2 Assume that

$$\overline{\lim_{t \to \infty}} \int_{t+\beta(t)}^{t} \left[\int_{\alpha(u)}^{\beta(u)} \exp\left(\int_{u+s}^{u} p(\sigma)\, d\sigma \right) d\eta(u,s) \right] du < e^{-1} \quad (2.9.8)$$

Then (2.9.1) has a nonoscillatory solution.

Proof: As before, we consider equation (2.9.3). We try to find a solution of the

$$z(t) = \exp\left(-\int_{t_0}^{t} \lambda(s)\, ds \right) \quad (2.9.9)$$

Putting (2.9.9) into (2.9.3) yields

$$-\lambda(t) + \int_{\alpha(t)}^{\beta(t)} \exp\left(\int_{t+s}^{t} \lambda(\sigma)\, d\sigma \right) \exp\left(\int_{t+s}^{t} p(\sigma)\, d\sigma \right) d\sigma(t,s) = 0 \quad (2.9.10)$$

Define the operator T as follows:

$$(T\lambda)(t) = \begin{cases} \displaystyle\int_{\alpha(t)}^{\beta(t)} \exp\left(\int_{t+s}^{t} \lambda(\sigma)\,d\sigma\right) \exp\left(\int_{t+s}^{t} p(\sigma)\,d\sigma\right) d\eta(t,s) & t \geq t_0 \\[4ex] \phi(t) & t_0 - \tau \leq t \leq t_0\,, \quad \inf_{t \geq t_0}(t + \alpha(t)) = t_0 - \tau \end{cases} \qquad (2.9.11)$$

T is a continuous operator defined on a space of continuous functions $C[t_0 - \tau, \infty)$ into itself. Our objective is to show that there exists a real-valued continuous function $\lambda(t)$ such that $\lambda(t)$ satisfies (2.9.10).

From (2.9.8), we can find $t_0 \in \mathbf{R}_+$ such that

$$e \int_{t+\alpha(t)}^{t} \left[\int_{\alpha(u)}^{\beta(u)} \exp\left(\int_{u+s}^{u} p(\sigma)\,d\sigma\right) d\eta(u,s)\right] du < 1 \qquad (2.9.12)$$

as $t \geq t_0$.

Let

$$y_0(t) = e \int_{\alpha(t)}^{\beta(t)} \exp\left(\int_{t+s}^{t} p(\sigma)\,d\sigma\right) d\eta(t,s)$$

and

$$y_n = Ty_{n-1} \qquad n = 1,\, 2,\, \ldots$$

Obviously, $y_n \in C$. Furthermore,

$$y_1 = Ty_0$$

$$= \int_{\alpha(t)}^{\beta(t)} \exp\left(\int_{t+s}^{t} e \int_{\alpha(\sigma)}^{\beta(\sigma)} \exp\left(\int_{\sigma+s}^{\sigma} p(\omega)\,d\omega\right) d\eta(\sigma,s)\,d\sigma\right) \exp\left(\int_{t+s}^{t} p(\sigma)\,d\sigma\right) d\eta(t,s)$$

$$\leq e \int_{\alpha(t)}^{\beta(t)} \exp\left(\int_{t+s}^{t} p(\sigma)\,d\sigma\right) d\eta(t,s) = y_0$$

for $t \geq t_0$. Let $0 \leq \phi(t) \leq y_0(t)$ on $[t_0 - \tau,\, t_0]$ in (2.9.11). Employing similar arguments, we see that $\{y_n\}$ is a nonincreasing sequence, that is,

$$0 \leq y_{n+1} \leq y_n \leq y_0 \qquad n = 1,\, 2,\, \ldots$$

Hence $\{y_n\}$ converges to a limit function $\lambda(t)$. Then Ty_n converges to $T\lambda$ according to the Lebesgue convergence theorem. Therefore $\lambda(t)$ satisfies (2.9.10), and (2.9.9) is a nonoscillatory solution of (2.9.3). Furthermore, we have proved that equation (2.9.1) has a nonoscillatory solution. The proof is complete.

Consider the equation

$$y'(t) + a(t)y(t) + a_1(t)y(t - \tau_1(t)) + \cdots + a_n(t)y(t - \tau_n(t)) = 0 \qquad (2.9.13)$$

where $a_i(t) \geq 0$, $\tau_i(t) > 0$ and $a(t)$ are locally integrable in t. This is a special case of $(2.9.1)$ since

$$\eta(t,s) = U(s - \tau_1(t))a_1(t) + \cdots + U(s - \tau_n(t))a_n(t)$$

where $U(\cdot)$ is a unit step function. Therefore we have the following corollary.

COROLLARY 2.9.1 If

$$\varliminf_{t\to\infty} \int_{t-\tau_{\min}(t)}^{t} \left[a_1(u) \exp \int_{u-\tau_1(u)}^{u} a(\sigma)\,d\sigma + \cdots + a_n(u) \exp \int_{u-\tau_n(u)}^{u} a(\sigma)\,d\sigma \right] du > e^{-1}$$
$$(2.9.14)$$

then every solution of $(2.9.13)$ oscillates. If

$$\varlimsup_{t\to\infty} \int_{t-\tau_{\max}(t)}^{t} \left[a_1(u) \exp \int_{u-\tau_1(u)}^{u} a(\sigma)\,d\sigma + \cdots + a_n(u) \exp \int_{u-\tau_n(u)}^{u} a(\sigma)\,d\sigma \right] du < \frac{1}{e}$$
$$(2.9.15)$$

then $(2.9.13)$ has a nonoscillatory solution.

Now we consider the first order distributed advanced type linear differential equation

$$y'(t) = a(t)y(t) + \int_{\alpha(t)}^{\beta(t)} y(t + s)\,d\eta(t,s) \qquad (2.9.16)$$

where a is locally integrable, $\beta(t) > \alpha(t) > 0$ are continuous. $\eta(t,s)$ satisfiel the conditions of Theorem $2.9.1$.

Using similar arguments we can prove the following results. We will state them without proofs.

THEOREM 2.9.3 Assume that

$$\varliminf_{t\to\infty} \int_{t}^{t+\alpha(t)} \left[\int_{\alpha(u)}^{\beta(u)} \exp\left(\int_{u}^{u+s} a(\sigma)\,d\sigma \right) d\eta(u,s) \right] du > e^{-1} \qquad (2.9.17)$$

Then every solution of $(2.9.16)$ oscillates.

THEOREM 2.9.4 Assume that

$$\overline{\lim_{t \to \infty}} \int_t^{t+\beta(t)} \left[\int_{\alpha(u)}^{\beta(u)} \exp\left(\int_u^{u+s} a(\sigma)\, d\sigma \right) d\eta\,(u,s) \right] du < e^{-1} \qquad (2.9.18)$$

then (2.9.16) has a nonoscillatory solution.

The equation

$$y'(t) = a(t)y(t) + a_1(t)y(t + \tau_1(t)) + \cdots + a_n(t)y(t + \tau_n(t)) \qquad (2.9.19)$$

where $a(t)$, $\tau_i(t) > 0$, $a_i(t) \geq 0$ are locally integrable in t. (2.9.13) is a special case of (2.9.16). Therefore from Theorems 2.9.3 and 2.9.4 we can obtain the same conclusions for (2.9.19).

2.10 NOTES

The oscillation of Eq. (2.1.3) is studied by Myskis in [189]. Ladas [150] obtained an integral condition for oscillation of (2.1.3) whenever $\tau(t) \equiv \tau > 0$. Tomaras [276] extended the result in [150] to the case with variable delay. Ladde [160] established the present form of condition (2.1.4) for the oscillation of (2.1.3). Then Ladas and Stavroulakis [151, 152] and Koplatadze and Canturija [125] studied differential inequalities (2.1.1) and (2.1.2) at almost the same time. Theorem 2.1.1 is based on the result in [125]. For Theorem 2.1.2 and Corollary 2.1.1 see [189] and [150]. Corollary 2.1.2 is a new result. Theorem 2.1.3 is a special case of Ladas, Lakshmikantham and Papadakis's result [146] which studies higher order differential equation with delay. A special case is treated in [111]. Theorem 2.1.4 is adapted from Ladde [162]. Lemma 2.1.1 is due to Driver [57]. Lemma 2.1.2 and Theorem 2.1.5 are from Ladas, Sficas, and Stavroulakis's paper [154]. Lemma 2.1.2 also is a special case of a Driver's general result [57]. Brauer [19] studied the relation between asymptotic behavior of solution of (2.1.32) and the delay τ whenever $p(t) \equiv p > 0$. Buchanan [20] studied the growth rate of oscillatory solution when $p(t) \equiv \pm 1$ with bounded delay. The results in Section 2.2 are based on Ladas, Sficas, and Stavroulakis's paper [155] but the present form of Theorem 2.2.2 is better than the original result in [155]. Theorem 2.3.1 and 2.3.2 are new. See also [189], Cooke [41], Lillo [168] and Driver [55] concerning equation (2.3.5). Lemmas 2.3.1 and 2.3.2, Theorem 2.3.3, Corollaries 2.3.1 and 2.3.2 are from Birkhoff [13]. Theorem 2.3.4 belongs to Birkhoff [14], while Theorem 2.3.5 is a new result. Theorems 2.3.6 and 2.3.7 are based on [14]. The oscillation of the advanced type equation (2.4.3) is studied by Zhang and Ding [302] and Zhang [301]. For related work see Anderson [1]. Kusano studied advanced type differential inequalities with constant coefficients [141]. Zhang in an unpublished paper (see Section 3) and Onose [212]

studied a general form of inequalities and equations (2.4.1) to (2.4.3).
Theorem 2.4.1 belongs to Zhang (see also Onose [212]). Theorem 2.4.2 is
taken from Zhang [301] but Theorem 2.4.3 is new. For Theorem 2.4.4 see
[302] and [141]. Theorem 2.4.5 and Corollaries 2.4.1 and 2.4.2 are new.
For a discussion of the asymptotic behavior of solution of advanced type
equations, see also Kato [111] and Heard [99]. Theorem 2.5.1 is a new
result. Lemma 2.6.1 and Theorem 2.6.1 are from Ladas and Stavroulakis
[156]. A more general result than Theorem 1.6.1 is obtained by Tramov
[277]. Theorem 2.6.2 is taken from Zhang [304] (see also [6]). Theorem
2.6.3 is a result of Zhang [304]; Theorem 2.6.4 is new (see also [304], [6],
[152], [160-162]). Theorems 2.6.5 to 2.6.7 are new. Theorem 2.7.1 is
from Ladde [161-162] and Ladas et al. [152] (see also [6] and [125]). Lemma
2.7.1 and Theorem 2.7.2 are from Arino, Gyori, and Jawhari's paper [6].
Theorem 2.7.3 belongs to Hunt and Yorke [101]. Theorem 2.7.4 is due to
Ladde [162] and Theorem 2.7.5 is based on Ladde's idea [160]. Theorem
2.8.1 is based on Tomaras's idea [273] (see also Lim [170] and Onose [207]).
Myskis [189] studied equation (2.9.1) systematically. Some assumptions
about (2.9.1) in the beginning of Section 2.9 are from [189]. Theorems
2.9.1 to 2.9.4 were obtained by Zhang (for related work see McCalla [185]).
Some survey papers on the oscillation of solutions of first order differential
equations with deviating arguments were written by Shevelo and Varekh
[235] and Zhang [304].

3

First Order Nonlinear Equations

3.0 INTRODUCTION

In this chapter, we study the oscillatory and nonoscillatory behavior of nonlinear first order ordinary differential equations with deviating arguments. Section 3.1 deals with oscillation and nonoscillation results for a single as well as several deviating arguments. Section 3.2 gives various results on differential inequalities with deviating arguments. These results are used to show the oscillatory and nonoscillatory behavior of ODEWDA. Section 3.3 is devoted to differential equations with mixed type of deviating arguments, while Section 3.4 investigates the oscillatory behavior of very general functional differential equations. Section 3.5 deals with positive solutions of superlinear equations. Sections 3.6 and 3.7 are devoted to oscillatory results for general nonlinear equations with and without forcing terms. Sections 3.8 and 3.9 consist of results relative to distributed type equations. Finally, Section 3.10 deals with oscillatory results on R.

3.1 EQUATIONS WITH DEVIATING ARGUMENTS

Let us consider a nonlinear delay differential equation.

$$y'(t) + p(t)f(y(g(t))) = 0 \tag{3.1.1}$$

Assume that f, p, and g in (3.1.1) satisfy the following conditions:

(i) $g \in C(R_+, R)$, $g(t) < t$ for $t \in R_+$, $g(t)$ is strictly increasing on R_+ and $\lim_{t \to \infty} g(t) = +\infty$;

(ii) $p(t)$ is locally integrable and $p(t) \geq 0$, a.e.;

(iii) $yf(y) > 0$ for $y \neq 0$, $f \in C(R, R)$, f is nondecreasing and

$$\lim_{y \to 0} \frac{y}{f(y)} = M < +\infty \tag{3.1.2}$$

Let us begin by proving the following lemma.

70

LEMMA 3.1.1 Assume that (i) holds. Let $\{t_n\}$ be a sequence defined by $t_n = g^{-1}(t_{n-1})$, t_0 being an arbitrary number. Then $t_n \to \infty$ as $n \to \infty$.

Proof   If the claim is false, let $t_n \to \beta < +\infty$. By continuity of g and g^{-1}, we have

$$\beta = \lim_{n \to \infty} t_n = \lim_{n \to \infty} g^{-1}(t_{n-1}) = g^{-1}(\beta) > \beta$$

which is a contradiction.

We are now in a position to prove the following result.

THEOREM 3.1.1 Assume that conditions (i), (ii), and (iii) hold. Assume further that

$$\overline{\lim_{t \to \infty}} \int_{g(t)}^{t} p(s)\, ds > M \tag{3.1.3}$$

where M is defined by (3.1.2). Then every solution of (3.1.1) is oscillatory.

Proof: For sufficiently large t*, we have

$$\int_{g(t)}^{t} p(s)\, ds > M + K \quad \text{for } t \ge t^* \tag{3.1.4}$$

where $K > 0$. Let $y(t)$ be a nonoscillatory solution of (3.1.1). Without loss of generality, assume that $y(t) > 0$ for $t \ge t_0 > g(t^*)$. Then

$$y'(t) = -p(t)f(y(g(t))) \le 0 \quad \text{for } t > t_1 = g^{-1}(t_0)$$

Thus y is nonincreasing and has a finite nonnegative limit α as $t \to \infty$. We claim that $\alpha = 0$. If not, let $\alpha > 0$ and $f(\alpha) > 0$. Let $s_0 = t_0$, and $s_n = g^{-1}(s_{n-1})$ such that

$$\int_{g(s_n)}^{s_n} p(s)\, ds > M + K \tag{3.1.5}$$

By integrating (3.1.1) from t_0 to s_n, we get

$$y(s_n) - y(t_0) = - \int_{t_0}^{s_n} p(s)f(y(g(s)))\, ds$$

This together with conditions (i) to (iii) and (3.1.5) yields

$$y(s_n) - y(t_0) \le -f(\alpha) \sum_{i=1}^{n} \int_{s_{i-1}}^{s_i} p(s)\, ds$$

$$= -f(\alpha) \sum_{i=1}^{n} \int_{g(s_i)}^{s_i} p(s)\, ds$$

$$< -f(\alpha) n (M + K)$$

Thus $y(s_n) \to -\infty$ as $n \to \infty$ which contradicts $y(t) \ge \alpha > 0$ for $t \ge s_0$ and $\alpha = 0$ as claimed. Therefore $y(t)$ is nonincreasing and converging to zero as $t \to +\infty$.

From (3.1.1), it follows that

$$y(t) - y(g(t)) + \int_{g(t)}^{t} p(s) f(y(g(s)))\, ds = 0$$

This implies

$$y(t) - y(g(t)) \left[1 - \frac{f(y(g(t)))}{y(g(t))} \int_{g(t)}^{t} p(s)\, ds \right] \le 0$$

and hence

$$\int_{g(t)}^{t} p(s)\, ds < \frac{y(g(t))}{f(y(g(t)))}$$

for sufficiently large t. Therefore

$$\overline{\lim_{t \to \infty}} \int_{g(t)}^{t} p(s)\, ds \le M$$

This is a contradiction to condition (3.1.3).

REMARK 3.1.1 Theorem 3.1.1 includes both the linear and sublinear cases of f. The function f is sublinear if $M = 0$ in (3.1.2). In case

$$\lim_{y \to 0} \frac{y}{f(y)} = +\infty \tag{3.1.6}$$

then f is said to be a generalized superlinear function. The superlinear case is more difficult to analyze as the following example shows.

EXAMPLE 3.1.1 Consider the delay differential equation

$$y'(t) + (t - \sqrt{t})^3 t^{-2} y^3 (t - \sqrt{t}) = 0 \qquad t \geq 2 \tag{3.1.7}$$

which possesses a nonoscillatory solution $y(t) = 1/t$, even though

$$\int_{g(t)}^{t} p(s)\ ds = \int_{t-\sqrt{t}}^{t} \frac{(s - \sqrt{s})^3}{s^2}\ ds \to +\infty \qquad t \to \infty$$

The next result is a variant of Theorem 3.1.1.

THEOREM 3.1.2 Assume that the hypotheses of Theorem 3.1.1 hold except that the relation (3.1.3) is replaced by

$$\lim_{t \to \infty} \int_{g(t)}^{t} p(s)\ ds > \frac{M}{e} \tag{3.1.8}$$

Then every solution of (3.1.1) oscillates.

Proof: Assume that there is a nonoscillatory solution $y(t) > 0$, $y(g(t)) > 0$ for $t \geq t_0 \geq 0$. So $y'(t) \leq 0$ and hence $y(t) \to \alpha \geq 0$ as $t \to \infty$. As in Theorem 3.1.1, $y(t) \to 0$ as $t \to \infty$.

There exists $t^* \in (g(t), t)$ such that

$$\int_{t^*}^{t} p(s)\ ds > \frac{M}{2e} \quad \text{and} \quad \int_{g(t)}^{t^*} p(s)\ ds > \frac{M}{2e} \tag{3.1.9}$$

Now integrating (3.1.1) from t^* to t,

$$y(t^*) - y(t) = \int_{t^*}^{t} p(s) f(y(g(s)))\ ds$$

$$\geq f(y(g(t))) \int_{t^*}^{t} p(s)\ ds > f(y(g(t))) \frac{M}{2e}$$

and from $g(t)$ to t^*, we obtain

$$y(g(t)) - y(t^*) = \int_{g(t)}^{t^*} p(s) f(y(g(s)))\ ds$$

$$\geq f(y(g(t^*))) \frac{M}{2e}$$

which implies

$$y(t^*) > f(y(g(t))) \frac{M}{2e} \geq \frac{f(y(g(t)))}{y(g(t))} f(y(g(t^*))) \left(\frac{M}{2e}\right)^2$$

and hence

$$\frac{y(g(t^*))}{y(t^*)} \leq \frac{y(g(t))}{f(y(g(t)))} \frac{y(g(t^*))}{f(y(g(t^*)))} \left(\frac{2e}{M}\right)^2 < \infty \qquad (3.1.10)$$

because of condition $(3.1.2)$.

Setting $w(t) = y(g(t))/y(t) \geq 1$, $\lim_{t\to\infty} w(t) = \ell \geq 1$, ℓ is finite, because of $(3.1.10)$. From $(3.1.1)$, we have

$$\ln w(t) = \int_{g(t)}^{t} p(s) \frac{f(y(g(s)))}{y(g(s))} w(s) \, ds$$

$$= w(\xi) \frac{f(y(g(\xi)))}{y(g(\xi))} \int_{g(t)}^{t} p(s) \, ds$$

where $g(t) < \xi < t$. Taking the limit inferior in the above equation, we obtain

$$\ln \ell \geq \frac{\ell}{M} \lim_{t\to\infty} \int_{g(t)}^{t} p(s) \, ds$$

But $\max_{\ell \geq 1} \ln \ell / \ell = 1/e$ and therefore

$$\frac{M}{e} \geq \lim_{t\to\infty} \int_{g(t)}^{t} p(s) \, ds$$

This is a contradiction, because $(3.1.8)$ holds.

REMARK 3.1.1 We did assume that $M \neq 0$ in $(3.1.8)$. In the case $M = 0$, this theorem is not applicable.

EXAMPLE 3.1.1 Consider the equation

$$y'(t) - \frac{2}{(e \ln \lambda)t} y(\lambda t) = 0 \qquad 0 < \lambda < 1 \qquad (3.1.11)$$

and note that

$$\int_{\lambda t}^{t} - \frac{2}{(e \ln \lambda)s} \, ds = \frac{2}{e}$$

Therefore $(3.1.11)$ satisfies the condition of Theorem 3.1.2, so all solutions of $(3.1.11)$ oscillate.

REMARK 3.1.2 Consider

$$y'(t) + p(t)f(y(t)) + F(t, y(t), y(g(t))) = 0 \qquad (3.1.12)$$

where $F \in C[R_+ \times R, R]$ and $F(t, u, v)u > 0$, for $u \cdot v > 0$. Then Theorem 3.1.2 is valid for (3.1.12).

REMARK 3.1.3 Theorems 3.1.1 and 3.1.2 can be extended to equations with several deviating arguments of the form

$$y'(t) + \sum_{i=1}^{n} p_i(t) f_i(y(g_i(t))) = 0 \tag{3.1.13}$$

Now we present a result for the advanced equation

$$y'(t) = \sum_{i=1}^{n} p_i(t) f_i(y(g_i(t))) \tag{3.1.14}$$

where $p_i(t) \geq 0$ and $g_i(t) > t$, $i \in I_n$ are continuous.

We present a sufficient condition for oscillation of (3.1.14). The proof is left to the reader.

THEOREM 3.1.3 For $n = 1$ if $uf_i(u) > 0$ for $u \neq 0$, $f_i(u)$ is nondecreasing in u,

$$\lim_{|u| \to \infty} \frac{u}{f_i(u)} = M_i > 0 \qquad i \in I_n \tag{3.1.15}$$

and

$$\lim_{t \to \infty} \int_{t}^{g_*(t)} \left(\sum_{i=1}^{n} p_i(s) \right) ds > \frac{M^*}{e} \tag{3.1.16}$$

then every solution of (3.1.14) oscillates, where $M^* = \max(M_1, \ldots, M_n)$ and $g_*(t) = \min(g_1(t), \ldots, g_n(t))$.

THEOREM 3.1.4 If condition (3.1.16) is replaced by

$$\overline{\lim_{t \to \infty}} \int_{t}^{g_*(t)} \left(\sum_{i=1}^{n} p_i(s) \right) ds > M^* \tag{3.1.17}$$

then the conclusion of Theorem 3.1.3 remains valid.

Proof: Otherwise, there is a nonoscillatory solution $y(t)$ and without loss of generality we assume that $y(t) > 0$. It is easy to see that $y(t) \to \infty$. From (3.1.14), we have

$$y(g_*(t)) - y(t) = \int_t^{g_*(t)} \sum_{i=1}^{n} p_i(s) f_i(y(g_i(s))) \, ds$$

$$\geq \sum_{i=1}^{n} f_i(y(g_i(t))) \int_t^{g_*(t)} p_i(s) \, ds$$

$$\geq \sum_{i=1}^{n} f_i(y(g_*(t))) \int_t^{g_*(t)} p_i(s) \, ds$$

and so

$$y(g_*(t)) \left[1 - \sum_{i=1}^{n} \frac{f_i(y(g_*(t)))}{y(g_*(t))} \int_t^{g_*(t)} p_i(s) \, ds \right] \geq y(t) > 0$$

Therefore

$$1 > \sum_{i=1}^{n} \frac{f_i(y(g_*(t)))}{y(g_*(t))} \int_t^{g_*(t)} p_i(s) \, ds$$

$$\geq \frac{1}{M^*} \overline{\lim_{t \to \infty}} \int_t^{g_*(t)} \sum_{i=1}^{n} p_i(s) \, ds$$

This is a contradiction. The proof is complete.

Next we present several results concerning the asymptotic behavior of (3.1.1). Let us set $g(t) = t - \tau(t)$ in (3.1.1).

THEOREM 3.1.5 Assume that p, $\tau \in C[R_+, R_+]$, $p(t) > 0$, $f \in C[R, R]$, $0 \leq \tau(t) \leq q$ and $yf(y) > 0$ for $y \neq 0$. If $\int^{\infty} p(t) \, dt = \infty$, then all nonoscillatory solutions of (3.1.1) tend to zero as $t \to \infty$.

Proof: Let $y(t) > 0$ be a nonoscillatory solution of (3.1.1) for sufficiently large t. Then $y'(t) < 0$ and it follows from the hypotheses that $\lim_{t \to \infty} y(t) = C \geq 0$ exists.

We show that $C = 0$. Otherwise, $C > 0$, and then there exists a $t^* \geq t_0$ such that $f(y(t - \tau(t))) \geq d > 0$ for $t \geq t^*$ and $f(C) \geq d > 0$. Thus $y'(t) \leq -p(t)d$ for $t \geq t^*$ which implies $y(t) \leq y(t^*) - d \int_{t^*}^{t} p(s) \, ds$. Hence $y(t)$ will become negative for sufficiently large t. This is a contradiction to the fact that $y(t) > 0$. Therefore $C = 0$. A similar argument holds if $y(t)$ is eventually negative.

For an advanced type equation

$$y'(t) = p(t)f(y(t + \tau(t)))$$ (3.1.18)

we have the following conclusions.

THEOREM 3.1.6 Assume that (3.1.18) satisfies the hypotheses of Theorem 3.1.5; then all nonoscillatory solutions of (3.1.18) tend to infinity as $t \to \infty$.

The proof of this theorem can be formulated on the basis of the proof of Theorem 3.1.5.

The following results provide sufficient conditions for asymptotic behavior of the bounded oscillatory solutions of (3.1.1).

THEOREM 3.1.7 Assume that p, $\tau \in C[R_+, R_+]$, $p(t) > 0$, $f \in C[R, R]$, $0 \le \tau(t) \le q$ and $yf(y) > 0$ for $y \ne 0$. If either $p(t) \to 0$ or $b(t)\tau(t) \to 0$ as $t \to \infty$, where $b(t) = \sup_{s \in [0, t]} p(s)$, then all bounded oscillatory solutions of (3.1.1) tend to zero as $t \to \infty$.

Proof: Let $y(t)$ be a bounded oscillatory solution. Assume that the conclusion of the theorem is false. This implies that there exists an $\epsilon > 0$ and sequences $\{t_n\}$, $\{t_n^*\} \to \infty$ such that for each n, either $y(t_n) = 0$, $y(t_n^*) = \epsilon$, $y'(t_n^*) \ge 0$ and $0 < y(t) < \epsilon$ whenever $t_n < t < t_n^* < t_{n+1}$, or $y(t_n) = 0$, $y(t_n^*) = -\epsilon$, $y'(t_n^*) \le 0$ and $0 > y(t) > -\epsilon$ whenever $t_n < t < t_n^* < t_{n+1}$. The following argument holds for both cases. We assume the former one. Integrating $y'(t)$ from t_n to t_n^*, we obtain

$$\epsilon = y(t_n^*) - y(t_n) \le \int_{t_n}^{t_n^*} p(s)|f(y(s - \tau(s)))|\, ds \le M \int_{t_n}^{t_n^*} p(s)\, ds$$ (3.1.19)

where M is a bound of $|f(y(\cdot))|$. Suppose $b(t)\tau(t) \to 0$ as $t \to \infty$. Since $b(t)$ is continuous and monotone increasing, we have

$$\epsilon \le M \int_{t_n}^{t_n^*} b(s)\, ds \le Mb(t_n^*)(t_n^* - t_n)$$

or

$$t_n^* - t_n \ge \frac{\epsilon}{Mb(t_n^*)}$$

Let n be chosen sufficiently large that $b(t_n^*)\tau(t_n^*) < \epsilon/M$. Then

$$\tau(t_n^*) < \frac{\epsilon}{Mb(t_n^*)} \le t_n^* - t_n$$

which implies $t_n < t_n^* - \tau(t_n^*) \leq t_n^*$. Thus $y(t_n^* - \tau(t_n^*)) > 0$ and we arrive at

$$y'(t_n^*) = -p(t_n^*)f(y(t_n^* - \tau(t_n^*))) < 0$$

This contradicts $y'(t_n^*) \geq 0$. Now, suppose $p(t) \to 0$ as $t \to \infty$. From (3.1.19) we get

$$\epsilon \leq M \int_{t_n}^{t_n^*} p(s)\,ds = Mp(\xi)(t_n^* - t_n)$$

or

$$t_n^* - t_n \geq \frac{\epsilon}{Mp(\xi)}$$

where $t_n < \xi < t_n^*$. Since $p(t) \to 0$, it follows that $p(\xi) \to 0$ as $n \to \infty$. Then n can be chosen sufficiently large that $t_n^* - t_n > q$. It follows that $t_n < t_n^* - \tau(t_n^*) \leq t_n^*$ and we obtain a contradiction as before. This completes the proof.

COROLLARY 3.1.1 Assume that $p, \tau \in C[R_+, R_+]$, $p(t) > 0$, $f \in C[R_+, R_+]$, $0 \leq \tau(t) \leq q$ and $yf(y) > 0$ for $y \neq 0$. Moreover, if $p(t)$ is bounded with $\int^\infty p(t)\,dt = \infty$ and $\tau(t) \to 0$ as $t \to \infty$, then all bounded solutions of (3.1.1) tend to zero as $t \to \infty$.

3.2 DIFFERENTIAL INEQUALITIES WITH DEVIATING ARGUMENTS

Consider the first order nonlinear delay differential inequalities

$$y'(t) + a(t)y(t) + p(t)f(y(t - \tau_1),\ \cdots,\ y(t - \tau_m)) \leq 0 \tag{3.2.1}$$

$$y'(t) + a(t)y(t) + p(t)f(y(t - \tau_1),\ \cdots,\ y(t - \tau_m)) \geq 0 \tag{3.2.2}$$

and the delay differential equation

$$y'(t) + a(t)y(t) + p(t)f(y(t - \tau_1),\ \cdots,\ y(t - \tau_m)) = 0 \tag{3.2.3}$$

where $p \in C[R_+, R_+]$, $a \in C[R_+, R]$, $\tau_i \in C^1[R_+, R]$, $\tau_i'(t) > 0$, $\tau_i(t) > 0$, and $t - \tau_i(t) \to \infty$ as $t \to \infty$, $i \in I_n$. Furthermore, suppose that

$$f \in C[R^m, R] \quad \text{and} \quad f(y_1, y_2, \cdots, y_m)y_1 > 0 \tag{A}$$

when y_i, $i \in I_m$, have the same sign. Then we can prove the following result.

THEOREM 3.2.1 Assume that (3.2.1) satisfies the above-mentioned conditions. In addition, suppose that

$$\liminf_{t \to \infty} \int_{t-\tau_i(t)}^{t} a(s)\, ds \geq k_i > -\infty \qquad i \in I_m \tag{3.2.4}$$

where k_i are constants, and there exist nonnegative numbers K, α_j, $j \in I_m$ such that

$$\sum_{i=1}^{m} \alpha_i = 1 \qquad K > 0 \tag{3.2.5}$$

$$|f(s_1, s_2, \cdots, s_m)| \geq K|s_1|^{\alpha_1} |s_2|^{\alpha_2} \cdots |s_m|^{\alpha_m} \tag{3.2.6}$$

hold for all $s \in R^m$, and

$$\liminf_{t \to \infty} \int_{t-\tau_*(t)}^{t} p(s)\, ds > \frac{1}{eKC} \tag{3.2.7}$$

where

$$C = \min_{1 \leq i \leq m} \exp k_i \quad \text{and} \quad \tau_*(t) = \min\,(\tau_1(t), \cdots, \tau_m(t)) \tag{3.2.8}$$

Then (3.2.1) has no eventually positive solution.

Proof: We shall show that the existence of an eventually positive solution leads to a contradiction. Suppose that $y(t)$ is a solution of (3.2.1) such that for sufficiently large constant T, $y(t) > 0$, $t \geq T$. Because of $t - \tau_i(t) \to \infty$ as $t \to \infty$, there exists $T_1 > T$ such that $t - \tau_i(t) \geq T$ for $t \geq T_1$, and hence

$$y(t - \tau_i(t)) > 0 \quad \text{for} \ \ t \geq T_1,\ i \in I_m$$

From (3.2.1) one gets

$$\exp\!\left(\int_T^t a(s)\, ds\right) (y'(t) + a(t)y(t) + p(t)f(y(t - \tau_1(t)), \cdots, y(t - \tau_m(t)))) \leq 0$$

for $t \geq T_1$, which implies

$$\left(y(t)\exp\!\left(\int_T^t a(s)\, ds\right)\right)' + \left(\exp\!\left(\int_T^t a(s)\, ds\right)\right) p(t)f(y(t - \tau_1(t)), \cdots, y(t - \tau_m(t))) \leq 0$$

$$\tag{3.2.9}$$

for $t \geq T_1$. By setting

$$z(t) = y(t) \exp \int_T^t a(s) \, ds$$

(3.2.9) reduces to

$$z'(t) + \left(\exp\left(\int_T^t a(s) \, ds\right)\right) p(t) f\left(\left(\exp\left(-\int_T^{t-\tau_1(t)} a(s) \, ds\right)\right) z(t - \tau_1(t)),\right.$$

$$\left. \ldots, \left(\exp\left(-\int_T^{t-\tau_m(t)} a(s) \, ds\right)\right) z(t - \tau_m(t))\right) \leq 0 \qquad (3.2.10)$$

for $t \geq T_1$. Since $z(t - \tau_i(t)) > 0$ and $f > 0$ for $t \geq T_1$, we see that $z'(t) < 0$ for $t \geq 2T_1$. Hence

$$z(t) < z(t - \tau_*(t)) \quad \text{for} \quad t \geq 2T_1$$

Setting $w(t) = z(t - \tau_*(t))/z(t)$, for $t \geq 2T_1$, we note that $w(t) > 1$. Dividing both sides of (3.2.10) by $z(t)$ and then integrating from $t - \tau_*(t)$ to t for $t \geq 2T_1$, we have

$$\ln z(t) - \ln z(t - \tau_*(t))$$

$$+ \int_{t-\tau_*(t)}^t \left[\left(\exp\left(\int_T^s a(u) \, du\right)\right) p(s) f\left(\left(\exp\left(-\int_T^{s-\tau_1(s)} a(u) \, du\right)\right) z(s - \tau_1(s)),\right.\right.$$

$$\left.\left. \ldots, \left(\exp\left(-\int_T^{s-\tau_m(s)} a(u) \, du\right)\right) z(s - \tau_m(s))\right) \middle/ z(s)\right] ds \leq 0$$

for $t \geq 3T_1$, from which we obtain

$$\ln w(t) \geq \int_{t-\tau_*(t)}^t \left[\left(\exp\left(\int_T^s a(u) \, du\right)\right) p(s) f\left(\left(\exp\left(-\int_T^{s-\tau_1(s)} a(u) \, du\right)\right) z(s - \tau_1(s)),\right.\right.$$

$$\left.\left. \ldots, \left(\exp\left(-\int_T^{s-\tau_m(s)} a(u) \, du\right)\right) z(s - \tau_m(s))\right) \middle/ z(s)\right] ds$$

for $t \geq 3T_1$. From (3.2.5), (3.2.6), and the nondecreasing property of $z(t)$, it follows that

$$\ln w(t)$$

$$\geq K \int_{t-\tau_*(t)}^{t} \left(\exp\left(\int_{T}^{s} a(u)\, du\right)\right) p(s) \left[\left(\exp\left(-\int_{T}^{s-\tau_1(s)} a(u)\, du\right)\right) z(s - \tau_1(s))\right]^{\alpha_1}$$

$$\cdots \left[\left(\exp\left(-\int_{T}^{s-\tau_m(s)} a(u)\, du\right)\right) z(s - \tau_m(s))\right]^{\alpha_m} \Big/ z(s)\, ds$$

$$\geq K \int_{t-\tau_*(t)}^{t} \left[\exp\left(\int_{T}^{s} a(u)\, du\right)\right]^{\alpha_1 + \cdots + \alpha_m} p(s) \left[\exp\left(-\int_{T}^{s-\tau_1(s)} a(u)\, du\right)\right]^{\alpha_1}$$

$$\cdots \left[\exp\left(-\int_{T}^{s-\tau_m(s)} a(u)\, du\right)\right]^{\alpha_m} \frac{z(s - \tau_*(s))}{z(s)}\, ds$$

$$\geq K \int_{t-\tau_*(t)}^{t} \left[\exp\left(\int_{s-\tau_1(s)}^{s} a(u)\, du\right)\right]^{\alpha_1} \left[\exp\left(\int_{s-\tau_2(s)}^{s} a(u)\, du\right)\right]^{\alpha_2}$$

$$\cdots \left[\exp\left(\int_{s-\tau_m(s)}^{s} a(u)\, du\right)\right]^{\alpha_m} p(s) \frac{z(s - \tau_*(s))}{z(s)}\, ds$$

$$\geq KC \int_{t-\tau_*(t)}^{t} p(s)w(s)\, ds \qquad \text{for } t \geq t_0 \geq 3T_1 \tag{3.2.11}$$

where t_0 is sufficiently large and C is as defined in (3.2.8). From (3.2.7), there exists a constant N such that

$$\int_{t-\tau_*(t)}^{t} p(s)\, ds \geq N > \frac{1}{eKC} \qquad \text{for } t \geq t_2 \geq t_0$$

where t_2 is sufficiently large. Therefore, for any t, $t \geq t_3 > t_2$, sufficiently large t_3, there exists a t^* such that $t^* - \tau_*(t^*) < t < t^*$, and

$$\int_{t}^{t^*} p(s)\, ds \geq \frac{N}{2}, \quad \int_{t^*-\tau_*(t^*)}^{t} p(s) \geq \frac{N}{2} \tag{3.2.12}$$

Now by integrating (3.2.10) from t to t^* and from $t^* - \tau_*(t^*)$ to t, and by using the argument that was used to obtain (3.2.11), we get

$$z(t) - z(t^*) \geq KC \int_t^{t^*} p(s)z(s - \tau_*(s))\, ds$$

$$\geq KC\, z(t^* - \tau_*(t^*)) \int_t^{t^*} p(s)\, ds$$

$$\geq \frac{NKC}{2}\, z(t^* - \tau_*(t^*)) \tag{3.2.13}$$

and

$$z(t^* - \tau_*(t^*)) - z(t) \geq KC \int_{t^*-\tau_*(t^*)}^{t} p(s)z(s - \tau_*(s))\, ds$$

$$\geq KC\, z(t - \tau_*(t)) \int_{t^*-\tau_*(t^*)}^{t} p(s)\, ds$$

$$\geq \frac{NKC}{2}\, z(t - \tau_*(t)) \tag{3.2.14}$$

From (3.2.13) and (3.2.14), we have

$$z(t) \geq \frac{NKC}{2}\, z(t^* - \tau_*(t^*)) \geq \frac{(NKC)^2}{4}\, z(t - \tau_*(t))$$

which implies that

$$\frac{4}{(NKC)^2} \geq \frac{z(t - \tau_*(t))}{z(t)} = w(t) \qquad \text{for } t \geq t_3$$

and therefore $w(t)$ is bounded.

Let

$$\liminf_{t \to \infty} w(t) = \ell$$

Then $\ell \geq 1$ and ℓ is finite. Taking liminf on both sides of (3.2.11), we get

$$\ln \ell \geq \ell\, KC\left(\liminf \int_{t-\tau_*(t)}^{t} p(s)\, ds\right)$$

Using the fact that $\max_{x \geq 1} (\ln x - ax) = -\ln a - 1$, the above relation reduces
to

$$\max_{\ell \geq 1} \left\{ \ln \ell - \ell\, KC \liminf \int_{t-\tau_*(t)}^{t} p(s)\, ds \right\} = -\ln\left(KC \liminf \int_{t-\tau_*(t)}^{t} p(s)\, ds\right) - 1$$

$$\geq 0$$

This last inequality implies

$$\ln \left(KC \, \liminf \int_{t-\tau_*(t)}^{t} p(s) \, ds \right) \leq -1$$

namely,

$$\liminf_{t \to \infty} \int_{t-\tau_*(t)}^{t} p(s) \, ds \leq \frac{1}{eKC}$$

which contradicts hypothesis (3.2.5). The proof is complete.

Employing similar arguments, one can prove the next result.

THEOREM 3.2.2 Assume that (3.2.2) satisfies the hypotheses of Theorem 3.2.1. Then (3.2.2) has no eventually negative solution.

From Theorem 3.2.1 and 3.2.2, it follows that the delay differential equation (3.2.3) has neither eventually positive nor eventually negative solutions.

COROLLARY 3.2.1 Under the hypotheses of Theorem 3.2.1, every solution of (3.2.3) is oscillatory.

EXAMPLE 3.2.1 The equation

$$y'(t) - y(t) + \exp\left(\frac{\pi}{2}\right) y\left(t - \frac{\pi}{2}\right) = 0 \tag{3.2.15}$$

satisfies the condition of Corollary 3.2.1, so every solution of (3.2.15) is oscillatory. In fact, $y(t) = (\exp t) \sin t$ is an oscillatory solution of (3.2.15).

REMARK 3.2.1 If condition (3.2.7) is replaced by

$$\limsup_{t \to \infty} \int_{t-\tau_*(t)}^{t} p(s) \, ds > \frac{1}{KC} \tag{3.2.16}$$

then Theorem 3.2.1 remains valid. This remark can be justified by using the method of proof used for Theorem 3.1.1.

REMARK 3.2.2 One can formulate results analogous to the above results with respect to the following form of first order differential inequality. Consider

$$y'(t) + a(t)y(t) + \sum_{i=1}^{n} p_i(t)f_i(y(t - \tau_1(t)), \ldots, y(t - \tau_m(t))) \leq 0 \tag{3.2.17}$$

Replace condition (3.2.7) by condition

$$\liminf_{t \to \infty} \int_{t-\tau_*(t)}^{t} \left(\sum_{i=1}^{n} k_i p_i(s) \right) ds > \frac{1}{eC} \tag{3.2.18}$$

Then the conclusion of Theorem 3.2.1 remains valid.

Using the above argument, we can obtain the following theorem.

THEOREM 3.2.3 Assume that all the hypotheses of Theorems 3.2.1 and 3.2.2 hold except that (3.2.4), (3.2.6), and (3.2.7) are replaced by

$$\liminf_{t \to \infty} \int_{t-\tau_i(t)}^{t} a(s) \, ds \geq k_i \geq 0 \qquad \text{for } i \in I_m \tag{3.2.19}$$

and

$$\liminf \int_{t-\tau_*(t)}^{t} p(s) \, ds > \frac{M}{Ce} \tag{3.2.20}$$

where

$$M = \limsup_{\substack{y_i \to 0 \\ 0 \leq i \leq m}} \frac{|y_1|^{\alpha_1} \cdots |y_m|^{\alpha_m}}{|f(y_1, \ldots, y_m)|} \tag{3.2.21}$$

and C, τ_* is defined by (3.2.8). Then the conclusions of Theorems 3.2.1, 3.2.2, and Corollary 3.2.1 remain true.

The above method can also be applied to the following type of advanced differential inequalities

$$y'(t) + a(t)y(t) - p(t)f(y(t + \tau_1(t)), \ldots, y(t + \tau_m(t))) \geq 0 \tag{3.2.22}$$

$$y'(t) + a(t)y(t) - p(t)f(y(t + \tau_1(t)), \ldots, y(t + \tau_m(t))) \leq 0 \tag{3.2.23}$$

$$y'(t) + a(t)y(t) - p(t)f(y(t + \tau_1(t)), \ldots, y(t + \tau_m(t))) = 0 \tag{3.2.24}$$

Let us next state a result whose proof can be formulated analogously.

THEOREM 3.2.4 Assume that p, $\tau_i \in C[R_+, R_+]$, p(t) > 0, $\tau_i(t) > 0$, $i \in I_m$, $a \in C[R_+, R]$, and f satisfies (A).
     Furthermore, assume that

$$\liminf_{t \to \infty} \int_t^{t+\tau_i(t)} (-a(s))\, ds = k_i > -\infty \qquad i \in I_m \tag{3.2.25}$$

where $k_i \in R$, and there exist nonnegative numbers K and α_j, $j \in I_m$, such that (3.2.5) and (3.2.6) hold, and

$$\liminf_{t \to \infty} \int_t^{t+\tau_*(t)} p(s)\, ds > \frac{1}{eKC} \tag{3.2.26}$$

where C, τ_* is defined by (3.2.8). Then (3.2.22) has no eventually positive solution; (3.2.23) has no eventually negative solution, and every solution of (3.2.24) is oscillatory.

EXAMPLE 3.2.2   The equation

$$y'(t) - 3\left[y\left(t + \frac{\pi}{2}\right)\right]^{1/3} [y(t + 2\pi)]^{2/3} = 0 \tag{3.2.27}$$

satisfies the conditions of Theorem 3.2.4 so every solution of (3.2.27) is oscillatory. In fact, the functions $y_1(t) = \cos^3 t$, $y_2(t) = \sin^3 t$ are oscillatory solutions of (3.2.27).

THEOREM 3.2.5   If $a(t) \leq 0$ in Theorem 3.2.4, then (3.2.6), (3.2.25), and (3.2.26) can be replaced by the condition

$$\lim_{t \to \infty} \int_t^{t+\tau_*(t)} p(s)\, ds > \frac{M}{e} \exp\left(- \lim_{t \to \infty} \int_t^{t+\tau_*(t)} (-a(s))\, ds\right) \tag{3.2.28}$$

where

$$M = \overline{\lim_{\substack{|y_i| \to \infty \\ 1 \leq i \leq m}}} \frac{|y_1|^{\alpha_1} |y_2|^{\alpha_2} \cdots |y_m|^{\alpha_m}}{|f(y_1, \ldots, y_m)|} \tag{3.2.29}$$

and the conclusions of Theorem 3.2.4 remain valid.

EXAMPLE 3.2.3   Consider the advanced type differential inequality

$$y'(t) - e^{-t} y(t) - e^{-2t} [y(t + 1)]^{1/3} \left[y\left(t + \frac{1}{2}\right)\right]^{2/3} \geq 0 \tag{3.2.30}$$

It does not satisfy conditions of Theorem 3.2.5. In fact, (3.2.30) has the positive solution $y(t) = e^{2t}$.

REMARK 3.2.3 Consider the mixed type differential equation

$$y'(t) + a(t)y(t) + p(t)f(y(t - \tau_1(t)), \cdots, y(t - \tau_m(t)))$$

$$+ q(t)F(y(t + g_1(t)), \cdots, y(t + g_m(t))) = 0 \qquad (3.2.31)$$

where a, p, τ_i, and f satisfy the conditions of Theorem 3.2.1, and $q(t) > 0$, $g_i(t) > 0$ are continuous and F satisfies condition (A). Then every solution of (3.2.31) oscillates under the conditions of Theorem 3.2.1. In fact, assume that there is a positive solution $y(t)$ of Theorem (3.2.31). Then (3.2.1) has a positive solution which contradicts the conclusion of Theorem 3.2.1. If there is a negative solution $y(t)$ of (3.2.31), it will contradict Theorem 3.2.2.

Consider the first order advanced type differential equation

$$y'(t) = f(t, y(\tau_1(t)), \cdots, y(\tau_m(t))) \qquad (3.2.32)$$

where $f \in C[R^+ \times R^m, R]$, $\tau_i(t) > t$ on $t \in R^+$ and $\tau_i \in C[R^+, R^+]$, $i \in I_m$.

THEOREM 3.2.6 Assume that there exists a function $a \in C[R^+, R^+]$ such that

$$f(t, y_1, \cdots, y_m) \operatorname{sgn} y_0 \geq a(t)|y_0| \qquad (3.2.33)$$

for $t \geq 0$, $|y_i| \geq |y_0|$, $y_i y_0 \geq 0$, $i = 1, 2, \cdots, m$, and

$$\lim_{t \to \infty} \int_t^{\tau_*(t)} a(s)\, ds > e^{-1} \qquad (3.2.34)$$

where $\tau_*(t) = \min(\tau_1(t), \cdots, \tau_m(t))$. Then every solution of (3.2.32) is oscillatory.

Proof: Otherwise, assume that there is a nonoscillatory solution $y(t)$. Without loss of generality, assume that $y(t) > 0$; then from (3.2.32) and (3.2.33), we obtain a first order advanced differential inequality

$$y'(t) - a(t)y(\tau_*(t)) \geq 0 \qquad (3.2.35)$$

This implies (3.2.35) has a positive solution $y(t)$. On the other hand, from Theorem 2.4.1, (3.2.35) has no eventually positive solution under condition (3.2.34). This contradiction establishes the theorem.

Let us consider a special case of (3.2.32)

$$y'(t) = p_1(t)y(\tau_1(t)) + \cdots + p_m(t)y(\tau_m(t)) \qquad (3.2.36)$$

where $p_i \in C[R^+, R^+]$, $\tau_i \in C[R^+, R^+]$, $\tau_i(t) > t$ on $t \in R^+$, $i \in I_m$.

COROLLARY 3.2.2 If

$$\varlimsup_{t \to \infty} \int_t^{\tau_*(t)} \left(\sum_{i=1}^m p_i(s) \right) ds > e^{-1} \qquad (3.2.37)$$

then every solution of (3.2.32) oscillates.

The following result provides a sufficient condition for existence of a nonoscillatory solution of (3.2.32).

THEOREM 3.2.7 Assume that there exists a function a(t) such that $a \in C[R^+, R^+]$ and

$$0 \le f(t, y_1, \cdots, y_m) \text{ sgn } y_0 \le a(t)|y_0| \qquad (3.2.38)$$

on $t \in R^+$, $|y_i| \le |y_0|$, $y_i y_0 \ge 0$, $i = 1, 2, \cdots, m$, and

$$\limsup_{t \to \infty} \int_t^{\tau^*(t)} a(s) \, ds < \frac{1}{e} \qquad (3.2.39)$$

where $\tau^*(t) = \max(\tau_1(t), \cdots, \tau_n(t))$. Then equation (3.2.32) has a nonoscillatory solution.

Proof: We denote by $C[t_0, +\infty)$ a locally convex space consisting of the set of all continuous functions. The topology of C is the topology of uniform convergence on every compact interval of $[t_0, +\infty)$. Let us define

$$S = \Big\{ y \in C[t_0, \infty) : y(t) \text{ is nondecreasing on } [t_0, +\infty)$$

$$\text{and } 1 \le y(t) \le \exp\left[e \int_{t_1}^t a(s) \, ds \right], \ t \in [t_1, +\infty),$$

$$y(t) \equiv 1 \text{ for } t \in [t_0, t_1], \text{ and}$$

$$y(\tau^*(t)) < e y(t) \text{ for } t > t_1 \Big\} \qquad (3.2.40)$$

We note that S is nonempty, convex, and a closed subset of $C[t_0, \infty)$. Define a map $F: S \to C[t_0, +\infty)$ as follows

$$(Fy)(t) = \begin{cases} \exp\left(\int_{t_1}^t \dfrac{f(s, y(\tau_1(s)), \cdots, y(\tau_m(s)))}{y(s)} \, ds \right) & t \in [t_1, +\infty) \\[2em] 1 & t \in [t_0, t_1] \end{cases} \qquad (3.2.41)$$

It is easy to see that $FS \subset S$. In fact,

$$\frac{f(s, y(\tau_1(s)), \cdots, y(\tau_m(s)))}{y(s)} = \frac{f(s, y(\tau_1(s)), \cdots, y(\tau_m(s)))}{y(\tau^*(s))} \frac{y(\tau^*(s))}{y(s)}$$

From (3.2.38) and (3.2.40), we observe that every element of FS is also an element of S. Thus $FS \subset S$.

It is obvious that the mapping F is continuous.

The functions of FS are equicontinuous on every $[t_0, T]$. According to the Arzela theorem we conclude that $\overline{FS}$ is a compact subset of S. By the Schauder-Tychonov fixed point theorem there exists a fixed point $y \in S$ such that

$$y = F(y)$$

This y is a nonoscillatory solution of (3.2.32).

COROLLARY 3.2.3 Assume that

$$\limsup_{t \to \infty} \int_t^{\tau^*(t)} \left(\sum_{i=1}^m p_i(s) \right) ds < \frac{1}{e} \tag{3.2.42}$$

Then (3.2.26) has a nonoscillatory solution.

The following theorem gives sufficient conditions concerning the oscillatory behavior of first order retarded differential equations of the type

$$y'(t) + f(t, y(\tau_1, (t)), \cdots, y(\tau_m(t))) = 0 \tag{3.2.43}$$

THEOREM 3.2.8 Assume that $\tau_i \in C[R_+, R]$, $\tau_i(t) < t$ and $\lim_{t \to \infty} \tau_i(t) = \infty$, $i \in I_m$. Suppose further that f satisfies the conditions of Theorem 3.2.6 and

$$\lim_{t \to \infty} \int_{\tau_*(t)}^t a(s) ds > e^{-1} \tag{3.2.44}$$

Then every solution of (3.2.43) oscillates. If f satisfies the conditions of Theorem 3.2.7 and

$$\limsup_{t \to \infty} \int_{\tau_*(t)}^t a(s) ds < \frac{1}{e} \tag{3.2.45}$$

then (3.2.43) has a nonoscillatory solution.

REMARK 3.2.4 Condition (3.2.39) (or (3.2.45)) guarantees that equation (3.2.32) (or (3.2.43)) has a nonoscillatory solution on $[t_0, \infty)$, where t_0 is sufficiently large. If (3.2.39) (or (3.2.45)) is replaced by the condition

$$\int_{t}^{\tau^*(t)} a(s)\, ds \le \frac{1}{e} \qquad t \in [t_0, \infty) \tag{3.2.46}$$

or

$$\int_{\tau_*(t)}^{t} a(s)\, ds \le \frac{1}{e} \qquad t \in [t_0, +\infty) \tag{3.2.47}$$

then the interval of existence of a nonoscillatory solution of (3.2.32) (or (3.2.43)) is $[t_0, +\infty)$.

EXAMPLE 3 2.4 Consider the first order advanced differential equation

$$y'(t) = \frac{2}{(e \ln 2)t}\, y^{1/3}(2t) y(3t) y^{1/3}(4t) \tag{3.2.48}$$

which satisfies condition (3.2.33) and

$$\int_{t}^{2t} a(s)\, ds = \frac{2}{e} > \frac{1}{e}$$

Then all solutions of (3.2.48) oscillate.

3.3 MIXED TYPE DIFFERENTIAL EQUATIONS

Consider two first order differential equations with mixed type deviating arguments:

$$y'(t) = \sum_{i=1}^{n} q_i(t) f_i(y(g_1(t)), \ldots, y(g_m(t))) + F(t, y(t), y(g_1(t)), \ldots, y(g_m(t))) \tag{3.3.1}$$

and

$$y'(t) + \sum_{i=1}^{n} q_i(t) f_i(y(g_1(t)), \ldots, y(g_m(t))) + F(t, y(t), y(g_1(t)), \ldots, y(g_m(t))) = 0 \tag{3.3.2}$$

where the following conditions are assumed to hold:

(a) q_i, $g_j \in C[[a,\infty), R]$, $q_i(t) \ge 0$, and $\lim_{t\to\infty} g_j(t) = \infty$, $i \in I_n$, $j \in I_m$; and there is at least one q_i which is different from zero;

(b) $f_i \in C[R^m, R]$, f_i is nondecreasing with respect to every element, and $u_1 f_i(u_1, \ldots, u_m) > 0$ as $u_1 u_j > 0$, $j \in I_m$;

(c) $F \in C[[a, \infty) \times R^{m+1}, R]$, and $u_0 F(t, u_0, u_1, \ldots, u_m) \geq 0$ for $u_0 u_i > 0$, $i \in I_m$.

THEOREM 3.3.1 Assume that conditions (a)-(c) hold. Further assume that each f_i, $i \in I_n$, satisfies

$$\int_M^\infty \frac{du}{f_i(u, \ldots, u)} < \infty \quad \text{and} \quad \int_{-M}^{-\infty} \frac{du}{f_i(u, \ldots, u)} < \infty \qquad M > 0 \qquad (3.3.3)$$

(This is one kind of superlinear condition on f_i.) Then all solutions of (3.3.1) are oscillatory if

$$\sum_{i=1}^n \int_\Omega q_i(t)\, dt = \infty \qquad (3.3.4)$$

where $\Omega_i = \{t \in [a, \infty) : g_i(t) > t\}$, the advanced part of $g_i(t)$, and $\Omega = \Omega_1 \cap \Omega_2 \ldots \cap \Omega_m$.

Proof: Let $y(t)$ be a nonoscillatory solution which is eventually positive. There is a $T > a$ such that $y(t) > 0$ and $y(g_i(t)) > 0$ for $t \geq T$, $i \in I_m$. By conditions (b) and (c), $f_i(y(g_1(t)), \ldots, y(g_m(t))) > 0$, $i \in I_n$, and $F(t, y(t), y(g_1(t)), \ldots, y(g_m(t))) \geq 0$ on $[T, \infty)$, and so from (a), $y'(t) > 0$ for $t \geq T$.

Let i be fixed. We divide (3.3.1) by $f_i(y(t), \ldots, y(t))$ and integrate it on $[T, T']$, $T' > T$. Using condition (c) and noting that $f_i(y(g_1(t)), \ldots, y(g_m(t))) \geq f_i(y(t), \ldots, y(t))$ for $t \in \Omega \cap [T, T']$, we then have

$$\int_T^{T'} \frac{y'(t)}{f_i(y(t), \ldots, y(t))}\, dt \geq \int_T^{T'} q_i(t) \frac{f_i(y(g_1(t)), \ldots, y(g_m(t)))}{f_i(y(t), \ldots, y(t))}\, dt$$

$$\geq \int_{\Omega \cap [T, T']} q_i(t)\, dt \qquad (3.3.5)$$

Letting $T' \to \infty$ in (3.3.5) and taking (3.3.1) into account, we find

$$\int_{\Omega \cap [T, \infty)} q_i(t)\, dt \leq \int_{y(T)}^{y(\infty)} \frac{du}{f_i(u, \ldots, u)} < \infty$$

Since i is arbitrary, this contradicts (3.3.4). Hence (3.3.1) cannot have eventually positive solutions. Similarly, one can prove that (3.3.1) does not possess eventually negative solutions.

THEOREM 3.3.2 Suppose that the conditions (a)-(c) hold and each f_i, $i \in I_n$ satisfies

$$\int_0^m \frac{du}{f_i(u, \ldots, u)} < \infty \quad \text{and} \quad \int_0^{-m} \frac{du}{f_i(u, \ldots, u)} < \infty \quad \text{for any } m > 0$$

$$(3.3.6)$$

(This is another kind of sublinear condition on f_i.) Then all solutions of (3.3.2) are oscillatory if

$$\sum_{i=1}^n \int_\Delta q_i(t) \, dt = \infty \tag{3.3.7}$$

where $\Delta_i = \{t \in [t_0, \infty) : t_0 \le g_i(t) \le t\}$, the retarded part of $g_i(t)$, and $\Delta = \Delta_1 \cap \Delta_2 \cdots \cap \Delta_m$.

Proof: By following the proof of Theorem 3.3.1 we assume that $y(t)$ is a positive solution of (3.3.2) such that $y(t) > 0$ and $y(g_i(t)) > 0$ for $t \ge t_0 > a$. From (3.3.2), $y'(t) < 0$, $t \ge T$. For any fixed i, and $t \in \Delta \cap [T, T']$, proceeding as in the proof of Theorem 3.3.1, we obtain from (3.3.2),

$$\int_T^{T'} \frac{-y'(t)}{f_i(y(t), \ldots, y(t))} \, dt \ge \int_T^{T'} q_i(t) \, \frac{f_i(y(g_1(t)), \ldots, y(g_m(t)))}{f_i(y(t), \ldots, y(t))} \, dt$$

$$\ge \int_{\Delta \cap [T, T']} q_i(t) \, dt$$

Letting $T' \to \infty$, using condition (3.3.6), we see that

$$\int_{\Delta \cap [T, \infty)} q_i(t) \, dt \le \int_{y(\infty)}^{y(T)} \frac{du}{f_i(u, \ldots, u)} < \infty \qquad i \in I_n$$

which contradicts (3.3.7). The proof is complete.

REMARK 3.3.1 If $g_i(t) > t$, $i \in I_m$ (resp. $g_i(t) < t$, $i \in I_m$), then condition (3.3.4) (resp. (3.3.7)) reduces to

$$\sum_{i=1}^n \int^\infty q_i(t) \, dt = \infty \tag{3.3.8}$$

Next we shall present sufficient conditions for the existence of nonoscillatory solutions of the following type of equations

$$y'(t) = \sum_{i=1}^n q_i(t) f_i(y(g_1(t)), \ldots, y(g_m(t))) \tag{3.3.9}$$

$$y'(t) + \sum_{i=1}^n q_i(t) f_i(y(g_1(t)), \ldots, y(g_m(t))) = 0 \tag{3.3.10}$$

THEOREM 3.3.3 Let conditions (a) and (b) hold. If

$$\sum_{i=1}^{n} \int^{\infty} q_i(t) \, dt < \infty \qquad\qquad (3.3.11)$$

then equations (3.3.9) and (3.3.10) have nonoscillatory solutions.

Proof: First, we prove the part with respect to (3.3.9). For any given constant $k > 0$, consider the integral equation

$$y(t) = k + \sum_{i=1}^{n} \int_{T}^{t} q_i(s) f_i(y(g_1(s)), \cdots, y(g_m(s))) \, ds \qquad (3.3.12)$$

where $T > a$ is chosen so that

$$\sum_{i=1}^{n} f_i(2k, \cdots, 2k) \int_{T}^{\infty} q_i(s) \, ds < k$$

Set $T_0 = \min_{1 \le i \le m} \inf_{t \ge T} g_i(t)$. Let $C[T_0, \infty)$ denote the locally convex space of all continuous functions $y \colon [T_0, \infty) \to R$ with the topology of uniform convergence on compact subintervals of $[T_0, \infty)$. Let $S = \{y \in C \colon k \le y(t) \le 2k, \ t \ge T_0\}$. We observe that S is a closed and convex subset of C. Define the operator $\Phi \colon S \to C$ by

$$\Phi y(t) = k + \sum_{i=1}^{n} \int_{T}^{t} q_i(s) f_i(y(g_1(s)), \cdots, y(g_m(s))) \, ds, \qquad t \ge T \quad (3.3.13)$$

$$\Phi y(t) = k, \qquad T_0 \le t \le T$$

It is easy to see that Φ maps S into S. Also $\{\Phi y\}$, $y \in S$, are uniformly bounded, and $|\Phi y(t_1) - \Phi y(t_2)| \le \Sigma_{1 \le i \le n} C_i \int_{t_1}^{t_2} q_i(s) \, ds$ so $\{\Phi y\}$, $y \in S$ are equicontinuous on any compact subintervals of $[T_0, +\infty)$. Therefore Φ maps S continuously into a compact subset of S. Consequently, by the Schauder-Tychonov fixed point theorem, Φ has a fixed point y in S.

Obviously, this fixed point $y = y(t)$ satisfies (3.3.12) for $t \ge T$. Clearly, $y(t)$ is a nonoscillatory solution of (3.3.9).

Similarly, a nonoscillatory solution of (3.3.10) is obtained as a solution to the integral equation

$$y(t) = 2k - \sum_{i=1}^{n} \int_{T}^{t} q_i(s) f_i(y(g_1(s)), \cdots, y(g_m(s))) \, ds \qquad (3.3.14)$$

By combining the results of Theorems 3.3.1 to 3.3.3, we may state the following results:

THEOREM 3.3.4 Suppose that (3.3.3) holds and that $g_i(t) > t$, $i \in I_m$. Then (3.3.8) is a necessary and sufficient condition for all solutions of (3.3.9) to be oscillatory.

THEOREM 3.3.5 Suppose that conditions (a)–(c), and (3.3.6) hold and that $g_i(t) < t$, $i \in I_m$. Then (3.3.8) is a necessary and sufficient condition for all solutions of (3.3.10) to be oscillatory.

EXAMPLE 3.3.1 Consider

$$y'(t) = \frac{|y(t+\sin t)|^{\alpha_1} \operatorname{sgn} y(t+\sin t)\, |y(t+\cos t)|^{\alpha_2} \operatorname{sgn} y(t+\cos t)}{t^{\beta}[\ln (t+\sin t)]^{\alpha_1}[\ln (t+\cos t)]^{\alpha_2}} \tag{3.3.15}$$

$t > 2\pi$, $\alpha_i > 0$, $\alpha = \alpha_1 + \alpha_2 > 1$, $\beta < 1$ are constants and $\Omega = \bigcup_{k=1}^{\infty} (2k\pi, 2k\pi + \pi/2)$. We note that

$$\int_{\Omega} \frac{dt}{t^{\beta}[\ln (t+\sin t)]^{\alpha_1}[\ln (t+\cos t)]^{\alpha_2}} = \sum_{k=1}^{\infty} \int_{2k\pi}^{2k\pi+\pi/2} \frac{dt}{t^{\beta}[\ln (t+\sin t)]^{\alpha_1}[\ln (t+\cos t)]^{\alpha_2}} = \infty \tag{3.3.16}$$

By Theorem 3.3.4 it follows that all solutions of (3.3.15) are oscillatory. If $\beta = 1$,

$$\int_{\Omega} \frac{dt}{t[\ln (t+\sin t)]^{\alpha_1}[\ln (t+\cos t)]^{\alpha_2}} = \sum_{k=1}^{\infty} \int_{2k\pi}^{2k\pi+\pi/2} \frac{dt}{t[\ln (t+\sin t)]^{\alpha_1}[\ln (t+\cos t)]^{\alpha_2}}$$

$$\leq \sum_{k=1}^{\infty} \int_{2k\pi}^{2k\pi+\pi/2} \frac{dt}{t\ln^{\alpha} t} < \infty$$

By Theorem 3.3.3, (3.3.15) has a nonoscillatory solution. In fact, $y(t) = \ln t$ is an unbounded nonoscillatory solution of (3.3.15).

REMARK 3.3.1 If $0 < \alpha \leq 1$ and $\beta = 1$, (3.3.16) holds, but (3.3.15) has a nonoscillatory solution $y = \ln t$. This example shows that the superlinear condition (3.3.3) is important in Theorem 3.3.4.

EXAMPLE 3.3.2 Consider

$$y'(t) = \frac{1}{2e}\, e^{-t}\, y(t+1)y^2\left(t + \frac{1}{2}\right) \tag{3.3.17}$$

We see that condition (3.3.8) is not satisfied. And so by Theorem 3.3.3 it should have a nonoscillatory solution. In fact, $y(t) = \exp(t/2)$ is such a solution of (3.3.17).

3.4 FUNCTIONAL DIFFERENTIAL EQUATIONS

We consider a scalar functional differential equation of the form

$$y'(t) = F(t, y_t) \tag{3.4.1}$$

Now we will state the meaning of the y_t. Let R be the real line. For any real $\beta > 0$, let $C(\beta)$ be the Banach space of real-valued continuous functions ϕ on $[-\beta, 0]$ with the usual sup norm. For $0 < \rho < \infty$, let $C_\rho(\beta) = \{\phi \in C(\beta) : \|\phi\| < \rho\}$. Let $T > 0$ and suppose y is a continuous function on $[-\beta, T]$. For any $t \in [0, T]$, let y_t denote a member of $C(\beta)$ defined by $y_t(s) = y(t+s)$, that is, y_t is the restriction of y on $[t - \beta, t]$ shifted back to $[-\beta, 0]$.

DEFINITION 3.4.1 Let $F : [0, \infty) \times C_\rho(\beta) \to R$, and $y'(t)$ denote the right-hand derivative of $y(t)$. Then it is said that (3.4.1) is a scalar delay differential equation. For a given initial point $t_0 \geq 0$ and an initial function $\phi_0 \in C_\rho(\beta)$, it is said that $y(t)$ is a solution of (3.4.1) at (t_0, ϕ_0) if there exists an $\epsilon > 0$ such that for $t_0 \leq t < t_0 + \epsilon$, $y_t \in C_\rho(\beta)$, $y(t)$ satisfies (3.4.1), and $y_{t_0} = \phi_0$.

It is well known [98] that if $F(t, \phi)$ is continuous, then (3.4.1) has a solution for each pair (t_0, ϕ_0). Without loss of generality, we shall assume that $t_0 = 0$ in what follows.

DEFINITION 3.4.2 For $\phi \in C(\beta)$, define

$$m(\phi) = \max(0, \sup_{s \in [-\beta, 0]} |\phi(s)|) \tag{3.4.2}$$

Let us first prove the following result.

THEOREM 3.4.1 Suppose that

$$-\alpha m(\psi) \leq F(t, \psi) \leq \alpha m(-\psi) \tag{3.4.3}$$

where $0 \leq \alpha\beta \leq e^{-1}$, $\psi \in C_\rho(\beta)$ and ψ has a constant sign (allowing zeros). Let $y(0, \phi)(t) \equiv y(t)$ be any solution of (3.4.1). Then

(i) $y(0, \phi)(t)$ converges to a finite limit monotonically as $t \to \infty$ and moreover

$$|y(0, \phi)(t)| \geq |\phi(0)| e^{-\alpha \Delta t} \tag{3.4.4}$$

where Δ is defined by $\Delta = \min\{|\lambda_1|, |\lambda_2|\}$, λ_1 and λ_2 are negative real roots of $\lambda e^{\alpha\beta\lambda} + 1 = 0$, $\phi \not\equiv 0$, $\phi \in C_\rho(\beta)$ with a constant sign (allowing zeros) and

$$|\phi(s)| \leq |\phi(0)| e^{-\alpha \Delta s} \qquad s \in [-\beta, 0] \tag{3.4.5}$$

(ii) In addition, suppose the following condition holds:
$\lim_{n \to \infty} F(t_n, \psi_n) \neq 0$, whenever $t_n \to \infty$ and $\psi_n \to \psi \neq 0$. Then
$\lim_{t \to \infty} y(0, \phi)(t) = 0$.

Proof: Let $\phi \in C_\rho(\beta)$ such that $\phi \not\equiv 0$ and

$$| \phi(s) | \leq | \phi(0) | e^{-\alpha \Delta s}$$

Without loss of generality, assume that $\phi(s) \geq 0$ on $[-\beta, 0]$. From the choice
of ϕ, $\phi(0) > 0$. Hence $y(t) > 0$ on $[0, \delta]$ for $0 < \delta < \beta$ and $\|y_t\| < \rho$. This
implies that $y'(t) = F(t, y_t) \leq 0$, and moreover $\|y_t\| \leq \phi(0)e^{-\alpha \Delta(t-\beta)}$.
We consider

$$z'(t) + \alpha z(t - \beta) = 0 \tag{3.4.6}$$

on $(0, \theta)$, where $\theta(s) = \phi(0)e^{-\alpha \Delta s}$ for $s \in [-\beta, 0]$. As in Section 2.1, if
$0 < \alpha \beta e \leq 1$, (3.4.6) has a nonoscillatory solution $z(t) = e^{-\alpha \Delta t}$, where
$\Delta = \min(|\lambda_1|, |\lambda_2|)$, λ_i are the real negative roots of $\lambda e^{\alpha \beta \lambda} + 1 = 0$, and so

$$z(t) = \phi(0)e^{-\alpha \Delta t} \qquad t \geq 0$$

is a solution of (3.4.6).
For $t \in [0, \delta]$

$$y'(t) \geq -\alpha \|y_t\| \geq -\alpha \phi(0)e^{-\alpha \Delta(t-\beta)}$$
$$\geq -\alpha z(t - \beta)$$
$$= z'(t)$$

Hence $y(t) \geq z(t) > 0$. This shows that $y(\delta) > z(\delta) > 0$ and hence, after
repeating the preceding steps a finite number of times, one can conclude
that $y(t) > z(t) > 0$ on $[0, \beta]$.
Define $f(t) = k_1 z(t) - y(t)$ on $[0, \beta]$, where

$$k_1 = \frac{y(\beta)}{z(\beta)} \geq 1$$

Observe that $f(\beta) = 0$. This together with (3.4.1) implies that

$$f'(t) = k_1 z'(t) - y'(t) \leq (k_1 - 1)z'(t) \leq 0$$

Hence $f(t) \geq 0$, on $t \in [0, \beta]$, which implies that

$$k_1 z(t - \beta) \geq y(t - \beta) \qquad \text{for } t \in [\beta, 2\beta]$$

Thus

$$k_1 z'(t) = -k_1 \alpha z(t - \beta) \leq -\alpha y(t - \beta) \tag{3.4.7}$$

Repeating the above procedure for $t \in [\beta, 2\beta]$ such that $y(t)$ is positive (for example as $t = \beta$, $y(\beta) > 0$), we get $y'(t) \leq 0$. Thus, for such a t

$$\|y_t\| = y(t - \beta) \tag{3.4.8}$$

From (3.4.7) and (3.4.8), we have

$$k_1 z'(t) \leq -\alpha \|y_t\| \leq F(t, y_t) = y'(t)$$

Therefore $0 < z(t) \leq k_1 z(t) \leq y(t)$ for $t \in [\beta, 2\beta]$. Proceeding inductively, we can show that

$$k_1, k_2 \cdots k_n z(t) \leq y(t) \quad \text{for } t \in [n\beta, (n+1)\beta]$$

where

$$k_n = \frac{y(n\beta)}{k_1 \cdots k_{n-1} z(n\beta)} \geq 1$$

Moreover, since $y(t)$ is monotonically decreasing for all t, there exists $\lim_{t \to \infty} y(t) = C < \phi(0)$.

Finally suppose that $\lim_{n \to \infty} F(t_n, \psi_n) \neq 0$ as $t_n \to \infty$, $\psi_n \to \psi \neq 0$. Since $y(t) \to C$ as $t \to \infty$, there exists a sequence $\{t_n\}$, as $t_n \to \infty$, $y'(t_n) \to 0$. Thus

$$F(t_n, y_{t_n}) \to 0$$

which implies that $y_{t_n} \to 0$ and hence $y(t_n) \to 0$. Therefore, by monotonicity, $\lim_{t \to \infty} y(t) = 0$. The proof is completed.

COROLLARY 3.4.1 Assume that the hypotheses of Theorem 3.4.1 hold. Then for $\tau \geq 0$,

$$|y(\tau)| \exp(-\alpha\Delta(t - \tau)) \geq (\leq) |y(t)| \qquad t \leq \tau(t \geq \tau) \tag{3.4.9}$$

Proof: Suppose ϕ is a nontrivial initial function such that

$$0 \leq \phi(s) \leq \phi(0) \exp(-\alpha \Delta s)$$

Let $\tau \in [n\beta, (n+1)\beta]$ and

$$k_\tau = \frac{y(\tau)}{k_1 \cdots k_n z(\tau)} \geq 1$$

where $z(t) = \phi(0) \exp(-\alpha\Delta t)$ for $t \geq 0$. As in Theorem 3.4.1, it follows that $f(t) = k_\tau k_1 \cdots k_n z(t) - y(t)$ is nonnegative on $[n\beta, \tau]$, and furthermore,

$$k_\tau k_1 \cdots k_n z(t) \geq (\leq) y(t) \qquad t \leq \tau \ (t \geq \tau) \tag{3.4.10}$$

and

$$y(\tau) = k_\tau k_1 \cdots k_n \, z(\tau) = k_\tau k_1 \cdots k_n \, \phi(0) \exp(-\alpha \, \Delta\tau)$$

Hence

$$k_\tau k_1 \cdots k_n \, z(t) = k_\tau k_1 \cdots k_n \, \phi(0) \exp(-\alpha \, \Delta t)$$

$$= y(\tau) \exp(-\alpha \, \Delta(t - \tau)) \qquad (3.4.11)$$

Combining (3.4.10) and (3.4.11), we obtain (3.4.9). The proof is complete.

EXAMPLE 3.4.1 Consider the equation

$$y'(t) = -a(t, \, y(t)) \, y(t - g(y(t))) \qquad (3.4.12)$$

where $g \in C[R, R_+]$ and $g(0) = 0$, $a \in C[R_+ \times R, R]$ such that $a(t, y) \le \alpha$. Choose $\beta > 0$ such that $\alpha\beta \le e^{-1}$. By continuity, there exists a $\rho > 0$ such that $|y| < \rho$ implies $g(y) < \beta$. Denoting the right-hand side of (3.4.12) by

$$F(t, \, \psi) = -a(t, \, \psi(0)) \, \psi(-g(\psi(0)))$$

We see that $F(t, \, \psi)$ satisfies condition (3.4.3). Thus, if $\phi(t)$ is any constant initial function such that $0 < |\phi(t)| \equiv C < \rho$ on $[-\beta, 0]$, then $y(t) \ne 0$ for $t \ge 0$.

EXAMPLE 3.4.2 Suppose that $a(\cdot, \cdot)$ and $g(\cdot, \cdot)$ are continuous functions such that $a(t, y) \le \alpha$, $0 \le g(t, y) \le \beta$ and $\alpha\beta \le e^{-1}$. If n is any positive odd integer, then solutions of

$$y'(t) = -a(t, \, y(t)) y^n(t - g(t, \, y(t))) \qquad (3.4.13)$$

with constant initial functions $\phi(t)$ of norm less than 1, have no zeros. For the special case $n = 1$, the bound on the norm may be dropped. Note, however, that if $\phi(t) \equiv 2$, the solution of $y'(t) = -y^3(t - e^{-1})$ is $y(t) = -8t + 2$ for $t \in [0, e^{-1}]$. Hence, $y(t)$ has a zero at $t = 1/4 < e^{-1}$.

COROLLARY 3.4.2 Suppose that $F(t, \, \psi)$ is of the form

$$F(t, \, \psi) = -a(t) \, \psi(\sigma(t, \, \psi)) \qquad (3.4.14)$$

where $0 \le a(t) \le \alpha$, $-\beta \le \sigma(t, \, \psi) \le 0$, and $\alpha\beta \le e^{-1}$. If $\phi(t)$ is any nontrivial initial function which has constant sign (allowing zeros) such that

$$|\phi(s)| \le |\phi(0)| \exp(-\alpha \, \Delta s) \qquad (3.4.15)$$

then $\lim_{t \to \infty} y(t) \exp \int_0^t a(s) \, ds$ exists and is finite, that is, solutions of (3.4.1) are asymptotically similar to those of $z'(t) = -a(t) z(t)$.

Proof: Suppose $\phi(t) \ge 0$. For $t \ge \beta$,

$$\frac{d}{dt}\left[y(t)\,\exp\int_0^t a(s)\,ds\right] = \exp\int_0^t a(s)\,ds\,[a(t)y(t) + y'(t)]$$

$$= a(t)\,\exp\int_0^t a(s)\,ds\,[y(t) - y(t - \sigma(t, y_t))] \le 0$$

Therefore

$$\lim_{t\to\infty} y(t)\,\exp\int_0^t a(s)\,ds$$

exists and is finite.

We are now in a position to prove results on oscillation.

THEOREM 3.4.2 Suppose that for any $\psi \in C[-\beta, -\nu]$ with constant sign (allowing zeros)

$$(\text{sgn }\psi)F(t, \psi) \le -\alpha \inf_{s\in[-\beta, -\nu]} \psi(s) \tag{3.4.16}$$

where $0 < \nu \le \beta$. If $\alpha\nu > e^{-1}$, then every solution of (3.4.1) oscillates.

Proof: Suppose that $y(t)$ is a nonoscillatory solution such that $y(t) > 0$ for $t \ge m \ge 0$. For $t \ge m + \beta$, $y_t > 0$, so $y'(t) = F(t, y_t) < 0$. Thus, for $t \ge m + 2\beta$

$$\inf_{s\in[-\beta, -\nu]} y_t(s) = y(t - \nu)$$

Hence

$$y'(t) \le -\alpha y(t - \nu)$$

has a positive solution $y(t)$. This is impossible according to Theorem 2.1.1. The proof is complete.

EXAMPLE 3.4.3 Consider

$$y'(t) + y(t - |y(t)|) = 0 \tag{3.4.17}$$

with initial function $\phi(t)$ such that $\phi(-1) = -1$ and $\phi(0) = 1$. Equation (3.4.17) has a solution $y(t) = t + 1$. This example shows that the delay must be bounded to ensure the oscillation of all solutions of (3.4.1) when $\alpha\nu > e^{-1}$.

REMARK 3.4.1 Condition (3.4.16) can be replaced by

$$(\text{sgn }\psi)F(t, \psi) \le -\alpha(t) \inf_{s\in[-\beta, -\nu]} |\psi(s)|$$

where $0 < \nu \le \beta$, $\alpha(t) > 0$ and

$$\lim_{t \to \infty} \int_{t-\nu}^{t} \alpha(s) \, ds > e^{-1} \tag{3.4.19}$$

and the conclusion of Theorem 3.4.2 remains valid.

Now we consider (3.4.1) with unbounded delay. We rewrite (3.4.1) as

$$y'(t) + F(t, y_t) = 0 \tag{3.4.20}$$

Assume that $g, r \in C[R_+, R]$ and they are monotonically increasing and satisfy

$$g(t) \le r(t) \le t \qquad t \ge 0$$

It is considered that g(t) represents the maximum retardation and r(t) the minimum retardation associated with the delay equation (3.4.20). For each fixed $t > 0$, the symbol y_t denotes a continuous function with domain $(-\infty, 0]$ such that its graph on $[g(t) - t, 0]$ coincides with the graph of $y(t)$ on the interval $[g(t), t]$.

THEOREM 3.4.3 We assume there exists a positive integrable function h and a time $T > 0$ such that for all $t > T$,

$$(\text{sgn } \phi) F(t, \phi) \ge h(t) | \phi(r(t)) | \tag{3.4.21}$$

for any $\phi \in \mu$, where $\mu = \{ \phi \in C(g(t), r(t)) : | \phi(t) |$ is monotone decreasing and $\phi(t)$ has constant sign $\}$. Further assume that for all large t, say $t \ge T$

$$\int_{r(t)}^{t} h(s) \, ds \ge 1 \tag{3.4.22}$$

Then all solutions of (3.4.20) are oscillatory.

Proof: It can be demonstrated that for any $T_0 \ge T$, a zero of y(t) must occur in the interval $(T_0, r^{-1}g^{-1}g^{-1}(T_0)]$. Let $T_1 = g^{-1}(T_0)$, $T_2 = g^{-1}(T_1)$ and $T_3 = r^{-1}(T_2)$. We prove this result by the method of contradiction. Assume that $y(t) > 0$ for all $t \in (T_0, T_3]$ (a similar proof holds for the case when $y(t) < 0$). This assumption implies that for $s \in (T_1, T_3]$, we have $y(s) > 0$ for $s \in (g(s), r(s))$ and hence by (3.4.21) $y'(t) \le 0$ indicating that $y(s)$ is monotone decreasing on $(T_1, T_3]$. Thus, for $s \in (T_2, T_3]$, $y(s)$ is monotone decreasing on the domain $[g(s), r(s)]$. Therefore $s \in (T_2, T_3]$ implies

$$y'(s) = -F(s, y_s) \le -h(s)y(r(s))$$

Integrating the foregoing inequality we have

$$y(t) \le y(T_2) - \int_{T_2}^{t} h(s)y(r(s)) \, ds \tag{3.4.23}$$

Now for $s \in (T_2, T_3]$, $r(s) \leq T_2$ and since $y(t)$ is monotone decreasing on $(T_1, T_3]$, we see that $y(r(s)) \geq y(T_2)$ for $s \in (T_1, T_3]$. Hence

$$y(t) \leq y(T_2) \left[1 - \int_{T_2}^{t} h(s)\, ds\right] \tag{3.4.24}$$

Setting $t = T_3$ in (3.4.24) and considering (3.4.22), we obtain $y(T_3) \leq 0$ in contradiction to the fact that $y(t) > 0$ on $(T_0, T_3]$, and so the theorem is valid.

COROLLARY 3.4.3 Consider

$$y'(t) + \sum_{i=1}^{n} p_i(t) y(g_i(t)) = 0 \tag{3.4.25}$$

where $p_i(t) > 0$ and p_i, $g_i \in C[R_+, R_+]$, $g_i(t) < t$, $i \in I_n$. Then, if

$$\int_{g^*(t)}^{t} \sum_{i=1}^{n} p_i(s)\, ds \geq 1 \tag{3.4.26}$$

for all large t, all solutions of (3.4.25) oscillate.

THEOREM 3.4.4 If under the conditions of Theorem 3.4.3 condition (3.4.22) is replaced by

$$\varliminf_{t \to \infty} \int_{r(t)}^{t} h(s)\, ds > e^{-1} \tag{3.4.27}$$

then all solutions of (3.4.20) oscillate.

Proof: As before, assume that there is a positive solution $y(t)$ and hence

$$y'(t) = -F(t, y_t) \leq -h(t) y(r(t)) \tag{3.4.28}$$

Then by Theorem 2.1.1, we conclude that the above inequality has no positive solution. This contradicts the assumption that $y(t) > 0$.

3.5 POSITIVE SOLUTIONS OF SUPERLINEAR EQUATIONS

We consider

$$y'(t) + p(t) |y(\tau(t))|^{\lambda} \operatorname{sgn} y(\tau(t)) = 0 \qquad t \geq 0 \tag{3.5.1}$$

where $\lambda > 0$ is constant, $p(t) \geq 0$ and $\tau(t)$ are continuous on R_+ and $\tau(t) \leq t$, with $\lim_{t \to \infty} \tau(t) = \infty$.

In the case $\lambda \leq 1$, we have obtained some criteria to guarantee that all solutions of (3.5.1) are oscillatory. Now we shall prove a result concerning the case $\lambda > 1$.

THEOREM 3.5.1 Assume that

$$\int_{\tau(t)}^{t} p(s)\,ds \leq \lambda^{\lambda/(1-\lambda)}\,(\lambda - 1)\int_{0}^{\tau(t)} p(s)\,ds \qquad \lambda > 1 \tag{3.5.2}$$

for all sufficiently large t. Then there exist positive numbers C and t_0 such that (3.5.1) has a solution $y(t)$ on $[t_0, +\infty)$ that satisfies the following inequality:

$$\left[C^{1-\lambda} + (\lambda - 1)\lambda^{\lambda/(\lambda-1)}\int_{t_0}^{t} p(s)\,ds \right]^{1/(1-\lambda)} \leq y(t) \leq \left[C^{1-\lambda} + (\lambda - 1)\int_{t_0}^{t} p(s)\,ds \right]^{1/(1-\lambda)}$$

$$\tag{3.5.3}$$

$t \geq t_0$

Proof: From (3.5.2), and $\tau(t) \leq t$, there exists $t_1 > 0$ and $C > 0$ such that

$$\int_{\tau(t)}^{t} p(s)\,ds \leq \lambda^{\lambda/(1-\lambda)}\left[C^{1-\lambda} + (\lambda - 1)\int_{t_0}^{\tau(t)} p(s)\,ds \right] \qquad t > \nu^*(t_0) \tag{3.5.4}$$

where

$$C^{1-\lambda} \geq \lambda^{\lambda/(\lambda-1)}\int_{0}^{\nu^*(t_0)} p(s)\,ds \tag{3.5.5}$$

$$\nu^*(t) = \sup\{s: \tau(s) < t\} \qquad t_0 = \nu^*(t_1)$$

Let $C[[t_1, +\infty), R]$ be the space of continuous functions with uniform convergence on every finite interval, and S be a subset of C which is defined by

$$S = \left\{ \begin{array}{l} y(t) - C = \text{const} \quad \text{for } t \subset [t_1, t_0), \\[2ex] y \in C[[t_0, +\infty), R]: \left[C^{1-\lambda} + (\lambda - 1)\lambda^{\lambda/(\lambda-1)}\int_{t_0}^{t} p(s)\,ds \right]^{1/(1-\lambda)} \leq y(t) \\[3ex] \qquad \leq \left[C^{1-\lambda} + (\lambda - 1)\int_{t_0}^{t} p(s)\,ds \right]^{1/(1-\lambda)} \quad \text{for } t \geq t_0\,, \\[3ex] \text{and } y(t) \leq y(\tau(t)) \leq \lambda^{1/(\lambda-1)} y(t) \quad \text{for } t \geq t_0 \end{array} \right. \tag{3.5.6}$$

Let

$$y_0(t) = \begin{cases} C & t \in [t_1, t_0] \\[2mm] \left[C^{1-\lambda} + (\lambda - 1) \displaystyle\int_{t_0}^{t} p(s)\, ds \right]^{1/(1-\lambda)} & t \geq t_0 \end{cases} \qquad (3.5.7)$$

It is easy to see that $y_0 \in S$ and S is nonempty closed convex subset of C. Define an operator $T : S \rightarrow C[[t_1, \infty), R]$ by

$$(Ty)(t) = \begin{cases} \left[C^{1-\lambda} + (\lambda - 1) \displaystyle\int_{t_0}^{t} p(s) \left[\dfrac{y(\tau(s))}{y(s)} \right]^{\lambda} ds \right]^{1/(1-\lambda)} & t \geq t_0 \\[4mm] C & t \in [t_1, t_0] \end{cases} \qquad (3.5.8)$$

From (3.5.6) and (3.5.8) we have

$$\left[C^{1-\lambda} + (\lambda - 1)\lambda^{\lambda/(\lambda-1)} \int_{t_0}^{t} p(s)\, ds \right]^{1/(1-\lambda)} \leq (Ty)(t)$$

$$\leq \left[C^{1-\lambda} + (\lambda - 1) \int_{t_0}^{t} p(s)\, ds \right]^{1/(1-\lambda)}$$

$$t \geq t_0$$

Obviously

$$(Ty)(\tau(t)) \geq (Ty)(t) \qquad t \geq t_0$$

$$\frac{(Ty)(\tau(t))}{(Ty)(t)} = \left[\frac{C^{1-\lambda} + (\lambda - 1) \displaystyle\int_{t_0}^{t} p(s) \left[\dfrac{y(\tau(s))}{y(s)} \right]^{\lambda} ds}{C^{1-\lambda} + (\lambda - 1) \displaystyle\int_{t_0}^{\tau(t)} p(s) \left[\dfrac{y(\tau(s))}{y(s)} \right]^{\lambda} ds} \right]^{1/(\lambda-1)}$$

$$\leq \left[1 + \frac{\lambda^{\lambda/(\lambda-1)} (\lambda - 1) \displaystyle\int_{\tau(t)}^{t} p(s)\, ds}{C^{1-\lambda} + (\lambda - 1) \displaystyle\int_{t_0}^{\tau(t)} p(s)\, ds} \right]^{1/(\lambda-1)} \leq \lambda^{1/(\lambda-1)} \qquad t \geq \nu^*(t_0)$$

and

$$\frac{(Ty)(\tau(t))}{(Ty)(t)} \leq \left[1 + \frac{(\lambda - 1) \displaystyle\int_{t_0}^{\nu^*(t_0)} p(s)\, ds}{C^{1-\lambda}} \right]^{1/(\lambda-1)} \leq \lambda^{1/(\lambda-1)} \qquad t \in [t_0, \nu^*(t_0))$$

This implies that

   $TS \subset S$

and it is easy to check that TS is compact. Hence there exists a fixed point $y \in S$ such that

   $(Ty)(t) = y(t) \qquad t \geq t_1$

by the Schauder–Tychonov fixed point theorem. From (3.5.8), this $y(t)$ satisfies (3.5.3). The proof is complete.

We consider the more general form

$$y'(t) + f(t, y(\tau(t))) = 0 \tag{3.5.9}$$

where f is continuous and $\tau(t)$ satisfies the conditions of Theorem 3.5.1. Then we have the following theorem.

THEOREM 3.5.2 Assume that

$$p_1(t)|y|^{\lambda} \leq f(t,y) \, \text{sgn} \, y \leq p_2(t)|y|^{\lambda} \tag{3.5.10}$$

where $\lambda > 1$, $p_i(t) \geq 0$ are continuous functions and for sufficiently large t

$$\int_{\tau(t)}^{t} p_2(s) \, ds \leq \lambda^{\lambda/(1-\lambda)} (\lambda - 1) \int_{0}^{\tau(t)} p_1(s) \, ds \tag{3.5.11}$$

Then there exist positive numbers C and t_0 such that equation (3.5.9) has a solution $y(t)$ on $[t_0, +\infty)$ which satisfies the following inequality

$$\left[C^{1-\lambda} + (\lambda - 1)\lambda^{\lambda/(\lambda-1)} \int_{t_0}^{t} p_2(s) \, ds \right]^{1/(1-\lambda)} \leq y(t)$$

$$\leq \left[C^{1-\lambda} + (\lambda - 1) \int_{t_0}^{t} p_1(s) \, ds \right]^{1/(1-\lambda)}$$

$$t \geq t_0$$

For example, equation (3.1.7) satisfies the conditions of Theorem 3.5.1, so there exists a nonoscillatory solution $y(t)$ satisfying (3.5.3). In fact $y(t) = 1/t$ is such a solution of (3.1.7).

3.6 GENERAL NONLINEAR EQUATIONS

In this section, we shall discuss some oscillation and nonoscillation theorems relative to the following type of equations:

$$y'(t) = \delta f(t, y(t), y(g_1(t)), \ldots, y(g_n(t))) \tag{3.6.1}$$

THEOREM 3.6.1 Assume that $\delta = 1$ or -1, and

(a) $f \in C[R_+ \times R^{n+1}, R]$ and $y_0 f(t, y_0, y_1, \ldots, y_n) \geq 0$, for $y_0 y_i > 0$, $i = 1, 2, \ldots, n$;

(b) $g_i \in C[R_+, R]$, $\lim_{t \to \infty} g_i(t) = \infty$, $i = 1, 2, \ldots, n$;

(c) $|f(t, \bar{y}_0, \bar{y}_1, \ldots, \bar{y}_n)| \leq |f(t, \bar{\bar{y}}_0, \bar{\bar{y}}_1, \ldots, \bar{\bar{y}}_n)|$ whenever $|\bar{y}_i| \leq |\bar{\bar{y}}_i|$, $\bar{y}_i \bar{\bar{y}}_i > 0$, $i = 0, 1, 2, \ldots, n$.

Then (3.6.1) has a nonoscillatory solution y with the property that $\lim_{t \to \infty} y(t) \neq 0$ if and only if there exists a nonzero constant c such that

$$\int^{\infty} |f(t, c, \ldots, c)| \, dt < \infty \tag{3.6.2}$$

Proof: For definiteness, we give a proof when $\delta = 1$. To prove sufficiency, we choose T such that

$$\int_T^{\infty} |f(t, c, \ldots, c)| \, dt \leq \frac{|c|}{2} \tag{3.6.3}$$

Set $T_0 = \min[T, \min\inf_{1 \leq i \leq n, t \geq T} g_i(t)]$. Let $CB[[T_0, +\infty), R)$ be the space of continuous bounded functions on $[T_0, \infty)$ in R. Define

$$S = \left\{ y \in CB : \frac{|c|}{2} \leq y(t) \, \text{sgn} \, c \leq |c|, \, t \geq T_0 \right\}$$

We note that S is a nonempty convex closed subset of $CB[T_0, \infty), R]$.

Consider the operator $\psi : S \to CB$ is defined by

$$\left\{ \begin{array}{ll} (\psi y)(t) = \dfrac{c}{2} + \displaystyle\int_T^t f(s, y(s), y(g_1(s)), \ldots, y(g_n(s))) \, ds & t \geq T \\[1em] (\psi y)(t) = \dfrac{c}{2} & T_0 \leq t \leq T \end{array} \right. \tag{3.6.4}$$

i) ψ maps S into itself.

In fact

$$\frac{|c|}{2} \leq \text{sgn} \, c(\psi y)(t) = \frac{|c|}{2} + \left| \int_T^t f(s, y(s), \ldots, y(g_n(s)) \, ds \right|$$

$$\leq \frac{|c|}{2} + \int_T^t |f(s, c, \ldots, c)| \, ds \leq |c| \qquad t \geq T$$

ii) ψ is continuous.

Let $\{y_n\} \subset S$ be a convergent sequence in CB, that is, $\lim_{n \to \infty} \| y_n - y \|$ = 0 for $y \in$ CB. Because S is a closed set, so $y \in S$.

Noting that

$$| \psi y_m - \psi y | = \left| \int_T^t (f(s, y_m(s), y_m(g_1(s)), \ldots, y_m(g_n(s))) \right.$$
$$\left. - f(s, y(s), y(g_1(s)), \ldots, y(g_n(s)))) \, ds \right|, \qquad t \geq T$$

and setting

$$G_m(s) = | f(s, y_m(s), y_m(g_1(s)), \ldots, y_m(g_n(s)))$$
$$- f(s, y(s), y(g_1(s)), \ldots, y(g_n(s))) |$$

we arrive at the estimate

$$\| \psi y_m - \psi y \| \leq \int_T^\infty G_m(s) \, ds$$

Because of continuity of f, it follows that $\lim_{m \to \infty} G_m(s) = 0$. From assumption (c), we obtain

$$G_m(s) \leq 2 | f(s, c, \ldots, c) |$$

Now, by applying the Lebesgue dominated convergence theorem, we conclude $\lim_{n \to \infty} \| \psi y_m - \psi y \| = 0$. That is, ψ is continuous.

iii) To show ψS is precompact, we observe

$$| (\psi y)(t_2) \quad (\psi y)(t_1) | = \left| \int_{t_1}^{t_2} f(s, y(s), \ldots, y(g_n(s))) \, ds \right|$$
$$\leq \int_{t_1}^{t_2} | f(s, c, \ldots, c) | \, ds$$

From (3.6.2), for any $\epsilon > 0$, there exists a T* such that $\int_{T*}^\infty | f(s, c, \ldots, c) | \, ds < \epsilon$. Therefore, for any $y \in S$, and $t_2 > t_1 > T*$, we have

$$| (\psi y)(t_2) - (\psi y)(t_1) | < \epsilon$$

For $T \leq t_1 < t_2 \leq T^*$, and for any given $\epsilon > 0$, by uniform continuity of f, one can find a $\delta > 0$ such that $|t_1 - t_2| < \delta$ implies

$$| (\psi y)(t_2) - (\psi y)(t_1)| \leq \int_{t_1}^{t_2} |f(s, c, \ldots, c)| \, ds < \epsilon$$

Therefore, $[T_0, +\infty)$ can be divided into finite intervals. On each of these subintervals, the oscillation of functions is less than ϵ. Thus $(\psi y)(t)$ is equicontinuous on $[T_0, +\infty)$. Hence ψS is precompact. By applying the Schauder fixed point theorem, we can conclude that there exists a $y \in S$ such that

$$y = \psi y$$

This y is a solution of (3.6.1). Furthermore, from the definition of ψ, sign of c, and condition (a), one can see that y(t) is a monotone function. Hence $\lim_{t \to \infty} y(t) = \text{const} \neq 0$.

To prove necessity, we assume that there exists a solution y(t) of (3.6.1) with the property that $\lim_{t \to \infty} y(t) \neq 0$, and without a loss in generality, assume that $\lim_{t \to \infty} y(t) = d > 0$. It then follows that there exists a T_1 such that $y(t) > d/2$, $y(g_1(t)) > d/2$, $\ldots$, $y(g_n(t)) > d/2$ for $t \geq T_1 \geq T$. From this and assumption (a), we note that y(t) is nondecreasing and moreover

$$y(t) = y(T_2) + \int_{T_2}^{t} f(s, y(s), \ldots, y(g_n(s))) \, ds$$

$$\geq y(T_2) + \int_{T_2}^{t} f\left(s, \frac{d}{2}, \ldots, \frac{d}{2}\right) ds \quad t \geq T_2$$

Set $T_1 = [\min T_2, \text{mininf}_{1 \leq i \leq n, t \geq T_2} g_i(t)]$. By taking the limit as $t \to \infty$, we conclude that the integral in the left converges. This completes the proof of the theorem.

REMARK 3.6.1 Equation (3.6.1) in Theorem 3.6.1 may be of the advanced type, retarded type, or mixed type.

EXAMPLE 3.6.1 Consider

$$y'(t) + \frac{(t-1)^3}{t^5} y^3(t-1) = 0 \qquad t \geq 2 \tag{3.6.5}$$

Equation (3.6.5) satisfies the conditions of Theorem 3.6.1. In fact, (3.6.5) has the solution $y(t) = 1 + 1/t$. Also,

$$y'(t) = \frac{(t-1)^3}{t^2(t-2)^3} y^3(t-1) \qquad t \geq 3$$

has the solution $y(t) = 1 - 1/t$.

The following result provides sufficient conditions for the oscillatory behavior of (3.6.1) with $\delta = 1$.

THEOREM 3.6.2 Assume that (a) and (b) of Theorem 3.6.1 are satisfied. Furthermore, suppose that there exists a continuous function H on R such that $yH(y) > 0$ for $y \neq 0$,

$$\int_M^\infty \frac{dy}{H(y)} < \infty, \qquad \int_{-M}^{-\infty} \frac{dy}{H(y)} < \infty \qquad \text{for} \ \ M > 0 \tag{3.6.6}$$

for every $\phi \in \Phi = \{\phi \in C[R,R] : |\phi| \text{ is monotone nondecreasing with positive sign}\}$, and

$$\int^\infty \frac{f(s, \phi(s), \phi(g_1(s)), \cdots, \phi(g_n(s)))}{H(\phi(s))} \, ds = \infty \tag{3.6.7}$$

holds. Then all solutions of (3.6.1) oscillate.

Proof: Without loss of generality, assume that $y(t) > 0$ and it is a solution of (3.6.1). Then there exists a T such that $y(t) > 0$, $y(g_i(t)) > 0$, $i \in I_n$, for $t \geq T$ and

$$\int_{y(T)}^{y(\infty)} \frac{dy}{H(y)} = \int_T^\infty \frac{f(s, y(s), y(g_1(s)), \cdots, y(g_n(s)))}{H(y(s))} \, ds$$

These relations give a contradiction to the assumption (3.6.6).

Now we consider a special case of (3.6.1), namely,

$$y'(t) = \sum_{i=1}^n q_i(t) f_i(y(g_i(t))) \tag{3.6.8}$$

COROLLARY 3.6.1 Assume that $g_i \in C(R_+, R)$, $\lim_{t \to \infty} g_i(t) = \infty$, $f_i \in C(R,R)$ is nondecreasing; $yf_i(y) > 0$ as $y \neq 0$, $q_i(t) \geq 0$, $i \in I_n$. Then

$$\sum_{i=1}^n \int^\infty q_i(s) \, ds < \infty \tag{3.6.9}$$

is a necessary and sufficient condition for (3.6.8) to have a bounded nonoscillatory solution.

3.7 NONLINEAR EQUATIONS WITH FORCING TERMS

In this section, we present oscillation and nonoscillation results for
nonhomogeneous differential equations with deviating arguments.

We consider

$$y'(t) + \sum_{i=1}^{n} p_i(t) f(y(g_i(t))) = q(t) y(t) + r(t) \qquad (3.7.1)$$

THEOREM 3.7.1 Assume that

(1) p_i, $r \in C[R_+, R]$, $p_i(t) \geq 0$, $i \in I_n$;

(2) $g_i \in C^1[R_+, R]$, $g_i(t) < t$, $\lim_{t \to \infty} g_i(t) = +\infty$, $g_i'(t) \geq 0$, $i \in I_n$;

(3) $q \in C[R_+, R]$;

(4) $f \in C[R, R]$, $yf(y) > 0$ for $y \neq 0$, $f(y)$ is nondecreasing and $f(xy) = f(x)f(y)$, for $x, y \in R$;

(5) $\limsup_{t \to \infty} \Sigma_{i=1}^{n} \int_{g^*(t)}^{t} p_i(s) f\left(\exp\left(\int_{C^*}^{g_i(s)} q(u)\, du\right)\right) \exp\left(-\int_{C^*}^{s} q(u)\, du\right) ds$

   $> M$, where $g^*(t) = \max_{i \in I_n} g_i(t)$, C^* is a constant, $M = \lim_{y \to 0}[y/f(y)]$;

(6) There exists a function $Q \in C^1[R_+, R]$ such that

$$Q'(t) = r(t) \exp\left(-\int_{C^*}^{t} q(u)\, du\right) \qquad t \geq 0$$

Then

(i) $\lim_{t \to \infty} Q(t) = 0$ implies that every solution $y(t)$ of (3.7.1) is either oscillatory or

$$\lim_{t \to \infty} y(t) \exp\left(-\int_{C^*}^{t} q(s)\, ds\right) = 0 \qquad (3.7.2)$$

(ii) If there exist constants q_1, q_2 and sequences $\{t_m'\}$, $\{t_m''\}$, such that

$\lim_{m \to \infty} t_m' = \lim_{m \to \infty} t_m'' = \infty$ and $Q(t_m') = q_1$, $Q(t_m'') = q_2$, $q_1 \leq q(t)$

$\leq q_2$, $t \geq 0$, then every solution $y(t)$ of (3.7.1) is oscillatory or such that

$$\lim_{t \to \infty} \left[y(t) \exp\left(-\int_{C^*}^{t} q(s)\, ds\right) - Q(t)\right] = -q_1 \text{ or } -q_2 \qquad (3.7.3)$$

Proof: Set $z(t) = y(t) \exp\left(-\int_{C*}^{t} q(s)\, ds\right)$, then using the assumptions (1)–(4), (3.7.1) becomes

$$z'(t) + \sum_{i=1}^{n} L_i(t) f(z(g_i(t))) = Q(t) \tag{3.7.4}$$

where

$$L_i(t) = p_i(t) f\left(\exp\left(\int_{C*}^{g_i(t)} q(s)\, ds\right)\right) \exp\left(-\int_{C*}^{t} q(s)\, ds\right)$$

and

$$Q(t) = r(t) \exp\left(-\int_{C*}^{t} q(s)\, ds\right)$$

We may suppose that $y(t)$ is a nonoscillatory solution of (3.7.1) and $y(t)$ is positive for sufficiently large t. In this case, $z(t)$ is also a nonoscillatory solution of (3.7.4) and $z(t)$ is positive for sufficiently large t. Set $\bar{y}(t) = z(t) - Q(t)$. Then $\bar{y}(t)$ satisfies

$$\bar{y}'(t) + \sum_{i=1}^{n} L_i(t) f(\bar{y}(g_i(t)) + Q(g_i(t))) = 0 \tag{3.7.5}$$

From (3.7.5), because of (1), (2), (4), and the fact that $z(t) = \bar{y}(t) + Q(t) > 0$, we see $\bar{y}'(t) < 0$, so that we have $\lim_{t \to \infty} \bar{y}(t) = c$, where c is a constant.

Suppose that the case (i) holds. If $c < 0$, then we get the contradiction that $z(t) < 0$ for sufficiently large t. If $c > 0$, then we obtain

$$z(g_i(t)) = \bar{y}(g_i(t)) + Q(g_i(t)) \geq \frac{c}{2} \qquad i \in I_n \tag{3.7.6}$$

for sufficiently large t. From (3.7.5), it follows that

$$\bar{y}'(t) + \sum_{i=1}^{n} L_i(t) f\left(\frac{c}{2}\right) \leq 0 \tag{3.7.7}$$

Integrating (3.7.7) from $g*(t)$ to t, we have

$$\bar{y}(t) - \bar{y}(g*(t)) + \left(\sum_{i=1}^{n} \int_{g*(t)}^{t} L_i(s)\, ds\right) f\left(\frac{c}{2}\right) \leq 0 \tag{3.7.8}$$

By taking the limsup of (3.7.8), as $t \to \infty$, we get a contradiction to (5). Hence we conclude that $c = 0$. From this, we see that

$$\lim_{t \to \infty} z(t) = \lim_{t \to \infty} y(t) \exp\left(-\int_{C^*}^{t} q(s)\, ds\right) = 0$$

Suppose that the case (ii) holds; put $s(t) = \bar{y}(t) + q_1$. Then we have

$$\lim_{t \to \infty} s(t) = \lim_{t \to \infty} (\bar{y}(t) + q_1) = c + q_1 \equiv d \quad (-\infty < d < \infty) \qquad (3.7.9)$$

If $d < 0$ in (3.7.9), then $\bar{y}(t) + q_1 < 0$, for sufficiently large t, say $t \geq t_1$. This leads to a contradiction to the fact that

$$\bar{y}(t'_s) + q_1 = \bar{y}(t'_s) + Q(t'_s) = z(t'_s) > 0 \qquad\qquad \text{for } t'_s > t_1$$

If d is positive, then

$$z(t) = \bar{y}(t) + Q(t) \geq \bar{y}(t) + q_1 = s(t) > \frac{d}{2} \qquad (3.7.10)$$

for sufficiently large t, say $t \geq t_2$. By using (3.7.5) and (3.7.10), we obtain, setting $s(t) = \bar{y}(t) + q_1$, that

$$s'(t) + \sum_{i=1}^{n} L_i(t) f(s(g_i(t))) \leq 0 \qquad (3.7.11)$$

has a positive solution. By the application of Theorem 3.1.1, this is a contradiction. The proof is complete.

REMARK 3.7.1   Condition (5) in Theorem 3.7.1 can be replaced by the condition

$$(5')\ \liminf_{t \to \infty} \sum_{i=1}^{n} \int_{g^*(t)}^{t} -p_i(s) f\left(\exp\left(\int_{C^*}^{g_i(s)} q(u)\, du\right) \exp\left(-\int_{C^*}^{s} q(u)\, du\right)\right) ds > \frac{M}{e}$$

This conclusion results from Section 3.1.

3.8   EQUATIONS WITH DISTRIBUTED TYPE
DEVIATING ARGUMENTS

We consider first order differential equations with distributed type deviating arguments of the form

$$y'(t) = \int_{\alpha(t)}^{\beta(t)} f(y(t + s))\, d\eta(t, s) \qquad (3.8.1)$$

$$y'(t) + \int_{-\beta(t)}^{-\alpha(t)} f(y(t + s))\, d\eta(t, s) = 0 \qquad (3.8.2)$$

Assume that $\eta(t, s)$ satisfies the conditions of Section 2.9 for existence of solutions.

THEOREM 3.8.1 Assume that

(i) $\eta(t, s)$ is nondecreasing in s for fixed t and nonnegative in t for fixed s;

(ii) $\beta(t) > \alpha(t) > 0$ are continuous;

(iii) $f \in C[R, R]$, $uf(u) > 0$ for $u \neq 0$, and $\lim_{u \to \infty} \dfrac{u}{f(u)} = M$;

(iv) $\displaystyle \lim_{t \to \infty} \int_t^{t+\alpha(t)} \left[\int_{\alpha(u)}^{\beta(u)} d\eta(u, s) \right] du > \dfrac{M}{e}$.

Then all solutions of (3.8.1) are oscillatory.

Proof: Otherwise, without loss of generality, assume that (3.8.1) has a nonoscillatory solution $y(t) > 0$ for $t \geq t_0$. From (i) and (iii), it follows that $y'(t) \geq 0$. Dividing both sides of (3.8.1) by $y(t)$ and integrating from t to $t + \alpha(t)$, we obtain

$$\ln \frac{y(t + \alpha(t))}{y(t)} = \int_t^{t+\alpha(t)} \left[\int_{\alpha(u)}^{\beta(u)} \frac{f(y(u + s))}{y(u)} d\eta(u, s) \right] du \qquad (3.8.3)$$

Let $\lim_{t \to \infty} y(t) = \ell$.

(a) If ℓ is finite, then $\lim_{t \to \infty} (f(y(t+s)))/y(t) = f(\ell)/\ell > 0$, for every $s > 0$, so there exists some $t_1 \geq t_0$ such that

$$\ln \frac{y(t + \alpha(t))}{y(t)} \geq \frac{1}{2} \frac{f(\ell)}{\ell} \int_t^{t+\alpha(t)} \left[\int_{\alpha(u)}^{\beta(u)} d\eta(u, s) \right] du \qquad t \geq t_1 \qquad (3.8.4)$$

Taking limits on both sides, as $t \to \infty$, we get

$$\lim_{t \to \infty} \int_t^{t+\alpha(t)} \left[\int_{\alpha(u)}^{\beta(u)} d\eta(u, s) \right] du = 0 \qquad (3.8.5)$$

which contradicts the hypothesis (iv).

(b) In the case of $\ell = \infty$, we set $w(t) = y(t + \alpha(t))/y(t) \geq 1$ and $\underline{\lim}_{t \to \infty} w(t) = d$. Here, there are two cases, namely, d is finite and $d = +\infty$.

First consider the case of d being a finite number. There exists a sequence $\{t_n\}$ so that $\lim_{n \to \infty} w(t_n) = d$. From (3.8.3),

$$\ln w(t_n) \geq w(\xi_n) \int_{t_n}^{t_n + \alpha(t_n)} \left[\int_{\alpha(u)}^{\beta(u)} \frac{f(y(u + s))}{y(u + s)} \, d\eta(u, s) \right] du$$

where $t_n \leq \xi_n \leq t_n + \alpha(t_n)$.

Taking limits on both sides of the above inequality we obtain

$$\frac{\ln d}{d} \geq \lim_{t \to \infty} \int_{\alpha}^{t + \alpha(t)} \left[\int_{\alpha(u)}^{\beta(u)} \frac{f(y(u + s))}{y(u + s)} \, d\eta(u, s) \right] du$$

Because of $\max_{d \geq 1} \ln d/d = 1/e$, and (iii), it follows that

$$\lim_{t \to \infty} \int_{\alpha}^{t + \alpha(t)} \left[\int_{\alpha(u)}^{\beta(u)} d\eta(u, s) \right] du \leq \frac{M}{e}$$

which contradicts the hypothesis (iv).

In the case of $d = +\infty$, there exists a sequence $\{t_n\}$ such that $\lim_{n \to \infty} w(t_n) = +\infty$ and $w(t) \geq w(t_n)$ for all $t \geq t_n$, so

$$\ln w(t_n) \geq w(t_n) \int_{t_n}^{t_n + \alpha(t_n)} \left[\int_{\alpha(u)}^{\beta(u)} \frac{f(y(u + s))}{y(u + s)} \, d\eta(u, s) \right] du$$

or

$$\frac{\ln w(t_n)}{w(t_n)} \geq \int_{t_n}^{t_n + \alpha(t_n)} \left[\int_{\alpha(u)}^{(u)} \frac{f(y(u + s))}{y(u + s)} \, d\eta(u, s) \right] du$$

Letting $n \to \infty$, the left hand side of the above inequality becomes zero because there exists limit $\lim_{n \to \infty} w(t_n) = +\infty$. Therefore, we have

$$\lim_{t \to \infty} \int_{t}^{t + \alpha(t)} \left[\int_{\alpha(u)}^{\beta(u)} d\eta(u, s) \right] du = 0$$

The proof is complete.

Similarly, we can present the following theorem whose proof can be formulated analogously to the proof of the previous theorem.

THEOREM 3.8.2 Assume that (i), (ii), and (iii) of Theorem 3.8.1 hold, but

$$\lim_{u \to 0} \frac{u}{f(u)} = M > 0 \tag{3.8.6}$$

$t - \beta(t) \to \infty$ as $t \to \infty$, and

$$\lim_{t \to \infty} \int_{t-\alpha(t)}^{t} \left[\int_{-\beta(u)}^{-\alpha(u)} d\eta(u, s) \right] du > \frac{M}{e} \tag{3.8.7}$$

Then all solutions of (3.8.2) are oscillatory.

COROLLARY 3.8.1 Consider the equation

$$y'(t) = \sum_{i=1}^{n} a_i(t)f(y(t + \tau_i(t))) \tag{3.8.8}$$

where $a_i \in C[R_+, R_+]$, $\tau_i \in C[R_+, R_+]$, $i \in I_n$, f satisfies condition (iii) of Theorem 3.8.1, and

$$\lim_{t \to \infty} \int_{t}^{t+\tau_*(t)} \left(\sum_{i=1}^{n} a_i(s) \right) ds > \frac{M}{e} \tag{3.8.9}$$

where $\tau_*(t) = \min(\tau_1(t), \cdots, \tau_n(t))$. Then every solution of (3.8.8) oscillates.

Proof: Let

$$\eta(t, s) = U(s - \tau_1(t))a_1(t) + \cdots + U(s - \tau_n(t))a_n(t) \tag{3.8.10}$$

where U(s) is a unit step function. It is obvious that this η satisfies the conditions that were outlined earlier. From this, (3.8.8) reduces to

$$y'(t) = \int_{\tau_*(t)}^{\tau^*(t)} f(y(t + s)) \, d\eta(t, s) \tag{3.8.11}$$

where $\tau^*(t) = \max(\tau_1(t), \cdots, \tau_n(t))$. By applying Theorem 3.8.1 for (3.8.11), the conclusion of Corollary 3.8.1 is established.

Similarly, we have the following corollary.

COROLLARY 3.8.2 Consider

$$y'(t) + \sum_{i=1}^{n} a_i(t)f(y(t + \tau_i(t))) = 0 \tag{3.8.12}$$

where $a_i \in C[R_+, R_+]$, $\tau_i(t) < 0$ are continuous, $i \in I_n$, the function f satisfies the conditions of Theorem 3.8.2, $\lim_{t \to \infty}(t - \tau(t)) = \infty$ and

$$\lim_{t \to \infty} \int_{t-\tau(t)}^{t} \left(\sum_{i=1}^{n} a_i(s) \right) ds > \frac{M}{e} \tag{3.8.13}$$

where $\tau(t) = \min(-\tau_1(t), \ldots, -\tau_n(t))$. Then every solution of (3.8.12) oscillates.

We note that (3.8.8) and (3.8.12) include linear equations with advanced and delayed arguments, respectively.

REMARK 3.8.1 The above results can be extended to more general functional differential equations of the type

$$y'(t) = \sum_{i=1}^{n} \int_{\alpha_i(t)}^{\beta_i(t)} f_i(y(t+s)) \, d\eta^i(t, s)$$

and

$$y'(t) + \sum_{i=1}^{n} \int_{-\beta_i(t)}^{-\alpha_i(t)} f_i(y(t+s)) \, d\eta^i(t, s) = 0$$

where f_i and $\eta^i(t, s)$ satisfy conditions of Theorems 3.8.1 and 3.8.2 respectively.

3.9 EQUATIONS WITH DISTRIBUTED TYPE
     DEVIATING ARGUMENTS (CONTINUED)

Let us consider the equation with a distributed type deviating argument

$$y'(t) = \int_{\alpha(t)}^{\beta(t)} f(t, y(t+s)) \, d\eta(t, s) \qquad (3.9.1)$$

where $\beta(t) > \alpha(t) \geq 0$ are continuous, and f and $\eta(t, s)$ satisfy certain regularity conditions to insure the existence of solutions.

DEFINITION 3.9.1 The function $f(t, y)$ which is continuous for $|y| < \infty$, $t \geq a$ and $yf(t, y) > 0$ for $y \neq 0$, $t \geq a$. (3.9.1) is said to be strongly superlinear if there exists a number $\sigma > 1$ such that for each fixed t, $f(t, y)/|y|^\sigma$ sgn y is nondecreasing in y for $y > 0$ and nonincreasing in y for $y < 0$.

THEOREM 3.9.1 Assume that

(i) $\eta(t, s)$ is nondecreasing in s for fixed t and nonnegative in t for fixed s.

(ii) $f(t, z)$ is strongly superlinear. $\alpha, \beta \in C[R_+, R_+]$ and $\beta(t) > \alpha(t)$ on R_+. Then

$$\int^{\infty} f(t, c) \int_{\alpha(t)}^{\beta(t)} d\eta(t, s)\, dt = \infty \cdot \operatorname{sgn} c, \qquad \text{for all } c \neq 0 \qquad (3.9.2)$$

is a necessary and sufficient condition for (3.9.1) to be oscillatory.

Proof: Let $y(t)$ be a nonoscillatory solution of (3.9.1). Suppose that $y(t) > 0$ for $t \geq t_1$. From (3.9.1), hypotheses on $f(t, z)$ and $\eta(t, s)$, we have that $y'(t) \geq 0$.

Dividing (3.9.1) by $y^{\sigma}(t)$ and integrating from $\alpha(t)$ to $\beta(t)$, we obtain

$$\frac{y'(t)}{y^{\sigma}(t)} = \int_{\alpha(t)}^{\beta(t)} \frac{f(t, y(t + s))}{y^{\sigma}(t)}\, d\eta(t, s)$$

$$\geq \int_{\alpha(t)}^{\beta(t)} \frac{f(t, y(t + s))}{y^{\sigma}(t + s)}\, d\eta(t, s)$$

$$\geq \frac{f(t, y(t + \alpha(t)))}{y^{\sigma}(t + \alpha(t))} \int_{\alpha(t)}^{\beta(t)} d\eta(t, s)$$

$$\geq \frac{1}{c^{\sigma}} f(t, c) \int_{\alpha(t)}^{\beta(t)} d\eta(t, s)$$

where $c = z(t_1) > 0$. By integrating the above inequality, we obtain

$$\int_{y(t_1)}^{y(\infty)} \frac{dy}{y^{\sigma}} \geq \frac{1}{c^{\sigma}} \int_{t_1}^{\infty} f(t, c) \int_{\alpha(t)}^{\beta(t)} d\eta(t, s)\, dt$$

which implies

$$\int^{\infty} f(t, c) \int_{\alpha(t)}^{\beta(t)} d\eta(t, s)\, dt < \infty$$

This contradicts (3.9.2)

A parallel argument holds if we assume that $y(t) < 0$ for $t \geq t_1$.

To prove necessity, suppose that

$$\int^{\infty} f(t, d) \int_{\alpha(t)}^{\beta(t)} d\eta(t, s)\, dt < \infty \qquad (3.9.3)$$

Without loss of generality, we assume that $d > 0$. The proof is based on a fixed point theorem of Chapter 1.

Let Y be the set of all nondecreasing functions $y(t)$ defined on $[T, \infty)$, and such that $d/2 \leq y(t) \leq d$ for every $t \geq T$. The set Y is considered endowed with the usual pointwise ordering $\leq$, that is, $y_1 \leq y_2 \Leftrightarrow y_1(t) \leq y_2(t)$

for every $t \geq T$. It is obvious that for every $A \subseteq Y$, sup A belongs to Y. It is enough to verify that sup A is nondecreasing on $[T, \infty)$. Assume that this is false. This implies that there exists $t_1 < t_2$ such that sup $A(t_1) >$ sup $A(t_2)$. Let $h =$ sup $A(t_1) -$ sup $A(t_2) > 0$. For any given $\epsilon > 0$, there exist $\bar{y} \in A$ such that $\bar{y}(t_1) \geq$ sup $A(t_1) - \epsilon$. On the other hand, $\bar{y}(t_2) \leq$ sup $A(t_2)$. Taking $\epsilon = h/2$, we see that $\bar{y}(t_1) \geq$ sup $A(t_1) - h/2 >$ sup $A(t_1) - h =$ sup $A(t_2) \geq \bar{y}(t_2)$, which is a contradiction, because $\bar{y} \in A$. Therefore, every $A \subseteq \bar{Y}$ has a least upper bound in Y.

We also consider the mapping F defined as follows:

$$y(t) = (Fy)(t) = \frac{d}{2} + \int_T^t \left[\int_{\alpha(u)}^{\beta(u)} f(u,\ y(u+s))\ d\eta\,(u,\,s) \right] du$$

The above integral is well defined on Y. This is because of the fact that $f(u,\ y(u+s))$ is nondecreasing in s and η is a function of bounded variation with respect to s. Choose $T_1 \geq T$ so large that

$$\int_{T_1}^{\infty} \left[f(u,\ d) \int_{\alpha(u)}^{\beta(u)} d\eta\,(u,\,s) \right] du < \frac{d}{2}$$

Thus $FY \subseteq Y$. Moreover F is obviously nondecreasing (with respect to the order of y). Consequently, by the fixed point theorem, there exists a $y \in Y$ such that $Fy = y$. The integral in the definition of the map F is continuous with respect to t and consequently y itself is continuous. It is obvious now that y is a nonoscillary solution of (3.9.1). The proof is complete.

COROLLARY 3.9.1 Equation (3.9.1) has a bounded nonoscillatory solution if and only if

$$\int^{\infty} f(t,\ c)\ \mathrm{sgn}\ c \int_{\alpha(t)}^{\beta(t)} d\eta\,(t,\,s)\ dt < \infty \qquad \text{for some } c \neq 0 \qquad (3.9.4)$$

Because $\alpha \geq 0$, (3.9.1) in Theorem 3.9.1 is of the advanced type. Now, we consider an equation with a mixed type of deviating argument

$$y'(t) = \int_{\alpha(t)}^{\beta(t)} f(t,\ y(t+s))\ d\eta\,(t,\,s) \qquad (3.9.5)$$

where $\beta(t) > \alpha(t)$; $\alpha(t)$ is permitted to be less than zero. We simply present a result similar to Theorem 3.9.1.

THEOREM 3.9.2 Assume that (i) and (ii) of Theorem 3.9.1 hold, and

$$\int^{\infty} f(t,\ c) \int_{\alpha^*(t)}^{\beta(t)} d\eta\,(t,\,s)\ dt = \infty \cdot \mathrm{sgn}\ c \qquad \text{for all } c \neq 0 \qquad (3.9.6)$$

where $\alpha^*(t) = \max(\alpha(t), 0)$. Then every solution of (3.9.5) is oscillatory.

We consider a special case

$$y'(t) = f(t, y(t + \tau(t))) \tag{3.9.7}$$

COROLLARY 3.9.2 Assume that $f(t, y)$ satisfies condition (ii) of Theorem 3.9.1 and $\tau(t) > 0$ is continuous. Then a necessary and sufficient condition for (3.9.7) to be oscillatory is that

$$\int^\infty f(t, c)\, dt = \infty \cdot \operatorname{sgn} c \quad \text{for all} \ c \neq 0 \tag{3.9.8}$$

Furthermore, (3.9.7) has a bounded nonoscillatory solution, if and only if

$$\int^\infty f(t, c)\, \operatorname{sgn} c\, dt < \infty \quad \text{for some} \ c \neq 0 \tag{3.9.9}$$

Proof: Set $\eta(t, s) = U(s - \tau(t))$, Then (3.9.1) reduces to (3.9.7). The conclusion of the corollary follows from Theorem 3.9.1.

Now we consider a retarded type equation

$$y'(t) + \int_{\alpha(t)}^{\beta(t)} f(t, y(t + s))\, d\eta(t, s) = 0 \tag{3.9.10}$$

where $\eta(t, s)$ and $f(t, y)$ are as defined before.

DEFINITION 3.9.2 Let $f \in C[R_+ \times R, R]$ and $yf(t, y) > 0$ as $y \neq 0$. Then $f(t, y)$ is said to be strongly sublinear if there exists a number $0 < \gamma < 1$ such that, for each t, $f(t, y)/|y|^\nu \operatorname{sgn} y$ is nonincreasing in y for $y > 0$ and nondecreasing in y for $y < 0$.

THEOREM 3.9.3 Assume that $\eta(t, s)$ satisfies the conditions of Theorem 3.9.1 and $f(t, y)$ is strongly sublinear. Then a necessary and sufficient condition for (3.9.10) to be oscillatory is that

$$\int^\infty f(t, c) \int_{\alpha(t)}^{\beta(t)} d\eta(t, s)\, dt = \infty \cdot \operatorname{sgn} c, \quad \text{for all} \ c \neq 0 \tag{3.9.11}$$

Proof: Let $y(t)$ be a nonoscillatory solution of (3.9.10). Suppose that $y(t) > 0$, and $y(t^*) > 0$, $t^* = \inf_{t \geq t_1}(t + \alpha(t))$, for $t \geq t_1$. It is obvious that $y'(t) \leq 0$. Dividing (3.9.10) by $y^\nu(t)$, we have

$$-\frac{y'(t)}{y^\nu(t)} = \int_{\alpha(t)}^{\beta(t)} \frac{f(t, y(t + s))}{y^\nu(t)}\, d\eta(t, s)$$

$$\geq \int_{\alpha(t)}^{\beta(t)} \frac{f(t,\ y(t+s))}{y^{\nu}(t+s)}\, d\eta(t,s)$$

$$\geq \frac{1}{c^{\nu}} \int_{\alpha(t)}^{\beta(t)} f(t,\ c)\, d\eta(t,s) \tag{3.9.12}$$

where $c = y(t^*) > 0$.

Integrating (3.9.12) from t_1 to ∞, we obtain

$$\int_{y(\infty)}^{y(t_1)} \frac{dy}{c^{\nu}} \geq \frac{1}{c^{\nu}} \int_{t_1}^{\infty} f(t,\ c) \int_{\alpha(t)}^{\beta(t)} d\eta(t,s)\, dt \tag{3.9.13}$$

We note that the integral in the left-hand side is convergent. Hence

$$\int^{\infty} f(t,\ c) \int_{\alpha(t)}^{\beta(t)} d\eta(t,s)\, dt < \infty \tag{3.9.14}$$

which contradicts (3.9.11). A similar argument holds if we assume that $y(t) < 0$ for $t \geq t_1$.

To prove necessity, suppose that

$$\int^{\infty} f\left(t,\ \frac{d}{2}\right) \int_{\alpha(t)}^{\beta(t)} d\eta(t,s)\, dt < \infty \tag{3.9.15}$$

for $d \neq 0$. By following the proof of Theorem 3.9.1 the rest of the proof can be formulated. We omit the details.

COROLLARY 3.9.3   Under the assumptions of Theorem 3.9.3, equation (3.9.10) has a bounded nonoscillatory solution if and only if

$$\int^{\infty} \left[f(t,\ c)\ \mathrm{sgn}\ c \int_{\alpha(t)}^{\beta(t)} d\eta(t,s) \right] dt < \infty \quad \text{for some } c \neq 0 \tag{3.9.16}$$

For the special case

$$y'(t) + f(t,\ y(t - \tau(t))) = 0 \tag{3.9.17}$$

we have the following result.

COROLLARY 3.9.4   Assume that $f(t,\ y)$ is strongly sublinear and $\tau(t) > 0$. Then (3.9.8) is a necessary and sufficient condition for (3.9.17) to be oscillatory. Furthermore, (3.9.17) has a bounded nonoscillatory solution, if and only if (3.9.9) holds.

Similar to Theorem 3.9.2, one can obtain a sufficient condition for oscillation of the mixed type equation

$$y'(t) + \int_{\alpha(t)}^{\beta(t)} f(t,\ y(t+s))\ d\eta(t,\ s) = 0 \tag{3.9.18}$$

Furthermore, the preceding discussion can be extended to the equation

$$y'(t) = \sum_{i=1}^{n} \int_{\alpha(t)}^{\beta(t)} f_i(t,\ y(t+s))\ d\eta^i(t,\ s) \tag{3.9.19}$$

3.10 OSCILLATION ON BOTH SIDES

Most of the work in the field of oscillation theory of ODEWDA discusses the case that the deviating argument $\tau(t)$ tends to $+\infty$ as $t \to \infty$. However, oscillation in both directions is also interesting.

EXAMPLE 3.10.1 Consider the equation

$$y'(t) + y\left(\frac{\pi}{2} - t\right) = 0 \qquad t \in R \tag{3.10.1}$$

which has a solution $y(t) = \cos t$ that is oscillatory in both directions.

The following theorem gives a sufficient condition for oscillation in both directions of a nonlinear differential equation with deviating argument of the form

$$y'(t) = f(t,\ y(t),\ y(\tau(t))) \qquad t \in R \tag{3.10.2}$$

where $f: R^3 \to R$, $\tau: R \to R$,

$$\tau(t) \to -\infty \qquad \text{as } t \to \infty$$

and $\tag{3.10.3}$

$$\tau(t) \to +\infty \qquad \text{as } t \to -\infty$$

DEFINITION 3.10.1 A function $y(t)$ is said to be a solution of (3.10.2) if it is defined on R and such that it satisfies the equation (3.10.2).

DEFINITION 3.10.2 A solution $y(t)$ of (3.10.2) is said to be oscillatory in both directions if there exist sequences $\{t_n\}$ and $\{t_n'\}$ in R such that $t_n \to \infty$, $t_n' \to -\infty$ as $n \to \infty$, and $y(t_n) = y(t_n') = 0$ for $n = 1,\ 2,\ \ldots$.

As before, we restrict our discussion to those solutions $y(t)$ of equation (3.10.2) which are not eventually identically equal to zero on any intervals $[T,\infty)$ and $(-\infty, T]$, T being any real number.

Next, consider the more general equation

$$y'(t) = \sum_{i=1}^{m} p_i(t) f_i(y(\tau_1(t)), \ldots, y(\tau_n(t))) \qquad t \in R \qquad (3.10.4)$$

THEOREM 3.10.1 Assume that

(1) $f \in C[R^n, R]$ and satisfies the relation $y_1 f_i(y_1, y_2, \ldots, y_n) > 0$ if $y_1 y_j > 0$, $j \in I_n$, for every $i \in I_m$, and

$$|f_i(\bar{y}_1, \bar{y}_2, \ldots, \bar{y}_n)| \leq |f_i(\bar{\bar{y}}_1, \bar{\bar{y}}_2, \ldots, \bar{\bar{y}}_n)|$$

if $|\bar{y}_j| \leq |\bar{\bar{y}}_j|$, $\bar{y}_j \bar{\bar{y}}_j > 0$, $j \in I_n$, $i \in I_m$;

(2) there exists a $T \geq 0$ such that $p_j(t)$, $\tau_i(t)$ are continuous for $|t| \geq T$, $i \in I_n$, $j \in I_m$, and $\tau_i(t)$ satisfies (3.10.3), $i \in I_n$. $p_j(t)$ are of the same sign for $j \in I_m$, all being either nonpositive or nonnegative,

$$\sum_{j=1}^{m} \int_T^{\infty} p_j(t)\, dt = (\infty) \cdot \operatorname{sgn} p_i \qquad (3.10.5)$$

and

$$\sum_{j=1}^{m} \int_{-\infty}^{-T} p_j(t)\, dt = (\infty) \cdot \operatorname{sgn} p_i \qquad (3.10.6)$$

Then every solution of (3.10.4) defined on $|t| \geq \bar{T} \geq T$ oscillates.

Proof: Otherwise, there exists a nonoscillatory solution $y(t)$ (even on one side, for example, as $t \to \infty$). This implies that there exists a $T' > 0$ such that $y(t)$ has the same sign for all $t \geq T'$. Without loss of generality, we assume that $y(t) > 0$ and $p_i(t) \geq 0$, for $t \geq T'$. Let $T''_i = \max_{t \geq T'} \tau_i(t)$, $\max_i T''_i = T''$, $\min_{t \leq T''} \tau_i(t) = T'''_i$, and $\min_{1 \leq i \leq n} T'''_i = T'''$. The relative position of T' and T''' on the real line can be arbitrary. If $T''' < T'$ then assume that $y(t) > 0$ on $T''' \leq t < T'$. Otherwise, we choose T' sufficiently large such that T''' is sufficiently large to guarantee that $y(t) > 0$ on $T''' \leq t < T'$.

If $t \leq T''$, then $\tau_i(t) \geq T'''$ and hence $y(\tau_i(t)) > 0$, $\forall i \in I_n$. Therefore, for $\forall i \in I_m$, $f_i > 0$, and $y'(t) \geq 0$ for $t \leq T''$. Now we discuss two possible cases:

Either

(i) $y(t) \geq 0$ for $t \leq T''$,

or

(ii) there exists $\bar{T} \leq T''$ such that $y(t) \leq 0$ for $t \leq \bar{T}$.

In the first case, from the definitions of T'' and T''', $\tau_i(t) \le T''$ as $t \ge T'''$. Therefore $y'(t) \ge 0$ as $t \ge T'''$. Hence $y(t) \ge y(T''')$, as $t > T'''$. Integrating (3.10.4) on (t, T''), $t < T''$, we have

$$y(T'') \ge y(T'') - y(t) = \sum_{i=1}^{m} \int_{t}^{T''} p_i(s) f_i(y(\tau_1(s)), \ldots, y(\tau_n(s)))\, ds$$

$$\ge \sum_{i=1}^{m} f_i(y(T''') \cdots y(T''')) \int_{t}^{T''} p_i(s)\, ds$$

Letting $t \to -\infty$, we obtain a contradiction to condition (3.10.5).

In the second case, there exists a $\bar{T} \le T''$ such that $y(t) < 0$ as $t \le \bar{T}$. We choose $\bar{\bar{T}} > T'$ such that $\max_{t > \bar{\bar{T}}} \tau_i(t) \le \bar{T}$, $i = 1, 2, \ldots, n$. Hence $y'(t) < 0$ as $t \ge \bar{\bar{T}}$. Integrating (3.10.4) on $(\bar{\bar{T}}, t)$ $t > \bar{\bar{T}}$, we have

$$-y(\bar{\bar{T}}) \le y(t) - y(\bar{\bar{T}}) = \sum_{i=1}^{m} \int_{\bar{\bar{T}}}^{t} p_i(s) f_i(y(\tau_1(s)), \ldots, y(\tau_n(s)))\, ds$$

$$\le \sum_{i=1}^{m} f_i(y(\bar{T}), \ldots, y(\bar{T})) \int_{\bar{\bar{T}}}^{t} p_i(s)\, ds$$

or

$$1 \ge -\frac{1}{y(\bar{\bar{T}})} \sum_{i=1}^{m} f_i(y(\bar{T}), \ldots, y(\bar{T})) \int_{\bar{\bar{T}}}^{t} p_i(s)\, ds$$

Letting $t \to +\infty$ we arrive at a contradiction. The proof is complete.

EXAMPLE 3.10.1 We consider the equation

$$y'(t) + y\left(\frac{\pi}{2} - t\right) = 0 \tag{3.10.7}$$

which satisfies all the conditions of Theorem 3.10.1. Therefore all solutions of (3.10.7) are oscillatory in both directions. In fact, $y = \cos t$ is an oscillatory solution.

EXAMPLE 3.10.2 We consider the equation

$$y'(t) = e^{2t} y(-t) \tag{3.10.8}$$

which satisfies condition (3.10.5), but it does not satisfy condition (3.10.6). Observe that (3.10.8) has a nonoscillatory solution $y(t) = e^t$.

3.11 NOTES

Lemma 3.1.1 and Theorem 3.1.1 are from Shreve's paper [240]. For
related work see Sficas and Staikos [226]. Theorem 3.1.2 is based on
Zhang et al. [303]. Theorems 3.1.3 and 3.1.4 are new. For a special case
of Theorem 3.1.4 see Tomaras [274]. Recently Kulenovic and Grammatikopoulos [127] studied the case of (3.1.1) with oscillating coefficients. Theorems 3.1.5 and 3.1.7 are based on Haddock [96]. Theorem 3.1.6 is new.
Arino and Seguier [5] studied the existence of an oscillatory solution going
to zero as $t \to \infty$ for some special nonlinear equations. The asymptotic
behavior of nonoscillatory solutions of (3.1.1) is discussed in Sficas [229],
Burton and Haddock [27], Driver [55], Cooke [41], Cooke and Yorke [42, 43].
For comparison theorems see Elbert [58]. Advanced type equations are
discussed in Anderson [1], Heard [99], Kato and McLeod [111], and Myskis
[189]. The work on oscillation relative to (3.2.1) to (3.2.3) was started by
Stavroulakis [259] and Koplatadze and Canturija [125]. Then Onose [213] and
Gyovi [95] studied this problem extensively. Theorems 3.2.1, 3.2.2,
3.2.4, and Corollary 3.2.1 are based on Onose's paper [213]. Theorems
3.2.3 and 3.2.5 are taken from [95]. Theorem 3.2.8 is adapted from [125].
Theorems 3.2.6 and 3.2.7 are new results whose arguments are based on
[125]. Corollaries 3.2.2 and 3.2.3 are due to Zhang [301]. Theorems 3.3.1
to 3.3.5 are generalizations of Kitamura and Kusuno's work [119]. For
related work see also [103] and [233]. Theorems 3.4.1 and 3.4.2, Corollaries 3.4.1 and 3.4.2 are from Winston [288]. Theorem 3.4.3 is due to
Burkowski and Ponzo [24]. Theorem 3.4.4 is a new result. Theorems 3.5.1
and 3.5.2 are from Koplatadze [123]. Theorems 3.6.1 and 3.6.2 are new.
For related work see [1] and [119]. Theorem 3.7.1 is from Onose [207]
(see also [273]). For an existence theorem for (3.8.1) see Myskis [189]
and Hale [98]. The results of Section 3.8 are taken from Zhang [307] (for
related work see [185] and [308]). The results of Section 3.9 are from
Zhang [306] (see also [119]). Theorem 3.10.1 is a generalization of work
in [232] and for related work see Fite [62], Sevelo and Vareh [235], and
Norkin [198].

4

Second Order Differential Equations

4.0 INTRODUCTION

Second order differential equations are most important in applications. Several phenomena in biological, physical, and engineering sciences can be described by a second order ordinary differential equation with or without deviating arguments. Thus, there is much literature regarding this kind of equation. Sections 4.1 to 4.4 concentrate on the unstable type equations. In Section 4.1, we classify solutions of linear equations. In Sections 4.2 and 4.3, we state certain important properties for second order linear unstable equations. In Section 4.4, we extend some of the results in Sections 4.1 to 4.3 to a class of nonlinear equations. Sections 4.5 and 4.6 discuss results on the oscillatory solutions of stable type second order differential equations with deviating arguments. In Section 4.7, we study oscillation results for the equations with deviating arguments of distributed type.

4.1 CLASSIFICATION OF SOLUTIONS
   OF LINEAR EQUATIONS

In this section, we classify solutions of initial value problem of the type

$$(r(t)y'(t))' = p(t)y(g(t)) \tag{4.1.1}$$

where p, q, $r \in C[R_+, R_+]$, $g(t) \leq t$ and $r(t) > 0$.

Initial conditions for (4.1.1) are the following:

$$y(s) = \phi(s) \quad \text{for} \quad s \in E_{t_0}, \ y(t_0) = y_0, \ y'(t_0) = y_0' \tag{4.1.2}$$

where $E_{t_0} = \{t_0\} \cup \{g(t) : g(t) < t_0, \ t > t_0\}, \quad \phi \in C(E_{t_0})$.

DEFINITION 4.1.1 Let S denote the set of all solutions of (4.1.1). We define the following subsets of S:

$$S^{+\infty} = \{y \in S: \lim_{t \to \infty} y(t) = \infty\},$$

$$S^{-\infty} = \{y \in S: \lim_{t \to \infty} y(t) = -\infty\},$$

$$S^{k} = \{y \in S: 0 < \lim_{t \to \infty} y(t) < \infty\},$$

$$S^{-k} = \{y \in S: -\infty < \lim_{t \to \infty} y(t) < 0\},$$

$$S^{0} = \{y \in S: y(t) \not\equiv 0 \text{ and } \lim_{t \to \infty} y(t) = 0 \quad \text{monotonically}\},$$

$$S^{\sim} = \{y \in S: y(t) \text{ is oscillatory}\}.$$

Let us now present some sufficient conditions for the qualitative behavior of the solutions of (4.1.1).

LEMMA 4.1.1 Assume that

(i) $p \geq 0$, $r > 0$ are continuous;

(ii) $g \in C[R_+, R_+]$, $g(t)$ is nondecreasing, $g(t) \leq t$, and $\lim_{t \to \infty} g(t) = \infty$;

(iii) $\lim_{t \to \infty} \int_{t_0}^{t} ds/r(s) = \infty$.

Then

(a) $\phi(t) \geq 0$ on E_{t_0} and $y_0' > 0$ imply $y(t, \phi, y_0') \in S^{\infty}$;

(b) $\phi(t) \leq 0$ on E_{t_0} and $y_0' < 0$ imply $y(t, \phi, y_0') \in S^{-\infty}$

Proof: Integrating (4.1.1) from t_0 to t, we have

$$r(t)y'(t) = r(t_0)y'(t_0) + \int_{t_0}^{t} p(s)y(g(s))\, ds \tag{4.1.3}$$

Dividing by $r(t)$ on both sides of (4.1.3) we get

$$y'(t) = \frac{r(t_0)y'(t_0)}{r(t)} + \frac{1}{r(t)} \int_{t_0}^{t} p(s)y(g(s))\, ds \tag{4.1.4}$$

which implies

$$y(t) = y(t_0) + r(t_0)y_0'(R(t) - R(t_0)) + \int_{t_0}^{t} \frac{1}{r(u)} \int_{t_0}^{u} p(s)y(g(s))\, ds\, du \tag{4.1.5}$$

where $R(t) = \int_{t_0}^{t} ds/r(s)$. From the second term of (4.1.5) we obtain the conclusion of the lemma.

LEMMA 4.1.2 Let the hypotheses (i) and (ii) of Lemma 4.1.1 be satisfied. Further assume that

$$\int^{\infty} (R(t) - R(s))p(s)\, ds = \infty \tag{4.1.6}$$

where $R(t) = \int_{t_0}^{t} ds/r(s)$.

Then

(a) $\phi(t) \geq 0$ on E_{t_0}, $\phi(t) \not\equiv 0$, and $y_0' \geq 0$ imply $y \in S^{+\infty}$;

and

(b) $\phi(t) \leq 0$ on E_{t_0}, $\phi(t) \not\equiv 0$, and $y_0' \leq 0$ imply $y \in S^{-\infty}$.

Proof: By computing the double integral in (4.1.5), we obtain the conclusion of the lemma.

LEMMA 4.1.3 Let the hypotheses of Lemma 4.1.1 hold. Further assume that $y_1(t)$ and $y_2(t)$ have the same initial function with $y_{10}' > y_{20}'$. Then $y_1(t) > y_2(t)$, $y_1'(t) > y_2'(t)$ on $t \geq t_0$ and $\lim_{t \to \infty} (y_1(t) - y_2(t)) = \infty$.

Proof: We consider $y(t) = y_1(t) - y_2(t)$ and note that $y(t)$ is a solution of equation (4.1.1) with initial function $\phi(t) \equiv 0$ and $y_0' = y_{10}' - y_{20}' > 0$. From Lemma 4.1.1 $y(t) \in S^{+\infty}$, and from (4.1.4), $y'(t) > 0$. The proof is complete.

THEOREM 4.1.1 Assume that the hypotheses of Lemma 4.1.1 are satisfied. Then for every initial function ϕ, (4.1.1) has no more than one bounded solution on $[t_0, \infty)$.

Proof: Suppose that the conclusion is false. Let $y_1(t)$ and $y_2(t)$ be bounded solutions with $y_{10}' > y_{20}'$. This implies that $|y_1(t) - y_2(t)|$ is bounded. On the other hand, by Lemma 4.1.3, $y_1 - y_2 \in S^{\infty}$. This contradiction establishes the theorem.

Before we formulate the decomposition of the parameter set R, we need to have a few notions and definitions.

We associate with $g(t)$ the function $\nu^*(t)$ which is defined on R as follows: For each $t \in [A, +\infty)$, let

$$\nu^*(t) = \sup\{\tau \in [A, \infty) : g(\tau) < t\}$$

If no such τ exists, then let $\nu^*(t) = A$ (the latter case is possible only if $t \leq A$).

The function $\nu^*(t)$ is characteristic of the duration of the aftereffect. It has the following properties [184]:

(1) $A \le \nu^*(t)$, $t \le \nu^*(t)$ for all t

(2) Let $\Delta(t) = t - g(t)$, $\Delta_0 = \sup_{t \ge A} \Delta(t)$ and $\delta_0 = \inf_{t \ge A} \Delta(t)$. Then $t + \delta_0 \le \nu^*(t) \le t + \Delta_0$ for $t \ge A - \Delta_0$

(3) If $t_1 < t_2 < \infty$, then $\nu^*(t_1) \le \nu^*(t_2)$

(4) The function $\nu^*(t)$ is continuous from the left, that is,

$$\lim_{t \to t_0-} \nu^*(t) = \nu^*(t_0)$$

(5) Assume that $\lim_{t \to \infty} g(t) = \infty$. This implies that the function $\nu^*(t)$ is bounded on any compact subinterval of $[t_0, \infty)$.

Now we fix ϕ and treat y_0' in (4.1.2) as a parameter, we decompose the set R of all the reals into the following subsets.

DEFINITION 4.1.2 Let $y(t, \phi(t), y_0')$ be a solution of (4.1.1) and (4.1.2). Define

$$K^{\infty} = \{y_0' \in R : y(t, \phi(t), y_0') \in S^{+\infty}\},$$

$$K^{-\infty} = \{y_0' \in R : y(t, \phi(t), y_0') \in S^{-\infty}\},$$

$$K^0 = \{y_0' \in R : y(t, \phi(t), y_0') \in S^0\},$$

and

$$K^{\sim} = \{y_0' \in R : y(t, \phi(t), y_0') \in S^{\sim}\}$$

LEMMA 4.1.4 Assume all the hypotheses of Lemma 4.1.1 hold. Then for any initial function $\phi \in C(E_{t_0})$, the sets K^{∞}, $K^{-\infty}$ are nonempty.

Proof: From hypothesis (ii), we note that $\nu^*(t)$ is defined on $[t_0, \infty)$. Set $M_t = \max_{s \in [t, \nu^*(t)]} p(s)$. Note that M_t exists for $t \ge t_0$. Further, we observe that

$$\min_{s \in [t, \nu^*(t)]} g(s) > -\infty$$

Without loss of generality, we assume that $\phi(t_0) \ge 0$.

First, we consider the case $\nu^*(t_0) > t_0$. Let $\|\phi\|_0 = \max_{E_{t_0}} |\phi(t)|$.

If $\|\phi\|_0 = 0$, then the assertion of the lemma follows from Lemma 4.1.1. Now suppose that $\|\phi\|_0 > 0$. One can find a number y_0' so large that the corresponding solution $y(t)$ will be positive on $(t_0, \nu^*(t_0)]$. Otherwise, there exists a T which is the infimum of the zeros of the solution $y(t)$ which lie on $(t_0, \nu^*(t_0)]$. From $\phi(t_0) \geq 0$, $y_0' > 0$, and by continuity of the solution and its derivative, it follows that $T > t_0$, $y(T) = 0$, and $y(t) > 0$ on (t_0, T), and that on (t_0, T), $y'(t)$ has at least one zero. But by (4.1.3), we have

$$r(t)y'(t) \geq r(t_0)y_0' - M_{t_0} \| \phi \|_0 (t - t_0) \qquad t_0 \leq t \leq T \qquad (4.1.7)$$

and for sufficiently large y_0', $y'(t) > 0$ ($t_0 \leq t \leq T$) which leads to a contradiction. From the above argument and (4.1.7), for sufficiently large y_0', we have $y'(\nu^*(t_0)) > 0$. After translating the initial point to the point $t = \nu^*(t_0)$ and using Lemma 4.1.1, we obtain the solution $y(t)$ of (4.1.1) with this value of $y_0' \in K$.

We shall now show the existence of a solution in $K^{-\infty}$.

Let $\phi(t_0) = 0$; in this case one can find values of $y_0' < 0$ so large in absolute value that the corresponding solution $y(t)$ will be negative on $(t_0, \nu^*(t_0)]$. The proof of this assertion is carried out in a manner similar to the proof in the preceding paragraph. To find a bound on the derivative, we use (4.1.3) and obtain the inequality

$$r(t)y'(t) \leq r(t_0)y_0' + M_{t_0} \| \phi \|_0 (t - t_0) \qquad t_0 \leq t \leq T \qquad (4.1.8)$$

From (4.1.8) and $y_0' < 0$ with $|y_0'|$ large, we have $y'(\nu^*(t_0)) < 0$. To prove $y \in S^{-\infty}$, it remains only to translate the initial point to the point $t = \nu^*(t_0)$, and refer to Lemma 4.1.1.

Finally, suppose that $\phi(t_0) > 0$. Let $\tilde{t}$ be the smallest zero of the solution $y(t)$ (if it exists) and let $t^* = \min \{\tilde{t}, \nu^*(t_0)\}$. First, we shall show that if $y_0' < 0$ and is sufficiently large in absolute value, then the corresponding solution has a negative derivative on $[t_0, t^*]$. Assume the contrary. Let t^0 be the infimum of those values of t for which $y'(t) \geq 0$. By the continuity of the derivative, it follows from the inequality $y_0' < 0$ that $y'(t^0) = 0$ and $y'(t) < 0$ for $t \in [t_0, t^0)$. But on this interval $0 < y(t) \leq \phi(t_0) \leq \| \phi \|_0$. From (4.1.8),

$$r(t)y'(t) \leq r(t_0)y_0' + M_{t_0} \| \phi \|_0 (t - t_0) \qquad t_0 \leq t \leq t^0$$

This relation implies that if y_0' is negative and sufficiently large in absolute value, then $y'(t^0) < 0$. This leads to a contradiction. Therefore, $y(t)$ has a negative derivative on $[t_0, t^*]$.

Now let us assume that $\nu^*(t_0) \leq \tilde{t}$; then for negative y_0' with large $|y_0'|$, we have

$$0 < y(t) \leq \| \phi \|_0 \qquad t \in [t_0, \nu^*(t_0)] \qquad (4.1.9)$$

This, together with (4.1.5), yields

$$y(t) \leq y_0 + y_0' r(t_0)(R(t) - R(t_0)) + \| \phi \|_0 M_{t_0} \int_{t_0}^{t} \frac{1}{r(u)} (u - t_0)\, du \qquad (4.1.10)$$

for $t \in [t_0, \nu^*(t_0)]$. But with increasing $|y_0'|$, the first zero of y(t) approaches the point $t = t_0$. This contradicts (4.1.9).

Thus the solution y(t) has at least one zero $\tilde{t} > t_0$ and $\tilde{t} < \nu^*(t_0)$. On $[t_0, \tilde{t}]$, $|y(t)| \leq \| \phi \|_0$, and the translation of the initial point to the point $t = \tilde{t}$ leads to the case already considered ($\tilde{\Phi}(\tilde{t}) = y(\tilde{t}) = 0$).

Finally, we consider the case $\nu^*(t_0) = t_0$. If $\phi(t_0) = 0$, then from Lemma 4.1.1 it follows that $y \in S^\infty$ if $y_0' > 0$ and $y \in S^{-\infty}$ if $y_0' < 0$.

Now suppose that $\phi(t_0) > 0$. It follows from Lemma 4.1.1 that y(t) belongs to S^∞ if $y_0' > 0$. We shall show that the existence of a solution in $S^{-\infty}$ whenever $\phi(t_0) > 0$. Let $\Delta(t) \not\equiv 0$ on each interval of the form $(t_0, t_0 + \epsilon)$. We consider the solution y(t), corresponding to the value $y_0' < 0$. By continuity of the derivative y'(t), there exists a point $\bar{t} > t_0$ such that on $[t_0, \bar{t}]$, $y(t) > 0$, $y'(t) < 0$, and $\Delta(\bar{t}) > 0$. We translate the initial point to the point $t = \bar{t}$. Since $\nu^*(t_0) = t_0$, then $E_{\bar{t}}^-$ is entirely contained in $[t_0, \bar{t}]$. From the fact that $\Delta(\bar{t}) > 0$, it follows that $\nu^*(\bar{t}) > \bar{t}$. On $E_{\bar{t}}^-$ the relation $|y(t)| \leq \phi(t_0)$ holds. By the arbitrary large choice of $|y_0'|$ and from (4.1.4), one can assert $|y'(\bar{t})|$ arbitrarily large. Now, by imitating the earlier proof for the case $\nu^*(t_0) > t_0$, the proof for $y \in K^{-\infty}$ follows immediately.

If $\Delta(t) \equiv 0$ on $[t_0, T]$ and $\Delta(t) \not\equiv 0$ on any interval of the form $(T, T + \epsilon)$, then, choosing $y_0' < 0$ sufficiently large in absolute value, it is possible to show that y'(t) takes on a value at t_0 smaller than any preassigned negative number. By following the above argument, we can show that the solutions belong to $S^{-\infty}$.

Finally, if $\Delta(t) \equiv 0$ on $[t_0, \infty)$, then by following the argument that was used in the case of $\nu^*(t_0) > t_0$, we can prove the existence of a point $\tilde{t}$ such that $y(\tilde{t}) = 0$ and $y'(\tilde{t}) < 0$. To complete the proof, it is sufficient to translate the initial time to the point $t = \tilde{t}$. The proof is therefore complete.

LEMMA 4.1.5 Under the hypotheses of Lemma 4.1.4 the sets $K^{+\infty}$ and $K^{-\infty}$ are open.

Proof: Let $y'(t_0) = y_0'$. Assume that the solution $y \in S^{+\infty}$. From the definition of $S^{+\infty}$, it follows that there exists an interval $[T, \nu^*(T)]$ on which y(t) and y'(t) are positive (or a time T such that $\nu^*(T) = T$). By continuous dependence of solutions and their derivatives on the initial conditions, all solutions whose derivatives at the initial time differ slightly from y_0', are positive and have positive derivatives in this interval (or at this point). If the initial point is translated to the point $t = \nu^*(T)$, then it follows from Lemma 4.1.1 that all those solutions belong to $S^{+\infty}$, i.e., $K^{+\infty}$ is open. Analogously, one can prove that the set $K^{-\infty}$ is also open.

THEOREM 4.1.2 Under the hypotheses of Lemma 4.1.4 the sets $K^{+\infty}$ and $K^{-\infty}$ are given by nonintersecting half lines $(-\infty, \alpha)$ and (β, ∞) $(\alpha \leq \beta)$. The set $F = R - (K^{+\infty} \cup K^{-\infty})$ is nonempty and consists of the interval $[\alpha, \beta]$ if $\alpha < \beta$, or the point α if $\alpha = \beta$.

Proof: The conclusion of the theorem follows immediately from Lemmas 4.1.3, 4.1.4, and 4.1.5.

THEOREM 4.1.3 If the set F consists of the interval $\lfloor \alpha, \beta \rfloor$ $(\alpha < \beta)$, then for every $y_0' \in F$, the corresponding solution is unbounded and oscillatory.

Proof: By the definition of the set F, the solutions corresponding to this set may be either bounded, or unbounded and oscillatory. In the case $\alpha < \beta$, we shall show that there are no bounded solutions. Let us assume the contrary, and let $y_1(t)$ be a bounded solution $(y_1'(t_0) = y_{10}')$. Let $y_{20}' \in F$ and $y_{20}' \neq y_{10}'$. From the definition of F and Theorem 4.1.1, it follows that the solution $y_2(t)$ with $y_2'(t_0) = y_{20}'$ is unbounded and oscillatory. On the other hand, by Lemma 4.1.2, $|y_1(t) - y_2(t)| \to \infty$ as $t \to \infty$, but since $y_1(t)$ is bounded, then $|y_2(t)| \to \infty$ as $t \to \infty$, i.e., $y_2(t)$ cannot be an oscillatory solution. A contradiction is obtained, which proves the theorem.

We note that if $g(t) \equiv t$, equation (4.1.1) may not have any oscillatory solutions. From Theorem 4.1.3 it follows that the possibility of an equation of the form (4.1.1) having oscillatory solutions when a retardation is present is intimately connected with the possibility of the presence of multiple zeros of nontrivial solutions of (4.1.1).

The following theorem shows us that if $y_0' \in F$, then either $y_0' \in K^0$ or $y_0' \in \tilde{K}$.

THEOREM 4.1.4 Assume that the hypotheses of Lemma 4.1.1 hold. Further assume $\int^{\infty} R(s)p(s)\,ds = \infty$. Then every bounded solution of (4.1.1) either belongs to S^0 or to $\tilde{S}$.

Proof: Let $y(t)$ be a bounded nonoscillatory solution of (4.1.1). Without loss of generality, we can assume that $y(t) > 0$, $t \geq t_0$. Then it follows that $(r(t)y'(t))' > 0$ for $t \geq t_0 + \tau$, where $\tau = \inf_{t \geq t_0} (t - g(t))$. That is, $r(t)y'(t)$ is increasing for large t, say $t \geq t_1$. If $ry' > 0$ for $t \geq t_1$ then in view of $R(t) \to \infty$, $y(t) \to \infty$ as $t \to \infty$. This is a contradiction to the boundedness of $y(t)$. Thus $r(t)y'(t) < 0$ for $t \geq t_1$. Obviously, limit $y(\infty) \geq 0$ and $r(\infty)y'(\infty) \leq 0$ exists. It is easy to see that $r(\infty)y'(\infty) = 0$. Otherwise, we would have $y(t) < 0$ for sufficiently large t.

We shall next show that $y(\infty) = 0$. From equation (4.1.1), it follows that

$$r(t)y'(t) = r(t_1)y'(t_1) + \int_{t_1}^{t} p(s)y(g(s))\,ds$$

and hence

$$r(t_1)y'(t_1) = -\int_{t_1}^{\infty} p(s)y(g(s))\,ds$$

From the foregoing equation, we obtain

$$y'(t) = \frac{r(t_1)y'(t_1)}{r(t)} + \frac{1}{r(t)} \int_{t_1}^{t} p(s)y(g(s))\,ds$$

and hence,

$$y(t) = y(t_1) - (R(t) - R(t_1)) \int_{t_1}^{\infty} p(s)y(g(s))\,ds + \int_{t_1}^{t} (R(t) - R(s))p(s)y(g(s))\,ds$$

$$= y(t_1) + \int_{t_1}^{t} (R(t_1) - R(s))p(s)y(g(s))\,ds - (R(t) - R(t_1)) \int_{t}^{\infty} p(s)y(g(s))\,ds$$

$$\leq y(t_1) + R(t_1)[r(t)y'(t) - r(t_1)y'(t_1)] - \int_{t_1}^{t} R(s)p(s)y(g(s))\,ds$$

$$\leq y(t_1) - R(t_1)r(t_1)y'(t_1) - \int_{t_1}^{t} R(s)p(s)y(g(s))\,ds$$

$$\leq y(t_1) - R(t_1)r(t_1)y'(t_1) - y(\infty) \int_{t_1}^{t} R(s)p(s)\,ds \to -\infty$$

as $t \to \infty$. This contradiction shows that $y(\infty) = 0$.

The case $y(t)$ is bounded and $y(t) < 0$ for sufficiently large t follows analogously.

COROLLARY 4.1.1 For any fixed initial function $\phi \in C$, under the conditions of Theorem 4.1.4,

$$R = K^{+\infty} \cup K^{-\infty} \cup K^{0} \cup K^{\sim}$$

and moreover, $K^{+\infty}$, $K^{-\infty}$, and $K^{0} \cup K^{\sim}$ are nonempty.

4.2 THE EXISTENCE OF SOLUTIONS OF A CLASS
OF BOUNDARY VALUE PROBLEMS

We shall in this section establish sufficient conditions for the unique solvability of the boundary value problem

$$y''(t) = p(t)y(g(t)) \tag{4.2.1}$$

$$y(t) = \phi(t) \quad \text{for } t \in [\tau_0, 0], \quad \lim_{t \to \infty} |y(t)| < \infty \tag{4.2.2}$$

where $p: R_+ \to R_+$ is summable on each finite segment of R, $g \in$ $C[R_+, R_+]$, $g(t) \leq t$, $\lim_{t \to \infty} g(t) = +\infty$, and $\phi \in C[[\tau_0, 0], R]$ where $\tau_0 = \inf\{g(t): t \in R_+\}$.

LEMMA 4.2.1   Let F be a set consisting of the intervals $[\alpha, \beta]$ $(\alpha < \beta)$. Then for $\nu \in (\alpha, \beta]$ there exists a number $\nu^* \in (\beta, \infty)$ such that

$$y(t, \phi(t), \nu^*) > |y(t, \phi(t), \nu)|$$

and

$$|y'(t, \phi(t), \nu^*)| > |y'(t, \phi(t), \nu)| \tag{4.2.3}$$

for $t \geq 1$ where $y(t, \phi(t), \nu^*)$ and $y(t, \phi(t), \nu)$ are solutions of the initial value problem (4.2.1).

Proof: Put

$$C_0 = \max\{|\phi(t)|: \tau_0 \leq t \leq 0\} + \max\{|y(t, \phi, \nu)| + |y'(t, \phi, \nu)|: 1 \leq t \leq \nu(1)\}$$

where $\nu(t) = \sup\{s: g(s) < t\}$ and $\nu^* = 2C_0 + C_0 \int_0^{\nu(1)} p(s)\, ds$.

We first prove that

$$y'(t, \phi(t), \nu^*) \geq 2C_0 \quad \text{for } 0 \leq t \leq \nu(1) \tag{4.2.4}$$

Assume the contrary, i.e., for some $t_0 \in (0, \nu(1)]$, $y'(t, \phi, \nu^*) > 0$ for $0 \leq t \leq t_0$, $y(t, \phi, \nu^*) \geq -C_0$ for $\tau_0 \leq t \leq t_0$, and $y'(t_0, \phi, \nu^*) < 2C_0$. But this is impossible because

$$y'(t_0, \phi, \nu^*) = \nu^* + \int_0^{t_0} p(s)y(g(s))\, ds \geq \nu^* - C_0 \int_0^{\nu(1)} p(s)\, ds = 2C_0$$

Therefore inequality (4.2.4) is proved.

From (4.2.4), we have

$$y'(1, \phi, \nu^*) > |y'(1, \phi, \nu)|$$

and

$$y(t, \phi, \nu^*) \geq 2C_0 t + \phi(0) > |y(t, \phi, \nu)| \quad \text{for } 1 \leq t \leq \nu(1) \tag{4.2.5}$$

By integrating (4.2.1), there result

$$y'(t, \phi, \nu^*) - y'(1, \phi, \nu^*) = \int_1^t p(s)y(g(s), \phi, \nu^*)\, ds$$

and

$$y'(t, \phi, \nu) - y'(1, \phi, \nu) = \int_1^t p(s)y(g(s), \phi, \nu)\, ds$$

By comparing these equations and using (4.2.5) the lemma follows.

THEOREM 4.2.1 Assume that p and g satisfy the hypotheses mentioned at the beginning of Section 4.2. Further assume that

$$\limsup_{t \to +\infty} \int_t^{\nu(t)} (s - t)p(s)\, ds < \infty \tag{4.2.6}$$

Then (4.2.1) together with (4.2.2) has a unique solution.

Proof: Theorem 4.1.3 implies that it is sufficient to prove that $\alpha = \beta$. Assume that $\alpha < \beta$. Then Lemma 4.2.1 implies that for each fixed $\nu \in (\alpha, \beta)$ there is $\nu^* > \beta$ such that (4.2.3) holds. We shall prove that, for large t,

$$y(t, \phi, \nu^*) \geq \left(1 + \frac{1}{M}\right) |y(t, \phi, \nu)| \tag{4.2.7}$$

where M is a positive number such that

$$M \geq \int_t^{\nu(t)} (s - t)p(s)\, ds \quad \text{for } t \geq 0 \tag{4.2.8}$$

By virtue of Theorem 4.1.3, the solution $y(t, \phi, \nu)$ is oscillatory. Let $t_0 \in [\nu(1), +\infty)$ be a zero of $y(t, \phi, \nu)$ and let $S_0 \in (t_0, +\infty)$ be any point that is not a zero of $y(t, \phi, \nu)$. For definiteness we assume that $y(S_0, \phi, \nu) > 0$. Then there are numbers $S_1 \in [t_0, S_0)$, $S_2 \in (S_0, \infty)$, and $\tilde{S} \in [S_1, S_2]$ for which

$$y(S_1, \phi, \nu) = y(S_2, \phi, \nu) = 0 \qquad y'(\tilde{S}, \phi, \nu) = 0$$

and

$$y(\tilde{S}, \phi\ \nu) \geq y(S, \phi, \nu) > 0 \quad \text{for } S_1 < S < S_2 \tag{4.2.9}$$

It is easily seen that $\tilde{S} < \nu(S_1)$, and by virtue of this inequality, (4.2.8), Lemma 4.2.1, and the relation

$$y(\tilde{S}, \phi, \nu) = -\int_{S_1}^{\tilde{S}} (S - S_1)p(s)y(g(s), \phi, \nu)\, ds$$

we conclude that

$$y(\tilde{S}, \phi, \nu) \leq My(S_1, \phi, \nu^*) \tag{4.2.10}$$

It follows from (4.2.9) and (4.2.10) that

$$y(S_0, \phi, \nu^*) = y(S_1, \phi, \nu^*) + y'(S_1, \phi, \nu^*)(S_0 - S_1) + \int_{S_1}^{S_0} (S_0 - S)p(s)y(g(s), \phi, \nu^*)\, ds$$

$$\geq \frac{1}{M} y(\tilde{S}, \phi, \nu) + y'(S_1, \phi, \nu)(S_0 - S_1) + \int_{S_1}^{S_0} (S_0 - S)p(s)y(g(s), \phi, \nu)\, ds$$

$$= \frac{1}{M} y(\tilde{S}, \phi, \nu) + y(S_0, \phi, \nu) \geq \left(1 + \frac{1}{M}\right) y(S_0, \phi, \nu)$$

Since S_0 is arbitrary, inequality (4.2.7) holds on the whole interval $[t_0, +\infty)$.

Let $t_1 \geq \nu(t_0)$ and $y'(t_1, \phi, \nu) = 0$. Relation (4.2.5) and (4.2.7) imply that

$$y'(t, \phi, \nu^*) = y'(t_1, \phi, \nu^*) + \int_{t_1}^{t} p(s)y(g(s), \phi, \nu^*)\, ds$$

$$> \left(1 + \frac{1}{M}\right) \int_{t_1}^{t} p(s)|y(g(s), \phi, \nu)|\, ds$$

$$\geq \left(1 + \frac{1}{M}\right)|y'(t, \phi, \nu)| \qquad t \geq t_1 \tag{4.2.11}$$

Starting from (4.2.7) and (4.2.11), we can apply induction to prove that there is a sequence of points $t_n \in R_+$ ($n = 1, 2, \ldots$) such that, for each positive integer n,

$$y(t, \phi, \nu^*) \geq \left(1 + \frac{1}{M}\right)^n |y(t, \phi, \nu)|$$

$$y'(t, \phi, \nu^*) \geq \left(1 + \frac{1}{M}\right)^n |y'(t, \phi, \nu)| \quad \text{for } t \geq t_n \tag{4.2.12}$$

and

$$\lim_{t \to \infty} \frac{y(t, \phi, \nu)}{y(t, \phi, \nu^*)} = 0 \tag{4.2.13}$$

Putting

$$\nu_0 = \frac{\nu + \alpha}{2} \qquad c = \frac{\nu - \nu_0}{\nu^* - \nu}$$

we conclude that $y(t, \phi, \nu_0) = (1 - c)y(t, \phi, \nu) + cy(t, \phi, \nu^*)$ and hence, in view of (4.2.13),

$$\lim_{t \to \infty} \frac{y(t, \phi, \nu_0)}{y(t, \phi, \nu^*)} - c < 0$$

But this is impossible because Theorem 4.1.3 implies that $y(t, \phi, \nu_0)$ is oscillating. This contradiction proves the theorem.

THEOREM 4.2.2   If (4.2.6) holds and

$$\int_0^\infty t\, p(t)\, dt = +\infty \tag{4.2.14}$$

then problem (4.2.1) and (4.2.2) has a unique solution y such that

$$\liminf_{t \to \infty} |y(t)| = 0 \tag{4.2.15}$$

Proof: Theorem 4.2.1 implies that it remains to prove that the solution y of (4.2.1) and (4.2.2) satisfies (4.2.15). Assuming the contrary, there are $\delta > 0$ and $t_0 > 0$ such that

$$|y(t)| \geq \delta \qquad y(t)y'(t) \leq 0 \qquad \text{for } t \geq t_0$$

Multiplying both sides of (4.2.1) by $(t - t_1)$ sign $y(t_1)$, where $t_1 = \nu(t_0)$ and integrating, we obtain

$$|y(t_1)| = (t - t_1)|y'(t)| + |y(t)| + \int_{t_1}^t (s - t_1)p(s)|y(s(s))|\, ds$$

for $t \geq t_1$. Hence

$$\int_{t_1}^\infty (s - t_1)p(s)\, ds \leq \frac{y(t_1)}{\delta}$$

which contradicts (4.2.14) and the theorem follows.

4.3 THE EXISTENCE OF BOUNDED OSCILLATORY SOLUTIONS

Consider the equation

$$y''(t) - y(t) = 0 \tag{4.3.1}$$

As we know, (4.3.1) has no oscillatory solution. On the other hand, consider the same equation with delay π, so that we have

$$y''(t) - y(t - \pi) = 0 \tag{4.3.2}$$

It is easy to check that $y_1 = \sin t$, $y_2 = \cos t$ are oscillatory solutions of (4.3.2).

Let us consider the general linear equation

$$y''(t) - p(t)y(t - \tau(t)) = 0 \qquad p(t) \geq 0 \qquad t \geq t_0 \geq 0 \tag{4.3.3}$$

We pose the problem: What conditions guarantee the presence of oscillatory solutions for (4.3.3)? This problem has no meaning when $\tau(t) \equiv 0$.

In this section, we obtain some sufficient conditions for every bounded solution to be oscillatory. Furthermore, certain asymptotic properties are also discussed. We begin with the following result.

THEOREM 4.3.1 Assume that the hypotheses (i), (ii), and (iii) of Lemma 4.1.1 are satisfied. Further, assume that

$$\limsup_{t \to \infty} \frac{1}{r(t)} \int_{g(t)}^{t} (\sigma - g(t))\, p(\sigma)\, d\sigma > 1 \tag{4.3.4}$$

Then bounded solutions of (4.1.1) are oscillatory.

Proof: Suppose not. Without loss of generality, let $y(t) > 0$ be a bounded solution of (4.1.1); as $t > T$, hence $(r(t)y'(t))' \geq 0$, i.e., ry' is nondecreasing.

(a) If $ry' > c > 0$, as $t \geq T_1 \geq T$, then $y' > c/r$. Integrating it we get $y(t)$ is unbounded, so this case is impossible.

(b) If $ry' \leq 0$, then $y' \leq 0$. Integrating (4.1.1) from s to t, we have

$$r(t)y'(t) - r(s)y'(s) = \int_{s}^{t} p(\sigma)y(g(\sigma))\, d\sigma$$

Then integrating from $g(t)$ to t we see that

$$r(t)y'(t)(t - g(t)) = \int_{g(t)}^{t} r(s)\, dy(s) + \int_{g(t)}^{t} [\sigma - g(t)]\, p(\sigma)y(g(\sigma))\, d\sigma$$

Hence

$$0 \geq r(t)y(t) - r(g(t))y(g(t)) - \int_{g(t)}^{t} y(s)\, dr(s) + \int_{g(t)}^{t} (\sigma - g(t))p(\sigma)y(g(\sigma))\, d\sigma$$

$$\geq r(t)y(t) - r(g(t))y(g(t)) - y(g(t))(r(t) - r(g(t)) + \int_{g(t)}^{t} (\sigma - g(t))p(\sigma)y(g(\sigma))\, d\sigma$$

$$\geq r(t)(y(t) - y(g(t)) + \int_{g(t)}^{t} (\sigma - g(t))p(\sigma)y(g(\sigma))\, d\sigma$$

or

$$0 \geq y(t) - y(g(t)) + \frac{1}{r(t)} \int_{g(t)}^{t} (\sigma - g(t))p(\sigma)y(g(\sigma))\, d\sigma$$

Dividing the above inequality by $y(g(t))$ and using the monotonicity of y, we get

$$0 \geq \frac{y(t)}{y(g(t))} + \left[\frac{1}{r(t)} \int_{g(t)}^{t} (\sigma - g(t))p(\sigma)\, d\sigma - 1\right]$$

Because of (4.3.4), we have arrived at a contradiction. The proof is complete.

COROLLARY 4.3.1 If $\tau \geq 0$, $p(t) \geq 0$ is continuous, and $\tau^2 p(t) \geq 2$ for $t \geq 0$, then bounded solutions of

$$y''(t) - p(t)y(t - \tau) = 0 \tag{4.3.5}$$

are oscillatory.

COROLLARY 4.3.2 If $k > 1$, $p(t) \geq 0$ is continuous, and

$$p(t) \geq \frac{2k^2}{((1 - k)t)^2} \tag{4.3.6}$$

for large t, then bounded solutions of

$$y''(t) - p(t)y\left(\frac{t}{k}\right) = 0 \tag{4.3.7}$$

are oscillatory.

EXAMPLE 4.3.1 The equation

$$\left(\frac{1}{t}y'\right)' - 4ty\sqrt{t^2 - \pi} = 0 \qquad t \geq 2 \tag{4.3.8}$$

satisfies the condition of Theorem 4.3.1. Therefore all bounded solutions are oscillatory. In particular, $y(t) = \cos t^2$ is a bounded oscillatory solution.

EXAMPLE 4.3.2 The equation

$$y''(t) - y(t - \tau) = 0 \qquad 0 \leq \tau \leq 2e^{-1} \tag{4.3.9}$$

does not satisfy the conditions of Theorem 4.3.1 as expected. (4.3.9) has a bounded nonoscillatory solution. Indeed, the characteristic equation

$$F(\lambda) = \lambda^2 - e^{-\lambda\tau} = 0 \tag{4.3.10}$$

has negative real root λ, and hence $y(t) = e^{\lambda t}$ is a bounded nonoscillatory solution.

REMARK 4.3.1 If we do not require $\int^{\infty} ds/r(s) = \infty$, but $r(t)$ is nondecreasing and (4.3.4) is satisfied, then the conclusion of Theorem 4.3.1 remains valid.

THEOREM 4.3.2 For equation (4.2.1), assume that the initial function $\phi(t) \not\equiv 0$ and there exists a nondecreasing g(t) such that

$$g(t) < t \qquad p(t) > 0 \quad \text{for } t \geq 0$$

$$\limsup_{t \to \infty} \int_{g(t)}^{t} [s - g(t)]p(s)\, ds > 1 \tag{4.3.11}$$

and (4.2.6) is satisfied. Then problem (4.2.1) and (4.2.2) has a unique solution which is oscillatory.

Proof: By Theorems 4.2.1 and 4.3.1, we obtain the conclusion of Theorem 4.3.2.

COROLLARY 4.3.1 Suppose that $\phi(t) \not\equiv 0$, $g(t)$ is nondecreasing, $g(t) < t$, $p(t) > 0$ are continuous, and

$$\Delta \leq \liminf (t - g(t)) \leq \limsup (t - g(t)) < +\infty$$

$$a_0 \leq \liminf p(t) \leq \limsup p(t) < +\infty$$

where Δ and a_0 are positive constants such that

$$\Delta^2 a_0 > 2 \tag{4.3.12}$$

Then problem (4.2.1) and (4.2.2) has a unique solution and this solution is oscillatory.

We consider the linear equation with several delays

$$y''(t) - \sum_{i=1}^{n} p_i(t)y(g_i(t)) = 0 \tag{4.3.13}$$

THEOREM 4.3.3 Assume that

(i) $p_i, g_i \in C[[0,\infty), R]$, $p_i \geq 0$, $i = 1, 2, \ldots, n$, and for some index i_0, $1 \leq i_0 \leq n$, $p_{i_0}(t) > 0$ for $t \geq 0$

(ii) $g_i(t) \leq t$ and $\lim_{t \to \infty} g_i(t) = \infty$ for $i = 1, 2, \ldots, n$

(iii) There exists a nonempty set of indices $K = \{k_1, k_2, \ldots, k_\ell\}$, $1 \leq k_1 < k_2 < \cdots < k_\ell < n$, such that for $t \geq t_0$, $g_k(t) < t$ and $g_k'(t) \geq 0$ for $k \in K$ and

$$\limsup_{t \to \infty} \sum_{k \in K} \int_{g^*(t)}^{t} [g_k(t) - g_k(s)]\, p_k(s)\, ds > 1 \tag{4.3.14}$$

where $g^*(t) \equiv \max_{k \in K} g_k(t)$.

Then every bounded solution of (4.3.13) is oscillatory.

Proof: Let $y(t)$ be a bounded nonoscillatory solution of (4.3.13). Then, without loss of generality, $y(t) > 0$, and because of condition on $g_i(t)$, there exists a $t_1 \geq t_0$ such that $y(g_i(t)) > 0$ for $t \geq t_1$ and $i = 1, 2, \ldots, n$. In view of Eq. (4.3.13), we have $y''(t) > 0$, $t \geq t_1$. Since $y(t) > 0$ and $y(t)$ is bounded, it follows that there exists a $t_2 \geq t_1$ such that $y'(t) < 0$, $t \geq t_2$. From these observations, we conclude that $y(t)$ is concave up and decreasing for $t \geq t_2$. Therefore, it lies above its tangent. That is, for $\tilde{t}, \tilde{s} \geq t_2$,

$$y(\tilde{t}) + y(\tilde{t})(\tilde{s} - \tilde{t}) \leq y(\tilde{s})$$

We note that $g_k(t) \to \infty$ as $t \to \infty$, so the above inequality implies that

$$y(g_k(t)) + y'(g_k(t))[g_k(s) - g_k(t)] \leq y(g_k(s))$$

for s, t sufficiently large, say $s, t \geq t_3 \geq t_2$, and for all $k \in K$. Multiplying the above inequality by $p_k(s)$ and summing up for all $k \in K$, we get

$$\sum_{k \in K} p_k(s) y(g_k(t)) + \sum_{k \in K} y'(g_k(t))[g_k(s) - g_k(t)] p_k(s)$$

$$\leq \sum_{k \in K} p_k(s) y(g_k(s)) \leq \sum_{k=1}^{n} p_k(s) y(g_k(s)) = y''(s)$$

Integrating the above inequality, with respect to s, from $g^*(t)$ to t, for t sufficiently large, we obtain

$$\sum_{k \in K} y(g_k(t)) \int_{g^*(t)}^{t} p_k(s) \, ds + \sum_{k \in K} y'(g_k(t)) \int_{g^*(t)}^{t} [g_k(s) - g_k(t)] p_k(s) \, ds$$

$$\leq y'(t) - y'(g^*(t))$$

Since $y'(t)$ increases in t and $g'(t) \geq 0$ this inequality, after some manipulation, becomes

$$\sum_{k \in K} y(g_k(t)) \int_{g^*(t)}^{t} p_k(s) \, ds - y'(g^*(t)) \left[\sum_{k \in K} \int_{g^*(t)}^{t} [g_k(t) - g_k(s)] p_k(s) \, ds - 1 \right]$$

$$\leq y'(t) \qquad\qquad\qquad\qquad\qquad\qquad\qquad\qquad (4.3.15)$$

Now, in view of the hypotheses (4.3.14), the left-hand side of (4.3.15) is nonnegative for sufficiently large t, while the right-hand side is negative. This contradiction proves our result.

EXAMPLE 4.3.3 Consider the equation

$$y''(t) - (K + 1) y(t - \pi) - K y(t) = 0 \qquad K \geq 0 \qquad\qquad (4.3.16)$$

and note that

$$\int_{t-\tau}^{t} (K + 1)(t - s) \, ds = \frac{K + 1}{2} \pi^2 > 1$$

By Theorem 4.3.3 every bounded solution of (4.3.16) is oscillatory. It is easily seen that Eq. (4.3.16) has the bounded oscillatory solutions $c_1 \cos t + c_2 \sin t$ for any real number c_1 and c_2.

REMARK 4.3.1 The result of Theorem 4.3.3 can be extended to a more general equation of the form

$$(r(t)y'(t))' = \sum_{i=1}^{n} p_i(t)y(g_i(t)) \qquad (4.3.17)$$

where $r(t) > 0$ and $\int^{\infty} d\sigma/r(\sigma) = \infty$.

4.4 CLASSIFICATION OF SOLUTIONS OF NONLINEAR EQUATIONS

We wish to extend, in this section, some results of Sections 4.1 and 4.3 to the nonlinear equation

$$y''(t) - f(t, y(t), y(g(t))) = 0 \qquad (4.4.1)$$

THEOREM 4.4.1 Assume that

(i) $f \in C[R^+ \times R \times R, R]$ and $f(t, u, v)$ is nondecreasing in u and v for fixed large t

(ii) $f(t, u, v)u > 0$ if $u \cdot v > 0$

(iii) $g \in C[R^+, R]$, $g(t) < t$, $g'(t) > 0$ and $\lim_{t \to \infty} g(t) = \infty$

(iv) For any constant $c \neq 0$,

$$\int^{\infty} f(s, g(s)c, g(s)c)\, ds = \pm\infty \qquad (4.4.2)$$

Then

$$S = S^{+\infty} \cup S^{-\infty} \cup S^{-k} \cup S^{k} \cup S^{0} \cup \tilde{S}$$

where $S^{+\infty} = \{y(t) \in S: \lim_{t \to \infty} y(t) = \infty,\ \lim_{t \to \infty} y'(t) = \infty\}$, $S^{k} = \{y(t) \in S: 0 < \lim_{t \to \infty} y(t) < \infty,\ \lim_{t \to \infty} y'(t) = 0\}$, and $y(t)$ is a solution of (4.4.1).

Proof: Let $y(t) \in S - \tilde{S}$.

Case 1. Let $y(t) > 0$ for large $t \geq t_1$. Then, in view of conditions (ii) and (iii), there exists $t_2 \geq t_1$ such that $y''(t) > 0$ for $t \geq t_2$. This implies that $y'(t)$ has a fixed sign for large t, say $t \geq t_3 \geq t_2$. First assume that $y'(t) > 0$ for $t \geq t_3$; we show that $y(t) \in S^{+\infty}$. In fact $\lim_{t \to \infty} y(t) = \infty$, $\lim_{t \to \infty} y'(t) = y'(\infty) > 0$, and $y'(\infty) < \infty$. By L'Hospital's rule, we get

$$\lim_{t \to \infty} \frac{y(g(t))}{g(t)} = \lim_{t \to \infty} y'(g(t)) = y'(\infty)$$

Thus, for arbitrarily large t, say $t \geq t_4 \geq t_3$,

$$\frac{y(g(t))}{g(t)} \geq \frac{y'(\infty)}{2} = c > 0 \tag{4.4.3}$$

and also

$$\frac{y(t)}{g(t)} \geq \frac{y'(\infty)}{2} = c > 0$$

in view of the fact that $y'(t) > 0$. Now, integrating (4.4.1) from t_4 to t and using (i), (4.4.2) and (4.4.3), we have

$$y'(t) = y'(t_4) + \int_{t_4}^{t} f(s, y(s), y(g(s))) \, ds$$

$$\geq y'(t_4) + \int_{t_4}^{t} f(s, g(s)c, g(s)c) \, ds$$

$$\to \infty \quad \text{as } t \to \infty$$

This contradiction proves that $y'(\infty) = \infty$. Hence, $y(t) \in \mathscr{S}^{+\infty}$.

Second, assume that $y'(t) < 0$ for $t \geq t_3$; then we will show that $y(t) \in S^0 \cup S^k$. Note that

$$\lim_{t \to \infty} y(t) = y(\infty) \quad \text{and} \quad \lim_{t \to \infty} y'(t) = y'(\infty)$$

exist and $y(\infty) \geq 0$, while $y'(\infty) \leq 0$. We show that

$$y'(\infty) = 0$$

Assume that $y'(\infty) < 0$. This implies $y'(t) < y'(\infty)$ for $t \geq t_3$, and hence $y(t) \leq y(t_3) + y'(\infty)(t - t_3) \to -\infty$ as $t \to \infty$, contradicting the fact that $y(t) > 0$ for $t \geq t_3$. Hence $y'(\infty) = 0$. This implies $y(t) \in S^0 \cup S^k$.

Case 2. Let $y(t) < 0$ for $t \geq t_1$. Then the proof of this case follows similarly to Case 1.

THEOREM 4.4.2 Assume that (4.4.1) satisfies the conditions (i), (ii), (iii) of Theorem 4.4.1. Furthermore, let $y(t)$ be a bounded solution of (4.4.1), with $|y(t)| \leq \beta$ for large t, and $\beta > 0$. Let us assume that there exists a function $G_\beta \in C[R^+, R^+]$ such that

$$z^2 G_\beta(t) \leq zf(t, x, z) \tag{4.4.4}$$

for $\operatorname{sgn} x = \operatorname{sgn} z$, $x \operatorname{sgn} x \leq z \operatorname{sgn} z \leq \beta$, and sufficiently large t. Further assume that

$$\limsup_{t \to \infty} \int_{g(t)}^{t} [g(t) - g(s)]G_\beta(s)\, ds > 1 \tag{4.4.5}$$

Then $y(t)$ is oscillatory.

Proof: Let $y(t)$ be a bounded nonoscillatory solution of (4.4.1), with bound β. Without loss of generality, assume $y(t) > 0$, and in view of (iii), $y(g(t)) > 0$ for $t \geq t_1$. This together with (4.4.1) and (ii) implies that $y''(t) > 0$, $t \geq t_1$. Since $y(t)$ is bounded, it follows that for some $t_2 \geq t_1$, $y'(t) < 0$, $t \geq t_2$. From these facts, we see that $y(t)$ is concave up and decreasing for $t \geq t_2$. Therefore, for $\tilde{t}, \tilde{s} \geq t_2$, we have

$$y(\tilde{t}) + y'(\tilde{t})(\tilde{s} - \tilde{t}) \leq y(\tilde{s})$$

This, together with the fact that $g(t) \to \infty$ as $t \to \infty$, yields

$$y(g(t)) + y'(g(t))[g(s) - g(t)] \leq y(g(s))$$

for sufficiently large s, $t \geq t_2$.

Now, multiplying the above inequality by $G_\beta(s)$ and using the relation (4.4.4) and the equation (4.4.1), we have

$$y(g(t))G_\beta(s) + y'(g(t))[g(s) - g(t)]G_\beta(s) \leq y(g(s))G_\beta(s) \leq y''(s)$$

Integrating this with respect to s from $g(t)$ to t, for large t, we obtain

$$y'(g(t)) \int_{g(t)}^{t} [g(s) - g(t)]G_\beta(s)\, ds \leq y'(t) - y'(g(t))$$

which implies that

$$y'(g(t)) \left\{ 1 - \int_{g(t)}^{t} [g(t) - g(s)]G_\beta(s)\, ds \right\} < y'(t) \tag{4.4.6}$$

Thus, in view of (4.4.5), we arrive at a contradiction to the fact that the left-hand side of (4.4.6) is nonnegative for large t, while the right-hand side is negative. A similar proof can be given if $y(t) < 0$. Hence the theorem is proved.

COROLLARY 4.4.1 Assume that (4.4.1) satisfies the conditions (i), (ii), and (iii) of Theorem 4.4.1. Furthermore, assume that for any $\beta > 0$, there exists a function $G_\beta \in C[R^+, R^+]$ such that (4.4.4) and (4.4.5) hold. Then every bounded solution of (4.4.1) is oscillatory.

COROLLARY 4.4.2 Consider

$$y''(t) - p_1(t)y(g(t)) - p_2(t)y(t) = 0 \tag{4.4.7}$$

where $p_1(t), p_2(t) \geq 0$ and are continuous on R^+, and

$$\overline{\lim_{t \to \infty}} \ \frac{1}{g(t)} \int^{t} [g(t) - g(s)] \, p_1(s) \, ds > 1 \tag{4.4.8}$$

Then every bounded solution of (4.4.7) is oscillatory.

Proof: For any $\beta > 0$, set $G_\beta(t) = p_1(t)$ and $f(t, u, v) = p_1(t)v + p_2(t)u$. Now it is easy to see that (4.4.7) fulfills all hypotheses of Theorem 4.4.2, in view of the condition (4.4.8). Hence, the conclusion of the corollary remains true.

EXAMPLE 4.4.1 Consider

$$y''(t) - y(t - \pi)[(k + 1) + ky^{2n}(t - \pi)] - ky(t)[1 + y^{2n}(t)] = 0 \tag{4.4.9}$$

$k \geq 0$, for any integer $n \geq 0$. For any $\beta > 0$, $G_\beta(t) = (k + 1)$ satisfies the condition (4.4.4). Then (4.4.5) reduces to

$$\int_{t-\pi}^{t} (k + 1)(t - s) \, ds = \frac{k + 1}{2} \pi^2 > 1$$

and by Theorem 4.4.2 every bounded solution of (4.4.9) is oscillatory. In fact, (4.4.9) has bounded oscillatory solutions $A \cos t + B \sin t$, where A and B are any arbitrary constants.

4.5 NONLINEAR EQUATIONS WITH $\int_{t_0}^{\infty} ds/r(s) = \infty$

We consider the nonlinear second order functional differential equation with deviating arguments

$$(r(t)y'(t))' + f(t, y(t), y(g(t)), y'(t), y'(h(t))) = 0 \tag{4.5.1}$$

Concerning equation (4.5.1) we have the following result.

THEOREM 4.5.1 Assume that

(a) $r \in C[R_+, R_+]$, $r(t) > 0$ for $t \geq t_0$, $t_0 \in R_+$, and

$$\lim_{t \to \infty} R(t) = \infty \tag{4.5.2}$$

where $R(t)$ is defined by $R(t) = \int_{t_0}^{t} ds/r(s)$

(b) $g, h \in C[R_+, R]$, $g(t) \leq t$, $\lim_{t \to \infty} g(t) = +\infty$

(c) $f \in C[R_+ \times R^4, R]$ and $uf(t, u, v, w, z) > 0$ for $u \cdot v > 0$, $t \geq t_0$

(d) There exist a constant β such that $0 < \beta < 1$ and

$$\int_{t_0}^{\infty} R^{\beta}(g(t)) \, \frac{|f(t, y(t), y(g(t)), y'(t), y'(h(t)))|}{|y(g(t))|^{\beta}} \, dt = +\infty \tag{4.5.3}$$

for every positive nondecreasing or negative nonincreasing function $y(t)$.

Then every solution of (4.5.1) oscillates.

Proof: Let $y(t)$ be a nonoscillatory solution of (4.5.1). Without loss of generality, we may suppose that $y(t) < 0$, $y(g(t)) < 0$ for $t \geq T$. From (4.5.1) and condition (c), we have $(r(t)y'(t))' > 0$. There are two possible cases.

Case (i): $r(t)y'(t) > 0$ for $t \geq t_1 \geq T$. This implies that

$$y'(t) \geq \frac{r(t_1)y'(t_1)}{r(t)} \qquad t \geq t_1$$

Integrating the above inequality, we obtain

$$y(t) - y(t_1) \geq r(t_1)y'(t_1) \int_{t_1}^{t} \frac{ds}{r(s)} \qquad \text{for } t \geq t_1 \tag{4.5.4}$$

Letting $t \to \infty$ in (4.5.4) we get $y(t) > 0$ which contradicts $y(t) < 0$.

Case (ii): $r(t)y'(t) < 0$ for $t \geq t_2 \geq T$. Then $y'(t) < 0$. From (4.5.1), we have

$$r(t)y'(t) \leq \int_{t}^{\infty} f(s, y(s), y(g(s)), y'(s), y'(h(s))) \, ds \qquad t \geq t_2 \tag{4.5.5}$$

We note that the integral on the right is convergent. From the monotone increasing property of ry', we get $r(g(t))y'(g(t)) \leq r(t)y'(t)$. From (4.5.5), we obtain

$$y'(g(t)) \leq \frac{1}{r(g(t))} \int_{t}^{\infty} f(s, y(s), y(g(s)), y'(s), y'(h(s))) \, ds$$

An integration of the above inequality yields

$$y(g(t)) \leq y(g(t)) - y(g(t_2))$$

$$\leq (R(g(t)) - R(g(t_2))) \int_{t}^{\infty} f(s, y(s), y(g(s)), y'(s), y'(h(s))) \, ds$$

Notice that both sides of the above inequality are negative. This, together with the above inequality, yields

$$(R(g(t)) - R(g(t_2)))^{\beta} |y(g(t))|^{-\beta} \leq \left| \int_{t}^{\infty} f(s, y(s), y(g(s)), y'(s), y'(h(s))) \, ds \right|^{-\beta} \tag{4.5.6}$$

Set

$$F(t) = -\int_{t}^{\infty} f(s, y(s), y(g(s)), y'(s), y'(h(s))) \, ds > 0$$

Multiplying the above inequality by $-f$ and integrating over $[t_3, t]$ for sufficiently large $t_3 \geq t_2$, we have

$$-\int_{t_3}^{t} (R(g(s)) - R(g(t_2)))^{\beta} |y(g(s))|^{-\beta} f(t, y(s), y(g(s)), y'(s), y'(h(s))) \, ds$$

$$\leq -\int_{t_3}^{t} F(s)^{-\beta} \, dF(s) = \frac{1}{1-\beta} [F(t_3)^{1-\beta} - F(t)^{1-\beta}] \qquad (4.5.7)$$

Letting $t \to \infty$ in the above inequality and using condition (a), one can conclude that the left side of (4.5.7) is convergent. This contradicts condition (4.5.3). The proof is therefore complete.

REMARK 4.5.1 From the proof of Theorem 4.5.1, we observe that the strict inequality in condition (c) can be relaxed. However, it is necessary to assume that $|F(t)| = \left| \int_{t}^{\infty} f \, ds \right| > 0$, for all nonoscillatory $y(t)$.

EXAMPLE 4.5.1 We consider

$$(r(t)y'(t))' + p(t)|y(g(t))|^{\beta} \operatorname{sgn} y(g(t)) = 0 \qquad (4.5.8)$$

where β is constant and $0 < \beta < 1$. To apply Theorem 4.5.1 to (4.5.8) we need to assume that $p(t) \geq 0$, $\int_{t}^{\infty} p(s) \, ds > 0$, and

$$\int_{}^{\infty} R(g(t))^{\beta} p(t) \, dt = \infty \qquad (4.5.9)$$

Then all solutions of (4.5.8) oscillate.

REMARK 4.5.2 Theorem 4.5.1 remains valid if the argument $g(t)$ is of mixed type, that is, it is advanced or retarded for certain values of t.

EXAMPLE 4.5.2 We consider

$$(ty'(t))' + p(t)y^{1/3}(\ln t)(1 + y'^2(\sqrt{t})) = 0 \qquad (4.5.10)$$

where $p(t) \geq 0$, $\int_{t}^{\infty} p(s) \, ds > 0$. According to Theorem 4.5.1, if

$$\int_{}^{\infty} (\ln \ln t)^{1/3} p(t) \, dt = \infty$$

then every solution to (4.5.10) oscillates.

In (4.5.1), if

$$f(t, y(t), y(g(t)), y'(t), y'(h(t))) = \sum_{i=1}^{n} p_i(t)|y(g(t))|^{\alpha_i} \operatorname{sgn} y(g(t)) \qquad (4.5.11)$$

then by Theorem 4.5.1, we have the following result.

COROLLARY 4.5.1 Assume that $r(t)$, $g(t)$ satisfy the conditions (a), (b) of Theorem 4.5.1, and $p_i \in C[R_+, R_+]$, $0 < \alpha_i < 1$, $i \in I_n$. Further assume that

$$\int^{\infty} R^{\alpha_k}(g(t)) P_k(t)\, dt = \infty \quad \text{for some } k \in I_n \tag{4.5.12}$$

Then every solution of (4.5.11) oscillates.

THEOREM 4.5.2 Assume that conditions (a), (b), and (c) of Theorem 4.5.1 hold. Further assume that there exists a positive number ϵ such that $0 < \epsilon < 1$ and

$$\int_{t_0}^{\infty} R^{1-\epsilon}(g(t)) \frac{f(t, y(t), y(g(t)), y'(t), y'(h(t)))}{y(g(t))}\, dt = \infty \tag{4.5.13}$$

for every positive nondecreasing or negative nonincreasing function $y(t)$. Then every solution of (4.5.1) oscillates.

Proof: In fact,

$$\int_{t_0}^{\infty} R^{1-\epsilon}(g(t)) \frac{|f(t, y(t), y(g(t)), y'(t), y'(h(t)))|}{|y(g(t))|^{1-\epsilon}}\, dt$$

$$= \int_{t_0}^{\infty} R^{1-\epsilon}(g(t)) \frac{|y(g(t))|}{|y(g(t))|^{1-\epsilon}} \frac{|f(t, y(t), y(g(t)), y'(t), y'(h(t)))|}{|y(g(t))|}\, dt \tag{4.5.14}$$

Since $|y(g(t))|^{\epsilon} \geq |y(g(t_0))|^{\epsilon} > 0$ for $t \geq T$, from (4.5.13) and (4.5.14) we obtain the relation (4.5.3). Then, from Theorem 4.5.1, we obtain Theorem 4.5.2.

Now we consider

$$(r(t)y'(t))' + \sum_{i=0}^{n} p_i(t) y^{2i+1}(g(t)) = 0 \tag{4.5.15}$$

COROLLARY 4.5.2 Assume that $r(t)$ and $g(t)$ satisfy the conditions (a) and (b) of Theorem 4.5.1, and $p_i \in C[R_+, R_+]$ for $i \in I_n$. Furthermore,

$$\int^{\infty} R^{1-\epsilon}(g(t))\left(\sum_{i=0}^{n} p_i(t)\right) dt = \infty \qquad 0 < \epsilon < 1 \tag{4.5.16}$$

Then every solution of (4.5.15) oscillates.

In fact, it is easy to see that (4.5.15) satisfies the conditions of Theorem 4.5.2.

In particular, for $n = 0$, (4.5.15) becomes

$$(r(t)y'(t))' + p(t)y(g(t)) = 0 \qquad\qquad (4.5.17)$$

and condition (4.5.16) becomes

$$\int^{\infty} R^{1-\epsilon}(g(t))p(t)\, dt = \infty \qquad 0 < \epsilon < 1 \qquad\qquad (4.5.18)$$

We note that ϵ cannot be equal to zero. In fact, consider the equation

$$y''(t) + \frac{1}{2\sqrt{2}}\,\frac{1}{t^2}\,y\!\left(\frac{t}{2}\right) = 0$$

which satisfies the condition $\int^{\infty} R(g(t))p(t)\, dt = \infty$, but it has a nonoscillatory solution $y(t) = t^{\frac{1}{2}}$.

THEOREM 4.5.3 Assume that (a), (b), and (c) of Theorem 4.5.1 hold. Further assume that there is a constant $\beta > 1$ such that

$$\int_{t_0}^{\infty} R(g(t)) \frac{|\,f(t, y(t), y(g(t)), y'(t), y'(h(t)))\,|}{|\,y(g(t))\,|^{\beta}}\, dt = \infty \qquad\qquad (4.5.19)$$

for every positive nondecreasing or negative nonincreasing function $y(t)$. Then every solution of (4.5.1) oscillates.

Proof: The method of proof is similar to Theorem 4.5.1.

COROLLARY 4.5.3 We consider

$$(r(t)y'(t))' + \sum_{i=1}^{n} p_i(t)y^{2i+1}(g(t)) = 0 \qquad\qquad (4.5.20)$$

Assume that $r(t)$, $g(t)$ satisfy conditions of Theorem 4.5.3. $p_i(t) \geq 0$ $(i = 1,\ 2,\ \ldots,\ n)$ and

$$\int^{\infty} R(g(t))\left(\sum_{i=1}^{n} p_i(t)\right) dt = \infty \qquad\qquad (4.5.21)$$

Then every solution of (4.5.20) oscillates.

REMARK 4.5.3 Condition (4.5.21) cannot be improved. (4.5.20), including the equation $y''(t) + p(t)y^{2n+1}(t) = 0$, $n \geq 1$, was discussed by Atkinson [7], but condition (4.5.21) is a necessary and sufficient condition for the oscillation of Atkinson's equation.

Now we consider two examples. One of them is

$$y''(t) + \frac{1}{4a^2t^2} y^3(t) = 0 \tag{4.5.22}$$

It is well known that every solution of (4.5.22) oscillates. The second equation,

$$y''(t) + \frac{1}{4a^2t^2} y^3(t^{1/3}) = 0 \tag{4.5.23}$$

has a nonoscillatory solution $y(t) = at^{\frac{1}{2}}$, but for the equation

$$y''(t) + \frac{1}{4a^2t^2} y^3(\lambda t) = 0 \qquad 0 < \lambda < 1 \tag{4.5.24}$$

every solution of (4.5.24) oscillates (according to Corollary 4.5.3).

These examples show that the order of the deviating argument g(t) is very important for the oscillation of the solutions. If g(t) is of the same order as t, then we can obtain a necessary and sufficient condition for the oscillation of a functional differential equation. The following results are based on the above idea.

THEOREM 4.5.4 Assume that conditions (a), (b), and (c) of Theorem 4.5.1 hold, $\lim_{t \to \infty} g'(t) = c$, $c > 0$, r(t) and r(g(t)) are of the same order if $t \to \infty$, and

$$\int^{\infty} R(t) \frac{|f(t, y(t), y(g(t)), y'(t), y'(h(t)))|}{|y(g(t))|^{\beta}} \, dt = \infty \tag{4.5.25}$$

for some $\beta > 1$ and every positive nondecreasing or negative nonincreasing function y(t). Then every solution of (4.5.1) oscillates.

Proof: Under the conditions of Theorem 4.5.4, R(t) and R(g(t)) are of the same order if $t \to \infty$, so we can obtain condition (4.5.19) from (4.5.25). The proof of the theorem follows by the application of Theorem 4.5.3.

We consider

$$(r(t)y'(t))' + p(t)(y(t) + y(g(t)))^{2n+1} = 0 \tag{4.5.26}$$

where n is a positive integer.

THEOREM 4.5.5 Assume that $p(t) \geq 0$, r(t), g(t) satisfy the conditions of Theorem 4.5.4. Then, a necessary and sufficient condition for (4.5.26) to be oscillatory is that

$$\int^{\infty} R(t)p(t) \, dt = \infty \tag{4.5.27}$$

Proof: The sufficiency of (4.5.27) follows immediately from Theorem 4.5.4.
To prove that (4.5.27) is necessary, we may suppose that there exists $M > 0$
such that

$$\int_{t_0}^{\infty} R(t)p(t)\,dt \leq M \tag{4.5.28}$$

Let $T_0 = \inf_{t \geq t_0} g(t)$. We introduce the Banach space $C[T_0, +\infty)$ of all bounded
continuous functions $y : [T_0, +\infty) \to R$ with norm

$$\|y\| = \sup\{|y(t)| : t \in [T_0, +\infty)\}$$

We consider subset Y of $C[T_0, +\infty)$, $Y = \{y \in C, a/2 \leq y(t) \leq a, a > 0\}$.
It is obvious that Y is a bounded, closed, and convex subset of $C[T_0, +\infty)$.

We define the operator ψ by

$$(\psi y)(t) = \begin{cases} \dfrac{a}{2} + \displaystyle\int_{t_0}^{t} R(\tau)(y(\tau) + y(g(\tau)))^{2n+1}\,d\tau \\[2ex] \qquad + R(t)\displaystyle\int_{t}^{\infty} p(\tau)(y(\tau) + y(g(\tau)))^{2n+1}\,d\tau \qquad t \geq t_0 \\[2ex] \dfrac{a}{2} \qquad\qquad\qquad\qquad\qquad\qquad\qquad T_0 \leq t \leq t_0 \end{cases} \tag{4.5.29}$$

(i) ψ maps Y into Y.

In fact, $(\psi y)(t) \geq a/2$, and

$$(\psi y)(t) \leq \frac{a}{2} + 2\int_{t_0}^{\infty} R(\tau)p(\tau)(y(\tau) + y(g(\tau)))^{2n+1}\,d\tau$$

$$\leq \frac{a}{2} + 2(2a)^{2n+1} M \leq a \qquad t \geq t_0$$

where we choose $a > 0$ such that $2(2a)^{2n+1}M \leq a/2$.

(ii) ψ is continuous.

Let $\{y_n\}$ be a sequence of elements of Y such that $\lim_{m \to \infty} \|y_m - y\| = 0$.
Since Y is closed, $y \in Y$ and

$$|(\psi y_m)(t) - (\psi y)(t)|$$

$$\leq 2\int_{t_0}^{\infty} R(\tau)p(\tau)|(y_m(\tau) + y_m(g(\tau)))^{2n+1} - (y(\tau) + y(g(\tau)))^{2n+1}|\,d\tau$$

$$= 2 \int_{t_0}^{\infty} G_m(\tau)\, d\tau \qquad t \geq t_0 \tag{4.5.30}$$

where

$$G_m(\tau) = R(\tau)\,|\,(y_m(\tau) + y_m(g(\tau)))^{2n+1} - (y(\tau) + y(g(\tau)))^{2n+1}\,|$$

Noting that $\lim_{m \to \infty} G_m(\tau) = 0$ and $\int_{t_0}^{\infty} G_m(\tau)\, d\tau \leq 2(2a)^{2n+1} \int_{t_0}^{\infty} R(\tau)p(\tau)\, d\tau$. From (4.5.30), we have

$$\| \psi y_m - \psi y \| \leq 2 \int_{t_0}^{\infty} G_m(\tau)\, d\tau \tag{4.5.31}$$

Applying the Lebesgue dominated convergence theorem, we obtain from (4.5.31) that $\lim_{m \to \infty} \| \psi y_m - \psi y \| = 0$, proving the continuity of ψ.

(iii) ψY is precompact.

It suffices to prove that ψY is equicontinuous on $[T_0, +\infty)$. Let $y \in Y$ and $t_1 > t_2 \geq t_0$. Then we have

$$(\psi y)(t_2) - (\psi y)(t_1) = \int_{t_1}^{t_2} R(\tau)p(\tau)(y(\tau) + y(g(\tau)))^{2n+1}\, d\tau$$

$$+ R(t_2) \int_{t_2}^{\infty} p(\tau)(y(\tau) + y(g(\tau)))^{2n+1}\, d\tau$$

$$- R(t_1) \int_{t_1}^{\infty} p(\tau)(y(\tau) + y(g(\tau)))^{2n+1}\, d\tau$$

so

$$|\,(\psi y)(t_2) - (\psi y)(t_1)\,| \leq 3(2a)^{2n+1} \int_{t_1}^{\infty} R(\tau)p(\tau)\, d\tau$$

Since the integral on the right-hand side tends to zero as $t_1 \to \infty$, given an $\epsilon > 0$, there is a $T > t_0$ such that for all $y \in Y$,

$$|\,(\psi y)(t_2) - (\psi y)(t_1)\,| < \epsilon \qquad \text{if } t_2 > t_1 \geq T \tag{4.5.32}$$

For $t_0 \leq t_1 < t_2 \leq T$, we have

$$| (\psi y)(t_2) - (\psi y)(t_1)| \le \int_{t_1}^{t_2} R(\tau)p(\tau)y(\tau) + y(g(\tau)))^{2n+1} \, d\tau$$

$$+ (R(t_2) - R(t_1)) \int_{t_2}^{\infty} p(\tau)(y(\tau) + y(g(\tau)))^{2n+1} \, d\tau$$

$$+ R(t_1) \int_{t_1}^{t_2} p(\tau)(y(\tau) + y(g(\tau)))^{2n+1} \, d\tau$$

$$= \int_{t_1}^{t_2} R(\tau)p(\tau)(y(\tau) + y(g(\tau)))^{2n+1} \, d\tau$$

$$+ (R(t_2) - R(t_1))\left[\int_{t_2}^{T} + \int_{T}^{\infty}\right] p(\tau)(y(\tau) + y(g(\tau)))^{2n+1} \, d\tau$$

$$+ R(t_1) \int_{t_1}^{t_2} p(\tau)(y(\tau) + y(g(\tau)))^{2n+1} \, d\tau \qquad (4.5.33)$$

but

$$\int_{T}^{\infty} p(\tau)(y(\tau) + y(g(\tau)))^{2n+1} \, d\tau \le (2a)^{2n+1} \int_{T}^{\infty} p(\tau) \, d\tau \le \frac{(2a)^{2n+1} M}{R(T)}$$

so, from (4.5.33), we obtain

$$| (\psi y)(t_2) - (\psi y)(t_1)|$$

$$\le 2(2a)^{2n+1} R(T) \int_{t_1}^{t_2} p(\tau) \, d\tau + (R(t_2) - R(t_1))(2a)^{2n+1}\left[\int_{t_0}^{T} p(\tau) \, d\tau + \frac{M}{R(T)}\right]$$

which shows that there is a $\delta > 0$ such that for all $y \in Y$,

$$| (\psi t)(t_2) - (\psi y)(t_1)| < \epsilon \quad \text{if } t_2 - t_1 < \delta \qquad (4.5.34)$$

In view of (4.5.32) and (4.5.34) we are able to decompose the interval $[t_0, +\infty)$ into a finite number of subintervals on each of which all functions ψy, $y \in Y$, have oscillations less than ϵ. Thus, ψY is precompact.

By Schauder's fixed point theorem there exists $y \in Y$ such that $y = \psi y$. Then, $y(t)$ is a nonoscillatory solution to the equation (4.5.26). This completes the proof.

COROLLARY 4.5.4 Under the conditions of Theorem 4.5.5, (4.5.26) has a bounded nonoscillatory solution $y(t)$, if and only if

$$\int_{t_0}^{\infty} R(t)p(t) \, dt < \infty \qquad (4.5.35)$$

Similarly, we can prove the following theorem.

THEOREM 4.5.6 Assume that $r(t)$, $g(t)$ of (4.5.20) satisfy the conditions of Theorem 4.5.4 and $p_i(t) > 0$, $i = 1, 2, \ldots, n$. Then a necessary and sufficient condition for (4.5.20) to be oscillatory is that

$$\int^{\infty} R(t)\left(\sum_{i=1}^{n} p_i(t) \right) dt = \infty \tag{4.5.36}$$

COROLLARY 4.5.5 Under the conditions of Theorem 4.5.6, (4.5.20) has a bounded nonoscillatory solution $y(t)$ if and only if

$$\int^{\infty} R(t)\left(\sum_{i=1}^{n} p_i(t) \right) dt < \infty \tag{4.5.37}$$

For the equation

$$(r(t)y'(t))' + p(t)y^{2n+1}(g(t)) = 0 \tag{4.5.38}$$

where n is a positive integer, if $p(t) \geq 0$ and $r(t)$, $g(t)$ of (4.5.38) satisfy the conditions of Theorem 4.5.6, then a necessary and sufficient condition for (4.5.37) to be oscillatory is that

$$\int^{\infty} R(t)p(t) \, dt = \infty \tag{4.5.39}$$

This is an extension of Atkinson's theorem.

Finally, we consider (4.5.1) under the condition

$$ap(t)\,\Omega(|y(g(t))|) \leq \frac{f}{y(g(t))} \leq bp(t)\,\Omega(|y(g(t))|) \tag{4.5.40}$$

where $b \geq a > 0$, $p(t) > 0$, $\Omega(0) = 0$, $\Omega(r)$ is a nondecreasing continuous function for $r > 0$, and

$$\int_{r_0}^{\infty} \frac{dr}{r\,\Omega(r)} < \infty \qquad r_0 > 0 \tag{4.5.41}$$

Apart from the hypotheses of Theorem 4.5.4 for $R(t)$, $g(t)$, assume that there is a constant $d > 2$ such that $R'(t) \leq dR(t)$.

THEOREM 4.5.7 Every solution of (4.5.1) is oscillatory if and only if

$$\int^{\infty} R(t)p(t) \, dt = \infty \tag{4.5.42}$$

Proof: The proof of sufficiency is similar to that of Theorem 4.5.1. To prove necessity of condition (4.5.42), we may suppose

$$\int_T^\infty p(t)R(t)\, dt \le M < \infty \tag{4.5.43}$$

Now we consider the integral equation

$$y(t) = \frac{c}{2} + \int_{T_0}^t (R(\tau) - R(T))f(\tau, y(\tau), y(g(\tau)), y'(\tau), y'(h(\tau)))\, d\tau$$

$$+ (R(t) - R(T)) \int_t^\infty f(\tau, y(\tau), y(g(\tau)), y'(\tau), y'(h(\tau)))\, d\tau \tag{4.5.44}$$

Let Z denote a set of all continuous functions z on $[T_0, +\infty)$ that take constant values on $[T_0, T]$ and have continuous first order derivatives on $[T, +\infty)$, where

$$T_0 = \min\left\{ \inf_{t \ge T} g(t),\ \inf_{t \ge T} h(t) \right\}$$

We define a seminorm on space Z by

$$\rho_\alpha(z) = \max_{t \in [T, \alpha]} |z'(t)| + |z(t)| \qquad \alpha \in (T, +\infty)$$

Then it is easy to see that Z is a locally convex space.

Sequence $\{z_n\}$ converges to z if and only if $\rho_\alpha(z_n - z) \to 0$ as $n \to \infty$ for each fixed $\alpha \in (T, +\infty)$. Now we define a set Y in the space Z composed of all $y \in Z$ which possess the following behavior:

(a) $|y(t) - c/2| \le c/4$ for $t \ge T_0$

(b) $|y'(t)| \le c/4$ for $t \ge T_0$

where $c > 0$ is a constant to be determined. Obviously, Y is a nonempty convex closed subset of Z.

We define an operator $S: Y \to Z$ by

$$z(t) = (Sy)(t) = \begin{cases} \dfrac{c}{2} + \displaystyle\int_T^t R(\sigma)f(\tau, y(\tau), \hat{y}(\tau), y'(\tau), \hat{y}'(\tau))\, d\tau \\[2em] \quad + R(t) \displaystyle\int_t^\infty f(\tau, y(\tau), \hat{y}(\tau), y'(\tau), \hat{y}'(\tau))\, d\tau \qquad t \ge T \\[2em] \dfrac{c}{2} + R(t) \displaystyle\int_T^\infty f(\tau, y(\tau), \hat{y}(\tau), y'(\tau), \hat{y}'(\tau))\, d\tau \qquad t \in [T_0, T] \end{cases}$$

where

$$\hat{y}(t) = \begin{cases} y(g(t)) & g(t) \geq T \\ y(T) & g(t) \leq T \end{cases} \qquad \hat{y}'(t) = \begin{cases} y'(h(t)) & h(t) \geq T \\ y'(T) & h(t) \leq T \end{cases}$$

Because of conditions (4.5.40) to (4.5.42), S is defined on Y.

Our objective is to show that S satisfies conditions of the Schauder–Tychonov fixed point theorem.

(i) S maps Y into itself.

In fact,

$$\left| z - \frac{c}{2} \right| \leq 2bc\,\Omega(c)\,M \leq dbc\,\Omega(c)\,M$$

We choose c such that $dbc\,\Omega(c)\,M \leq c/4$; then $|z - c/2| \leq c/4$ for all $y \in Y$, $t \geq T$, and

$$|z'(t)| \leq dbc\,M\Omega(c) \leq \frac{c}{4} \qquad t \geq T$$

Thus, S maps Y into itself.

(ii) $\overline{SY}$ is a compact subset of Y.

In fact, for all $y \in Y$, $t_2 > t_1 > T$,

$$|z'(t_2) - z'(t_1)| = \left| R'(t_2) \int_{t_2}^{\infty} f \, d\tau - R'(t_1) \int_{t_1}^{\infty} f \, d\tau \right|$$

$$\leq |R'(t_2) - R'(t_1)| \int_{t_2}^{\infty} f \, d\tau + R'(t_1) \left| \int_{t_1}^{t_2} f \, d\tau \right|$$

$$\leq |R'(t_2) - R'(t_1)|\, bc\,\Omega(c) \frac{M}{R(T)} + dbc\,\Omega(c) \left| \int_{t_1}^{t_2} R(\tau)p(\tau) \, d\tau \right|$$

Therefore $z'(t)$, $z \in SY$, is equicontinuous on any bounded interval $[T, \alpha]$. Thus, using Ascoli's theorem, any sequence $\{z_k\}$ in $\overline{SY}$ includes a subsequence $\{w_k\}$ such that the sequence $\{w'_k\}$ is uniformly convergent on any $[T, \alpha]$. From (a), (b), sequence $\{w_k(t)\}$ is bounded. Thus there exists a subsequence $\{v_k\}$ of $\{w_k\}$ such that $\{v_k(t)\}$ is convergent.

Therefore, for any sequence $\{z_k\}$ in $\overline{SY}$, there is a subsequence $\{v_k\}$ such that $\lim_{k \to \infty} v_k = v \in Y$ according to the topology in Z.

We have just proved the fact that $\overline{SY}$ is a compact subset of Y.

(iii) S is a continuous operator.

In fact, let $\{y_k\} \in Y$ be an elementary sequence such that

$\lim_{m \to \infty} \| y_m - y \| = 0$. Since Y is closed, so $y \in \bar{Y}$. Obviously, $\lim_{k \to \infty} \hat{y}_k(s) = \hat{y}(s)$, $\lim_{k \to \infty} \hat{y}_k'(s) = \hat{y}'(s)$ for $s \geq T$, and $\{y_k\}$ and $\{y_k'\}$ are uniformly convergent on any $[T, \alpha]$.

Now we want to prove that $\{z_k\} = \{Sy_k\}$ also has this convergence property.

In fact,

$$z_k'(t) = R'(t) \int_t^\infty f(\tau, y_k, \hat{y}_k, y_k', \hat{y}_k') \, d\tau \qquad t \geq T$$

Since f is continuous then

$$\lim_{k \to \infty} f(\tau, y_k, \hat{y}_k, y_k', \hat{y}_k') = f(\tau, y, \hat{y}, y', \hat{y}') \qquad \tau \geq T$$

and

$$\left| \int_t^\infty f \, d\tau \right| \leq bc\Omega(c) \int_t^\infty p(\tau) \, d\tau$$

Using Lebesgue's convergence theorem, we obtain

$$\lim_{k \to \infty} z_k'(t) = z'(t) \qquad t \geq T$$

In the same way, we can obtain $\lim_{k \to \infty} z_k(t) = z(t)$, $t \geq T$.

Now we want to prove that the convergence is uniform on $[T, \alpha]$, for every fixed $\alpha \in (T, +\infty)$. In fact,

$$| z_n'(t) - z'(t) | \leq d \int_T^\infty G_n(\tau) \, d\tau \qquad t \geq T$$

where

$$G_n(\tau) = R(\tau) | f(\tau, y_n(\tau), \hat{y}_n(\tau), y_n'(\tau), \hat{y}_n'(\tau))$$

$$- f(\tau, y(\tau), \hat{y}(\tau), y'(\tau), \hat{y}'(\tau)) |$$

so

$$\max_{[T, \alpha]} | z_n'(t) - z'(t) | \leq d \int_T^\infty G_n(\tau) \, d\tau$$

but $\lim_{n \to \infty} G_n(\tau) = 0$ and $G_n(\tau) \leq 2bc\Omega(c)R(\tau)p(\tau)$.

Using the Lebesgue dominated convergence theorem, we obtain

$$\lim_{n \to \infty} \max_{[t, \alpha]} | z_n'(t) - z'(t) | = 0$$

This fact shows that $\{z_k\} = \{Sy_k\}$ is convergent according to the topology in Z. This proves the continuity of S.

From the preceding observations, we see that Schauder-Tychonov's fixed point theorem can be applied to the operators. Let $y \in Y$ be a fixed point of S. Then $y(t)$ is a bounded nonoscillatory solution of (4.5.1). This completes the proof of Theorem 4.5.7.

COROLLARY 4.5.6   Under the conditions of Theorem 4.5.7, equation (4.5.1) has a bounded nonoscillatory solution if and only if

$$\int^{\infty} R(t)p(t) \, dt < \infty$$

REMARK 4.5.4   Indeed, Theorem 4.5.7 yields two conclusions, i.e., if the inequality on the left of condition (4.5.41) holds, then the condition (4.5.42) is sufficient for every solution of (4.5.1) to be oscillatory. If the inequality on the right of condition (4.5.40) holds, then condition (4.5.43) is sufficient for (4.5.1) to have a bounded nonoscillatory solution.

4.6 NONLINEAR EQUATIONS WITH $\int^{\infty} ds/r(s) < \infty$

In this section, we shall discuss the case $\int^{\infty} dt/r(t) < \infty$ relative to equation (4.5.1). For simplicity, we restrict our discussion to the equation

$$(r(t)y'(t))' + f(y(g(t)), t) = 0 \tag{4.6.1}$$

DEFINITION 4.6.1   Equation (4.6.1) is called:

(i) <u>Superlinear</u> if, for each fixed t, $f(y,t)/y$ is nondecreasing in y for $y > 0$ and nonincreasing in y for $y < 0$

(ii) <u>Strongly superlinear</u> if there exists a number $\sigma > 1$ such that, for each fixed t, $f(y,t)/|y|^{\sigma} \,\mathrm{sgn}\, y$ is nondecreasing in y for $y > 0$ and nonincreasing in y for $y < 0$

(iii) <u>Sublinear</u> if, for each fixed t, $f(y,t)/y$ is nonincreasing in y for $y > 0$ and nondecreasing in y for $y < 0$

(iv) <u>Strongly sublinear</u> if there exists a number $\tau < 1$ such that, for each t, $f(y,t)/|y|^{\tau}\,\mathrm{sgn}\, y$ is nonincreasing in y for $y > 0$ and nondecreasing in y for $y < 0$.

LEMMA 4.6.1   Assume that

(a) $r(t)$ is positive continuous for $t \geq \alpha$ and $\int^{\infty} dt/r(t) < \infty$

(b) $g(t)$ is continuous for $t \geq \alpha$ and $g(t) \leq t$, $\lim_{t \to \infty} g(t) = \infty$

(c) $f(y, t)$ is continuous for $|y| < \infty$, $t \geq \alpha$, and $yf(y, t) > 0$ for $y \neq 0$, $t \geq \alpha$.

If $y(t)$ is a positive solution of (4.6.1), then it is bounded above and satisfies

$$y(t) \geq -r(t)y'(t)\rho(t) \tag{4.6.2}$$

for all sufficiently large t, where $\rho(t) = \int_t^\infty ds/r(s)$ for all $t > 0$.

Proof: Suppose that there is a solution $y(g(t)) > 0$ for $t \geq t_1$. From (4.6.1), we have $(r(t)y'(t))' \leq 0$, $t \geq t_1$, i.e., $r(t)y'(t) \leq r(t_1)y'(t_1)$, for $t \geq t_1$. Dividing this inequality by $r(t)$ and integrating over $[t_1, t]$, we get

$$y(t) - y(t_1) \leq r(t_1)y'(t_1) \int_{t_1}^t r^{-1}(s) \, ds < \infty \tag{4.6.3}$$

i.e., $y(t)$ is bounded above. Letting $t \to \infty$, we have

$$y(t_1) \geq -r(t_1)y'(t_1)\rho(t_1)$$

for sufficiently large t_1. The proof is complete.

THEOREM 4.6.1 Assume that (4.6.1) satisfies the assumptions (a) to (c) of Lemma 4.6.1 and furthermore that it is either superlinear or sublinear. A necessary and sufficient condition for (4.6.1) to have a nonoscillatory solution which is asymptotic to a nonzero constant is

$$\int^\infty \rho(t) |f(c, t)| \, dt < \infty \quad \text{for some } c \neq 0 \tag{4.6.4}$$

Proof: Necessity: Let $y(t)$ be a nonoscillatory solution of (4.6.1) with $y(t) \to c \neq 0$ as $t \to \infty$. Without loss of generality, we may assume that $c > 0$. Hence there are positive constants c_1, c_2, and t_1 such that

$$c_1 \leq y(g(t)) \leq c_2 \qquad t \geq t_1 \tag{4.6.5}$$

By multiplying equation (4.6.1) by $\rho(t)$ and integrating over $[t_1, t]$, we have

$$r(t)y'(t)\rho(t) + y(t) - r(t_1)y'(t_1)\rho(t_1) - y(t_1) + \int_{t_1}^t \rho(s)f(y(g(s)), s) \, ds = 0$$

which implies, in view of (4.6.2),

$$\int_{t_1}^\infty \rho(s)f(y(g(s)), s) \, ds < \infty \tag{4.6.6}$$

From (4.6.5) and (4.6.6), it follows that $\int^\infty \rho(s)f(c_1, s) \, ds < \infty$ whenever (4.6.1) is superlinear and $\int^\infty \rho(s)f(c_2, s) \, ds < \infty$ whenever (4.6.1) is sublinear.

Sufficiency: Based on the superlinearity or sublinearity of (4.6.1) we choose $a = c/2$ or $a = c$. Consider the integral equation

$$y(t) = \begin{cases} a + \rho(t) \int_T^t f(y(g(s)), s)\, ds + \int_t^{\infty} \rho(s)f(y(g(s)), s)\, ds & \text{if } t \geq T \\[3mm] a + \int_T^{\infty} \rho(s)f(y(g(s)), s)\, ds & \text{if } t < T \end{cases} \tag{4.6.7}$$

where T is chosen so large that

$$\int_T^{\infty} \rho(t)f(c, t)\, dt < a \tag{4.6.8}$$

It is clear that a solution of (4.6.7) is a solution of equation (4.6.1).

Let $\tau = \inf[g(t): t \geq T]$ and denote by $C[\tau, \infty)$ the linear space of all continuous functions $y: [\tau, \infty) \to R$ such that

$$\sup\{\rho(t)\,|\,y(t)\,| : t \geq \tau\} < \infty$$

We define

$$\|y\| = \sup\{\rho(t)\,|\,y(t)\,| : t \geq \tau\} \qquad y \in C[\tau, \infty)$$

Then we can easily see that $y \to \|y\|$ is a norm for which $C[\tau, \infty)$ is a Banach space. Consider the set $\bar{Y}$ of all functions $y \in C[\tau, \infty)$ satisfying $a \leq y(t) \leq 2a$ on $[\tau, \infty)$. Clearly, $\bar{Y}$ is a bounded, closed, and convex subset of $C[\tau, \infty)$.

Let us now define the operator Φ by

$$(\Phi y)(t) = \begin{cases} a + \rho(t) \int_T^t f(y(g)s)), s)\, ds + \int_t^{\infty} \rho(s)f(y(g(s)), s)\, ds & t \geq T \\[3mm] a + \int_T^{\infty} \rho(s)f(y(g(s)), s)\, ds & \tau \leq t \leq T \end{cases} \tag{4.6.9}$$

We shall show that Φ is continuous and maps $\bar{Y}$ into a compact subset of $\bar{Y}$.

(i) Φ maps $\bar{Y}$ into $\bar{Y}$: If $y \in \bar{Y}$, then $(\Phi y)(t) \geq a$, $t \geq \tau$, and

$$(\Phi y)(t) \leq a + \int_T^{\infty} \rho(s)f(y(g(s)), s)\, ds \leq a + \int_T^{\infty} \rho(s)f(c, s)\, ds \leq 2a$$

for $t \geq \tau$.

(ii) Φ is continuous: Let $\{y_n\} \subset \bar{Y}$ be a sequence converging to y: $\lim_{n \to \infty} \|y_n - y\| = 0$. Since $\bar{Y}$ is closed, $y \in \bar{Y}$. By the definition of Φ we see that

$$|(\Phi y_n)(t) - (\Phi y)(t)| \leq \int_T^\infty \rho(s)G_n(s)\ ds \qquad t \geq \tau$$

where

$$G_n(s) = |f(y_n(g(s)), s) - f(y(g(s)), s)|$$

Hence we find

$$\|\Phi y_n - \Phi y\| \leq \rho(T) \int_T^\infty \rho(s)G_n(s)\ ds \qquad\qquad (4.6.10)$$

But $\lim_{n \to \infty} G_n(s) = 0$, which is a consequence of the convergence $y_n \to y$ in $C[\tau, \infty)$, and the fact

$$\rho(s)G_n(s) \leq \rho(s)[f(y_n(g(s)), s) + f(y(g(s)), s)]$$

$$\leq 2\rho(s)f(c, s) \qquad s \geq T$$

From (4.6.8) and the Lebesgue dominated convergence theorem we obtain

$$\lim_{n \to \infty} \int_T^\infty \rho(s)G_n(s)\ ds = 0$$

Consequently, from (4.6.10), $\lim_{n \to \infty} \|\Phi y_n - \Phi y\| = 0$, proving the continuity of Φ.

(iii) $\Phi\bar{Y}$ is compact: According to a theorem of Levitan [167] it suffices to show that, for any given $\epsilon > 0$, the interval $[\tau, \infty)$ can be divided into a finite number of subintervals in such a way that the oscillations on each subinterval of all functions $\rho\Phi y$, $y \in \bar{Y}$, are less than ϵ. Since each Φy is constant on $[\tau, T]$, we need only to examine the behavior of $\rho\Phi y$ on the interval $[T, \infty)$. For $t_2 > t_1 \geq T$ we see that

$$(\rho\Phi y)(t_2) - (\rho\Phi y)(t_1)$$

$$= a(\rho(t_2) - \rho(t_1)) + \rho^2(t_2) \int_T^{t_2} f(y(g(s)), s)\ ds - \rho^2(t_1) \int_T^{t_1} f(y(g(s)), s)\ ds$$

$$+ \rho(t_2) \int_{t_2}^\infty \rho(s)f(y(g(s)), s)\ ds - \rho(t_1) \int_{t_1}^\infty \rho(s)f(y(g(s)), s)\ ds$$

$$= a(\rho(t_2) - \rho(t_1)) + (\rho^2(t_2) - \rho^2(t_1)) \int_T^{t_2} f(y(g(s)), s)\, ds$$

$$+ \rho^2(t_1) \int_{t_1}^{t_2} f(y(g(s)), s)\, ds - (\rho(t_2) - \rho(t_1)) \int_{t_2}^{\infty} \rho(s) f(y(g(s)), s)\, ds$$

$$- \rho(t_1) \int_{t_1}^{t_2} \rho(s) f(y(g(s)), s)\, ds \tag{4.6.11}$$

Hence

$$|(\rho \Phi y)(t_2) - (\rho \Phi y)(t_1)| \le a(\rho(t_2)) - \rho(t_1)) + \rho(t_2) \int_T^{t_2} \rho(s) f((y(g(s)), s)\, ds$$

$$+ \rho(t_1) \int_T^{t_1} \rho(s) f(y(g(s)), s)\, ds + \rho(t_2) \int_{t_2}^{\infty} \rho(s) f(y(g(s)), s)\, ds$$

$$+ \rho(t_1) \int_{t_1}^{\infty} \rho(s) f(y(g(s)), s)\, ds$$

$$\le 2a\rho(t_1) + 2\rho(t_1) \int_T^{\infty} \rho(s) f(y(g(s)), s)\, ds \le 4a\rho(t_1)$$

Since $\rho(t_1) \to 0$ as $t_1 \to \infty$, we conclude from the above inequalities that there exists $t^* > T$ such that for all $y \in \bar{Y}$,

$$|(\rho \Phi y)(t_2) - (\rho \Phi y)(t_1)| < \epsilon \qquad t_2 > t_1 \ge t^*$$

Let $T \le t_1 < t_2 \le t^*$. Then from (4.6.11),

$$|(\rho \Phi y)(t_2) - (\rho \Phi y)(t_1)| \le a|\rho(t_2) - \rho(t_1)| + |\rho^2(t_2) - \rho^2(t_1)| \int_T^{t^*} f(y(g(s)), s)\, ds$$

$$+ \rho^2(T) \int_{t_1}^{t_2} f(y(g(s)), s)\, ds + |\rho(t_2) - \rho(t_1)| \int_T^{\infty} \rho(s) f(y(g(s)), s)\, ds$$

$$+ \rho(T) \int_{t_1}^{t_2} \rho(s) f(y(g(s)), s)\, ds$$

$$\leq 2a\,|\rho(t_2) - \rho(t_1)| + 2a|\rho^2(t_2) - \rho^2(t_1)|\int_T^{t*} f(c, s)\, ds$$

$$+ 2a\rho^2(T)\int_{t_1}^{t_2} f(c, s)\, ds + 2a\rho(T)\int_{t_1}^{t_2}\rho(s)f(c, s)\, ds$$

where in view of the uniform continuity of $\rho(t)$ and $f(c, t)$ on $[T, t*]$ we see that there exists $\delta > 0$ such that for all $y \in \bar{Y}$,

$$|(\rho\Phi y)(t_2) - (\rho\Phi y)(t_1)| < \epsilon \qquad \text{if} \quad t_2 - t_1 < \delta$$

This permits us to divide $[T, t*]$ into a finite number of subintervals on each of which every $\rho\Phi y$, $y \in \bar{Y}$, has oscillation less than ϵ. Thus it follows that $\Phi\bar{Y}$ is a compact subset of $\bar{Y}$.

From the preceding considerations we see that the fixed point theorem of Schauder can be applied to the operator Φ. Let $y \in \bar{Y}$ be a fixed point of Φ. Then, by (4.6.9), $y(t)$ satisfies the integral equation (4.6.7) for $t \geq T$, and since

$$y'(t) = -r^{-1}(t)\int_T^t f(y(g(s)), s)\, ds < 0$$

$y(t)$ decreases to a positive number in $[a, 2a]$ as t grows to infinity. Theorem 4.6.1 is thereby proved.

THEOREM 4.6.2 Assume that (4.6.1) satisfies the assumptions (a) to (c) of Lemma 4.6.1 and furthermore it is either superlinear or sublinear. A necessary and sufficient condition for (4.6.1) to have a nonoscillatory solution which is asymptotic to $a\rho(t)$ as $t \to \infty$ for $a \neq 0$ is that

$$\int^{\infty} f(c\rho(g(t)), t)\, dt < \infty \qquad \text{for some } c \tag{4.6.12}$$

Proof: An argument similar to the proof of Theorem 4.6.1 can be formulated. However, a Banach space $C[\tau, \infty)$ with a norm defined by

$$\|y\| = \sup\{\rho^{-1}(t)|y(t)| : t \geq \tau\} < \infty$$

is needed in this case.

THEOREM 4.6.3 Assume that conditions (a), (b), and (c) of Lemma 4.6.1 hold and let equation (4.6.1) be strongly superlinear. A sufficient condition for (4.6.1) to be oscillatory is that

$$\int^{\infty} f(c\rho(t), t)\, dt = \infty \qquad \text{for all } c > 0 \tag{4.6.13}$$

Proof: Let $y(t)$ be a positive nonoscillatory solution of (4.6.1). The derivative $y'(t)$ is of one sign for all large t.

<u>Case 1</u>. $y'(t) \geq 0$. There are positive constants c_1, t_1 such that $y(g(t)) \geq c_1$ for $t \geq t_1$. Integrating (4.6.1) from t_1 to t and then letting $t \to \infty$, we see that

$$\int_{t_1}^{\infty} f(y(g(s)), s)\, ds < \infty$$

This implies $\int^{\infty} f(c_1 \rho(s), s)\, ds < \infty$.

<u>Case 2</u>. $y'(t) \leq 0$. By Lemma 4.6.1 and the decreasing character of $y(t)$ we obtain

$$y(g(t)) \geq -r(t)y'(t)\rho(t) \geq c_2\rho(t) \qquad t \geq t_2$$

where

$$c_2 = -r(t_2)y'(t_2) > 0$$

$$\frac{f(y(g(t)), t)}{y^{\sigma}(g(t))} \geq \frac{f(c_2\rho(t), t)}{[c_2\rho(t)]^{\sigma}} \qquad t \geq t_2$$

From the above inequality and superlinearity of f, we have

$$\{-[-r(t)y'(t)]^{-\sigma+1}\}' = (\sigma - 1)[-r(t)y'(t)]^{-\sigma} f(y(g(t)), t)$$

$$= (\sigma - 1)(-r(t)y'(t))^{-\sigma} y^{\sigma}(g(t)) \frac{f(y(g(t)), t)}{y^{\sigma}(g(t))}$$

$$\geq (\sigma - 1)(-r(t)y'(t))^{-\sigma}(-r(t)y'(t)\rho(t))^{\sigma} \frac{f(y(g(t)), t)}{y^{\sigma}(g(t))}$$

$$= (\sigma - 1)\rho^{\sigma}(t) \frac{f(c_2\rho(t), t)}{(c_2\rho(t))^{\sigma}} = \frac{(\sigma - 1)}{c_2^{\sigma}} f(c_2\rho(t), t)$$

Integrating the above inequality we have

$$\frac{(\sigma - 1)}{c_2^{\sigma}} \int_{t_2}^{t} f(c_2\rho(s), s)\, ds \leq [-r(t_2)y'(t_2)]^{-(\sigma-1)} - [-r(t)y'(t)]^{-(\sigma-1)}$$

which yields $\int_{t_2}^{\infty} f(c_2\rho(t), t)\, dt < \infty$. This is a contradiction to (4.6.13). Thus Theorem 4.6.3 is proved.

From Theorem 4.6.2, it follows that if equation (4.6.1), whether superlinear or sublinear, is oscillatory, then

$$\int_{t_2}^{\infty} f(c\rho(g(t)), t)\, dt = \infty \qquad \text{for all } c > 0 \tag{4.6.14}$$

In some cases conditions (4.6.13) and (4.6.14) are equivalent. For example consider

(i) $r(t) = ct(\log t)^{\alpha}$, $g(t) = t^{\beta}$ or $g(t) = \nu t$, where $c > 0$, $\alpha > 1$, $0 < \beta < 1$, and $0 < \nu < 1$

(ii) $r(t) = ct^{\rho}$, $g(t) = \nu t$, where $c > 0$, $\rho > 1$, and $0 < \nu < 1$

(iii) $r(t) = ce^{qt}$, $g(t) = t - \tau(t)$, $0 \leq \tau(t) \leq M$, where $c > 0$, $q > 0$, and $M > 0$.

In these cases (4.6.13) is a necessary and sufficient condition for the oscillation of equation (4.6.1).

In general, there is a gap between conditions (4.6.13) and (4.6.14).

EXAMPLE 4.6.1 Consider the delay equation

$$[t^3 y'(t)]' + t[y(t^{1/3})]^3 = 0 \tag{4.6.15}$$

It satisfies condition (4.6.14), but does not satisfy condition (4.6.13). In fact, this equation has a nonoscillatory solution $y(t) = 1/t$.

Similarly, we have the following theorem.

THEOREM 4.6.4 Assume that conditions (a), (b), and (c) of Lemma 4.6.1 hold and let equation (4.6.1) be strongly sublinear. A sufficient condition for (4.6.1) to be oscillatory is that

$$\int^{\infty} \rho(t)f(c, t)\, dt = \infty \qquad \text{for all } c > 0 \tag{4.6.16}$$

By combining Theorems 4.6.1 and 4.6.4, we have the following theorem.

THEOREM 4.6.5 Assume that the conditions (a), (b), and (c) of Lemma 4.6.1 hold and let equation (4.6.1) be strongly sublinear. A necessary and sufficient condition for (4.6.1) to be oscillatory is that (4.6.16) be valid.

4.7 EQUATIONS WITH DEVIATING ARGUMENT

We consider

$$(r(t)y'(t))' + \int_0^{\beta(t)} f(y(t - s))\, d\eta(t, s) = 0, \quad \beta(t) > 0, \quad t \geq a \geq 0 \tag{4.7.1}$$

with initial condition $y(s) = \phi(s)$ for $s \in (-\infty, a]$ and $y'(a) = A$.

In this section our purpose is to present criteria for all solutions of (4.7.1) to be oscillatory.

THEOREM 4.7.1 Assume that

(i) $\eta(\cdot,\cdot)$ satisfies the conditions in Section 3.8

(ii) $r \in C^1[a,\infty)$ and $r(t) > 0$

(iii) $0 < \beta(t) < t$, $\beta \in C^1[a,\infty)$, $\beta'(t) \leq 1$, and $\lim_{t\to\infty}(t - \beta(t)) = \infty$

(iv) $f \in C(-\infty,\infty) \cap C^1(R - \{0\})$, $yf(y) > 0$, and $f'(y) \geq 0$ for $y \neq 0$.

Further assume that $\int^\infty dt/r(t) = \infty$, and there exist two positive functions $\rho \in C^2(0,\infty)$, $\phi \in C^1(0,\infty)$ with the following properties:

$$\rho'(t) \geq 0, \quad (r(t)\rho'(t))' \leq 0, \quad \phi'(y) \geq 0 \tag{4.7.2}$$

$$\int_\delta^{+\infty} \frac{dy}{f(y)\,\phi(y)} < \infty, \quad \int_{-\delta}^{-\infty} \frac{dy}{f(y)\,\phi(y)} < \infty \quad \text{for some } \delta > 0 \tag{4.7.3}$$

$$\int_{T_2}^{\infty} \frac{\rho(g(t))}{\phi(AR_T(g\,t)))} \mathop{V}_{0}^{\beta(t)} \eta(t,s)\, dt = \infty \tag{4.7.4}$$

for any $T > a$, $T_2 > T$, $A > 0$ constant, $g(t) = t - \beta(t)$, and $R_T(t) = \int_T^t ds/r(s)$, where $\mathop{V}_0^{\beta(t)} \eta(t,s)$ is the variation of the function $\eta(t,s)$ in $s \in [0, \beta(t)]$ for each $t \in R_+$. Then all solutions of (4.7.1) are oscillatory.

Proof: Suppose there exists a nonoscillatory solution $y(t)$ of (4.7.1). Without loss of generality we may assume that $y(t - s) > 0$, $s \in [0, \beta(t)]$, for all $t \geq T$. From (4.7.1), $(r(t)y'(t))' \leq 0$ for $t \geq T$. Since $\int^\infty dt/r(t) = \infty$, it follows that $y'(t) \geq 0$ for $t \geq T_1 \geq T$. We then have

$$r(t)y'(t) \leq r(T_1)y'(T_1) \qquad t \geq T_1$$

or

$$y'(t) \leq \frac{r(T_1)y'(T_1)}{r(t)}$$

Hence

$$y(g(t)) \leq AR_{T_1}(g(t)) \qquad t \geq T_2$$

where T_2 is a sufficiently large number such that $g(t) > T_1$ for $t \geq T_2$.

Multiplying (4.7.1) by $\rho(g(t))/f(y(g(t)))\,\phi(AR_{T_1}(g(t)))$ and integrating on $[T_2, t]$, we obtain

$$\int_{T_2}^{t} \frac{\rho(g(s))(r(s)y'(s))'}{f(y(g(s)))\,\phi(AR_{T_1}(g(s)))}\,ds$$

$$+ \int_{T_2}^{t} \frac{\rho(g(s))}{f(y(g(s)))\,\phi(AR_{T_1}(g(s)))}\left[\int_{0}^{\beta(s)} f(y(s-\sigma))\,d\eta(s,\sigma)\right] ds = 0$$

Integrating the first integral by parts, we get

$$\frac{\rho(g(t))r(t)y'(t)}{f(y(g(t)))\,\phi(AR_{T_1}(g(t)))} - \int_{T_2}^{t} (r(s)y'(s))[\rho'(g(s))g'(s)f(y(g(s)))\,\phi(AR_{T_1}(g(s)))$$

$$- \rho(g(s))[f(y(g(s)))\,\phi(AR_{T_1}(g(s)))]']/(f(y(g(s)))\,\phi(AR_{T_1}(g(s))))^{-2})\,ds$$

$$+ c + \int_{T_2}^{t} \frac{\rho(g(s))}{\phi(AR_{T_1}(g(s)))} \int_{0}^{\beta(s)} d\eta(s,\sigma)\,ds \leq 0$$

where c is a constant.

We consider the first integral in the above inequality:

$$\int_{T_2}^{t} \frac{r(s)y'(s)\rho'(g(s))g'(s)}{f(y(g(s)))\,\phi(AR_{T_1}(g(s)))}\,ds \leq \int_{T_2}^{t} \frac{r(g(s))y'(g(s))\rho'(g(s))g'(s)}{f(y(g(s)))\,\phi(AR_{T_1}(g(s)))}\,ds$$

$$\leq r(g(T_2))\rho'(g(T_2)) \int_{y(g(T_2))}^{y(g(t))} \frac{dy}{f(y)\,\phi(y)} < \infty$$

For the second integral,

$$\int_{T_2}^{t} \frac{r(s)y'(s)\rho(g(s))[f(y(g(s)))\,\phi(AR_{T_1}(g(s)))]'}{[f(y(g(s)))\,\phi(AR_{T_1}(g(s)))]^2}\,ds \geq 0$$

But condition (4.7.4) shows that the third integral will tend to infinity as $t \to \infty$ and this implies that

$$\frac{\rho(g(t))r(t)y'(t)}{f(y(g(t)))\,\phi(AR_{T_1}(g(t)))} \to -\infty \qquad \text{as } t \to \infty$$

which contradicts the fact that $y'(t) \geq 0$ for all $t \geq T_1$. The proof is complete.

COROLLARY 4.7.1 We consider

$$(r(t)y'(t))' + \sum_{i=1}^{n} a_i(t)y(t - \tau_i(t)) = 0 \qquad\qquad (4.7.5)$$

where $\tau_i(t) \leq t$; $\tau_i'(t) \leq 1$, $\lim_{t\to\infty}(t - \tau_i(t)) = \infty$; $a_i(t) \geq 0$ is integrable;

$r(t) > 0$ is continuous and $\int^{\infty} dt/r(t) = \infty$; and

$$\int_{T}^{\infty} R_T^{1-\epsilon}(g(t)) \left(\sum_{i=1}^{n} a_i(t) \right) dt = \infty \qquad 1 \geq \epsilon > 0 \qquad (4.7.6)$$

where $g(t) = \max (t - \tau_1(t), \cdots, t - \tau_n(t))$. Then all solutions of (4.7.5) are oscillatory.

Proof: In fact, we take

$$\eta(t, s) = \sum_{i=1}^{n} U(s - \tau_i(t)) a_i(t) \qquad (4.7.7)$$

Equation (4.7.5) becomes (4.7.1) whenever $f(y) = y$. We set $\rho(t) = R_T(t)$ and $\phi(y) = y^\epsilon$, $1 \geq \epsilon > 0$. By applying Theorem 4.7.1, the conclusion of Corollary 4.7.1 follows.

COROLLARY 4.7.2 We consider

$$(r(t)y'(t))' + a(t)|y(g(t))|^\alpha \ \text{sgn} \ y(g(t)) = 0 \qquad \alpha > 1 \qquad (4.7.8)$$

where $r(t)$, $g(t)$, and $a(t)$ satisfy the conditions of Corollary 4.7.1 and

$$\int_{T}^{\infty} R_T(g(t)) a(t) \ dt = \infty \qquad \text{for any } T > 0 \qquad (4.7.9)$$

Then all solutions of (4.7.8) are oscillatory.

Proof: Applying Theorem 4.7.1 to the particular case

$$\eta(t, s) = U(s - g(t)) a(t), \quad f(y) = |y|^\alpha \, \text{sgn} \, y, \quad \rho(t) = R_T(t), \quad \phi(y) \equiv 1$$

the conclusion remains valid.

THEOREM 4.7.2 Assume that the conditions (i) to (iv) of Theorem 4.7.1 hold and $\int^{\infty} ds/r(s) < \infty$. In addition, assume that there exists a positive function $\sigma(t) \in C^2(0, \infty)$ with the properties:

$$\sigma'(t) \leq 0, \quad (r(t)\sigma'(t))' \geq 0$$

$$\int^{\infty} \frac{dt}{\sigma(t) r(t)} = \infty \qquad (4.7.10)$$

$$\int^{\infty} \sigma(t) \ \overset{\beta(t)}{\underset{0}{V}} \ \eta(t, s) \ dt = \infty \qquad (4.7.11)$$

and

$$\int_{\pm 0}^{\pm \delta} \frac{dy}{f(y)} < \infty \qquad (4.7.12)$$

for some $\delta > 0$. Then all solutions of (4.7.1) are oscillatory.

Proof: Otherwise, there is a positive solution $y(t - s) > 0$, $s \in [0, \beta(t)]$ for $t \geq t_1$. It follows that $r(t)y'(t)$ is nonincreasing for $t \geq t_1$ and so $y'(t)$ is eventually of constant sign. We multiply (4.7.1) by $\sigma(t)/f(y(g(t)))$ and integrate from t_1 to t, where $g(t) = t - \beta(t)$. This process yields

$$\frac{\sigma(t)r(t)y'(t)}{f(y(g(t)))} + \int_{t_1}^{t} \frac{\sigma(s)r(s)y'(s)[f(y(g(s)))]'}{f[y(g(s))]^2} \, ds$$

$$= c + \int_{t_1}^{t} \frac{r(s)y'(s)\sigma'(s)}{f(y(g(s)))} \, ds - \int_{t_1}^{t} \frac{\sigma(s)}{f(y(g(s)))} \int_{0}^{\beta(s)} f(y(s - s_1) \, d\eta(s, s_1) \, ds$$

$$\tag{4.7.13}$$

where c is a constant.

First we consider the case $y'(t) \geq 0$ for $t \geq t_1$. The integral on the left side in (4.7.13) is nonnegative. The first integral on the right side is nonpositive, and

$$\int_{t_1}^{t} \frac{\sigma(s)}{f(y(g(s)))} \int_{0}^{\beta(s)} f(y(s - s_1)) \, d\eta(s, s_1) \, ds \geq \int_{t_1}^{t} \sigma(s) \int_{0}^{\beta(s)} d\eta(s, s_1) \, ds \to \infty$$

as $t \to \infty$. This together with (4.7.13) leads to a contradiction.

Second, we consider the case $y'(t) \leq 0$. We multiply (4.7.1) by $\sigma(t)/f(y(t))$ and integrate from t_1 to t, which gives rise to an expression similar to (4.7.13) with $g(t) \equiv t$. We see that

$$\int_{t_1}^{t} \frac{r(s)y'(s)\sigma'(s)}{f(y(s))} \, ds \leq r(t_1)\sigma(t_1) \int_{t_1}^{t} \frac{y'(s) \, ds}{f(y(s))} < \infty$$

and

$$\int_{t_1}^{t} \frac{\sigma(s)}{f(y(s))} \int_{0}^{\beta(s)} f(y(s - s_1)) \, d\eta(s, s_1) \, ds \geq \int_{t_1}^{t} \sigma(s) \int_{0}^{\beta(s)} d\eta(s, s_1) \, ds \to \infty$$

as $t \to \infty$. Hence we can choose $t_2 \geq t_1$ so that the right side of (4.7.13) is less than -1, i.e.,

$$1 + \int_{t_2}^{t} \frac{\sigma(s)r(s)y'(s) \, f(y(s)) \, '}{[f(y(s))]^2} \, ds \leq \frac{\sigma(t)r(t)(-y'(t))}{f(y(t))}$$

$$\tag{4.7.14}$$

for $t \geq t_2$. Multiplying both sides of the above inequality by

$$- \frac{f(y(t)) \, '}{f(y(t))} \left[1 + \int_{t_2}^{t} \frac{\sigma(s)r(s)y'(s)[f(y(s))]'}{[f(y(s))]^2} \, ds \right]^{-1} \geq 0$$

and integrating from t_2 to t we obtain

$$\ln \frac{f(y(t_2))}{f(y(t))} \leq \ln\left[1 + \int_{t_2}^{t} \frac{\sigma(s)r(s)y'(s)\,[f(y(s))]'}{[f(y(s))]^2}\,ds\right] \tag{4.7.15}$$

From (4.7.14) and (4.7.15), we have

$$f(y(t_2)) \leq \sigma(t)r(t)(-y'(t))$$

Hence

$$y(t) - y(t_2) \leq -f(y(t_2)) \int_{t_2}^{t} \frac{ds}{\sigma(s)r(s)} \longrightarrow -\infty$$

as $t \to \infty$. This contradicts the fact that $y(t) > 0$. The proof is complete.

COROLLARY 4.7.3 Consider

$$(r(t)y'(t))' + a(t)\,|y(g(t))|^{\alpha}\,\mathrm{sgn}\,y(g(t)) = 0 \qquad 0 < \alpha < 1 \tag{4.7.16}$$

Assume that $r(t) > 0$ and is continuous and $\int^{\infty} dt/r(t) < \infty$; $a(t) \geq 0$ is integrable, and

$$\int^{\infty} \rho(t)a(t)\,dt = \infty \tag{4.7.17}$$

where $\rho(t) = \int_{t}^{\infty} ds/r(s)$. Then all solutions of (4.7.16) oscillate.

Proof: We apply Theorem 4.7.2 to the particular case where $\eta(t, s) =$ $U(s - g(t))a(t)$, $f(y) = |y|^{\alpha}\,\mathrm{sgn}\,y$, $0 < \alpha < 1$, and $\sigma(t) = \rho(t) = \int_{t}^{\infty} ds/r(s)$.

Now we consider an equation with advanced argument of the form

$$(r(t)y'(t))' + \int_{\alpha(t)}^{\beta(t)} f(t,\ y(t + s))\,d\eta(t, s) = 0 \tag{4.7.18}$$

where $0 \leq \alpha(t) \leq \beta(t)$ are continuous, f and η satisfy the conditions of Section 3.8.

THEOREM 4.7.3 Assume that $r(t) > 0$ is continuous, $\int^{\infty} dt/r(t) < \infty$, and $\eta(t, s)$ satisfies the conditions of Section 3.8. Assume also that f is either superlinear or sublinear and

$$\int^{\infty} \rho(t)\,|f(t, c)|\,\mathop{V}_{\alpha(t)}^{\beta(t)}\,\eta(t, s)\,dt = \infty \qquad \text{for all } c \neq 0 \tag{4.7.19}$$

where $\rho(t) = \int_{t}^{\infty} ds/r(s)$, and $\mathop{V}_{\alpha(t)}^{\beta(t)}\,\eta(t, s)$ is the variation of the function

$\eta(t, s)$ in $s \in [\alpha(t), \beta(t)]$ for every $t \in R_+$. Then all nonoscillatory solutions of (4.7.18) tend to zero at $t \to \infty$.

Proof: Let $y(t)$ be a nonoscillatory solution of (4.7.18). Without loss of generality, we may suppose that $y(t) > 0$ for $t \geq t_1$. By imitating the method of proof of Lemma 4.6.1, we can find positive numbers t_1, a_1, a_2 such that

$$y(t) \geq -r(t)y'(t)\rho(t) \quad \text{for } t \geq t_1 \tag{4.7.20}$$

and

$$a_1\rho(t) \leq y(t) \leq a_2 \quad \text{for } t \geq t_1 \tag{4.7.21}$$

Multiplying (4.7.18) by $\rho(t)$ and integrating from t_1 to t, we get

$$\rho(t)r(t)y'(t) + y(t) - \rho(t_1)r(t_1)y'(t_1) - y(t_1) + \int_{t_1}^{t} \rho(s) \int_{\alpha(s)}^{\beta(s)} f(s, y(s+\sigma))\, d\eta(s,\sigma)\, ds = 0 \tag{4.7.22}$$

Assume that

$$\int_{t_1}^{t} \rho(s) \int_{\alpha(s)}^{\beta(s)} f(s, y(s+\sigma))\, d\eta(s,\sigma)\, ds \to \infty \tag{4.7.23}$$

We will prove that (4.7.23) is impossible. Indeed, from (4.7.23) and (4.7.22), we have $\rho(t)r(t)y'(t) \to -\infty$ as $t \to \infty$. Hence, there exist numbers t_2 and $M > 0$ such that $t_2 \geq t_1$ and

$$\rho(t)y'(t)r(t) \leq -M \quad \text{for } t \geq t_2$$

Dividing the above inequality by $\rho(t)r(t)$ and integrating gives

$$y(t) - y(t_2) \leq -M \int_{t_2}^{t} (\rho(s)r(s))^{-1}\, ds = M \ln\left[\frac{\rho(t)}{\rho(t_2)}\right]$$

Since $\rho(t) \to 0$ as $t \to \infty$, it follows that $y(t) \to -\infty$ as $t \to \infty$. This contradicts the positivity of $y(t)$. Consequently, (4.7.23) cannot hold, which implies that

$$\int_{t_1}^{\infty} \rho(s) \int_{\alpha(s)}^{\beta(s)} f(s, y(s+\sigma))\, d\eta(s,\sigma)\, ds < \infty \tag{4.7.24}$$

Letting $t \to \infty$ on (4.7.22), we obtain

$$\lim_{t \to \infty} (\rho(t)r(t)y'(t) + y(t))$$

which is finite. We claim that $\lim_{t \to \infty} y(t)$ exists. In fact, if it is false, then there exist numbers ξ and η such that

$$\liminf_{t\to\infty} y(t) \leq \xi < \eta \leq \limsup_{t\to\infty} y(t)$$

We can choose an increasing sequence $\{\tau\}_{n=1}^{\infty}$ with the following properties:

$$\lim_{n\to\infty} \tau_n = \infty \qquad y'(\tau_n) = 0 \qquad n = 1, 2, \ldots \tag{4.7.25}$$

$$y(\tau_{2n-1}) < \xi \qquad y(\tau_{2n}) > \eta \qquad n = 1, 2, \ldots \tag{4.7.26}$$

In view of (4.7.25), the limit

$$\lim_{n\to\infty} [\rho(\tau_n)r(\tau_n)y'(\tau_n) + y(\tau_n)] = \lim_{n\to\infty} y(\tau_n)$$

exists. This, however, is a contradiction to (4.7.26). Therefore, the $\lim_{n\to\infty} y(t)$ exists.

Our purpose is to prove that $y(\infty) = 0$. If $y(\infty) > 0$, then there are positive numbers c_1, c_2 such that

$$c_1 \leq y(g(t)) \leq c_2 \qquad \text{for } t \geq t_1 \tag{4.7.27}$$

where $g(t) \geq t$ is any continuous function. Using (4.7.24) and (4.7.27) we obtain

$$\int_{t_1}^{\infty} \rho(t) \int_{\alpha(t)}^{\beta(t)} f(t, c_1) \, d\eta\,(t, \sigma) \, dt < \infty$$

or

$$\int_{t_1}^{\infty} \rho(t) \int_{\alpha(t)}^{\beta(t)} f(t, c_2) \, d\eta\,(t, \sigma) \, dt < \infty$$

whenever f is superlinear or sublinear, respectively. This contradiction shows that $y(\infty) = 0$. Thus the proof is complete.

THEOREM 4.7.4 Besides (4.7.19), assume that (4.7.18) satisfies the conditions of Theorem 4.7.3. Then a necessary condition for (4.7.18) to have a nonoscillatory solution y(t) having the property

$$\lim_{t\to\infty} y(t)/\rho(t) = \text{const} \neq 0 \tag{4.7.28}$$

is that

$$\int^{\infty} |f(t, c\rho(t + \beta(t)))| \quad \overset{\beta(t)}{\underset{\alpha(t)}{V}} \eta\,(t, s) \, dt < \infty \qquad \text{for some } c \neq 0 \tag{4.7.29}$$

and a sufficient condition is that

$$\int_{\alpha(t)}^{\infty} |f(t, c\rho(t + \alpha(t)))| \overset{\beta(t)}{\underset{\alpha(t)}{V}} \eta(t, s) \, dt < \infty \qquad \text{for some } c \neq 0 \qquad (4.7.30)$$

whenever f is superlinear.

Proof: Necessity: Let $y(t)$ be a nonoscillatory solution with the property $\lim_{t \to \infty} y(t)/\rho(t) = \beta \neq 0$. Without loss of generality, we may suppose that $\beta > 0$. This implies that one can find positive numbers t_1, c_1, c_2 such that

$$c_1 \rho(g(t)) \leq y(g(t)) \leq c_2 \rho(g(t)) \qquad \text{for } t \geq t_1 \qquad (4.7.31)$$

since $y(t) \to 0$ as $t \to \infty$, so $y'(t) \leq 0$ eventually.

Choose t_1 sufficiently large so that $y'(t) \leq 0$ for $t \geq t_1$ and (4.7.20) holds. This implies

$$-r(t)y'(t) \leq c_2 \qquad \text{for } t \geq t_1$$

Integrating (4.7.18) from t_1 to t and using the above inequality, we obtain

$$\int_{t_1}^{\infty} \int_{\alpha(t)}^{\beta(t)} f(t, \, y(t + s)) \, d\eta(t, s) \, dt < \infty \qquad (4.7.32)$$

From (4.7.32) and (4.7.31), we have

$$\int_{t_1}^{\infty} f(t, \, c_1 \rho(t + \beta(t))) \overset{\beta(t)}{\underset{\alpha(t)}{V}} \eta(t, s) \, dt < \infty$$

whenever f is superlinear.

Sufficiency: Suppose (4.7.30) holds with $c > 0$. Choose $T > 0$ so large that

$$\int_{T}^{\infty} f(t, \, c\rho(t + \alpha(t))) \overset{\beta(t)}{\underset{\alpha(t)}{V}} \eta(t, s) \, dt < \frac{c}{4} \qquad (4.7.33)$$

We consider the integral equation

$$y(t) = \frac{c}{2} \rho(t) + \rho(t) \int_{T}^{t} \int_{\alpha(s)}^{\beta(s)} f(s, \, y(s + \sigma)) \, d\eta(s, \sigma) \, ds$$

$$+ \int_{t}^{\infty} \rho(s) \int_{\alpha(s)}^{\beta(s)} f(s, \, y(s + \sigma)) \, d\eta(s, \sigma) \, ds \qquad (4.7.34)$$

We define the linear space $c_\rho[T, \infty)$ of all continuous functions $y: [T, \infty) \to R$ such that

$$\|y\|_\rho = \sup \{\rho(t)^{-1} |y(t)| : t \geq T\} < \infty$$

It is clear that $c_\rho[T,\infty)$ is a Banach space with norm $\|\cdot\|_\rho$. We define an operator Φ by

$$(\Phi y)(t) = \frac{c}{2}\rho(t) + \rho(t)\int_T^t \int_{\alpha(s)}^{\beta(s)} f(s,\, y(s+\sigma))\, d\eta(s,\sigma)\, ds$$

$$+ \int_t^\infty \rho(s) \int_{\alpha(s)}^{\beta(s)} f(s,\, y(s+\sigma))\, d\eta(s,\sigma)\, ds$$

and seek a fixed point of Φ in the set $Y = \{y \in c_\rho[T,\infty) : (c/2)\rho(t) \le y(t) \le c\rho(t),\ \text{for } t \ge T\}$, which is a bounded, convex, and closed subset of $c_\rho[T,\infty)$. For this purpose we shall show that Φ is continuous and maps Y into a compact subset of Y.

(i) Φ maps Y into Y: If $y \in Y$, then clearly $(\Phi y)(t) \ge (c/2)\rho(t)$, $t \ge T$, and

$$(\Phi y)(t) \le \frac{c}{2}\rho(t) + \rho(t)\int_T^\infty \int_{\alpha(s)}^{\beta(s)} f(s,\, y(s+\sigma))\, d\eta(s,\sigma)\, ds$$

$$\le \frac{c}{2}\rho(t) + \rho(t)\int_T^\infty \int_{\alpha(s)}^{\beta(s)} f(s,\, c\rho(s+\sigma))\, d\eta(s,\sigma)\, ds$$

$$\le \frac{c}{2}\rho(t) + \rho(t)\int_T^\infty f(s,\, c\rho(s+\alpha(s)))\int_{\alpha(s)}^{\beta(s)} d\eta(s,\sigma)\, ds$$

$$\le c\rho(t) \qquad t \ge T$$

(ii) Φ is continuous: Let $\{y_n\}$ be a sequence of elements of Y such that $\lim \|y_n - y\|_\rho = 0$. Since Y is closed, $y \in Y$ and

$$|(\Phi y_n)(t) - (\Phi y)(t)| \le \rho(t)\int_T^\infty \int_{\alpha(s)}^{\beta(s)} |f(s,\, y_n(s+\sigma)) - f(s,\, y(s+\sigma))|\, d\eta(s,\sigma)\, ds$$

$$(4.7.35)$$

It follows that

$$\|\Phi y_n - \Phi y\|_\rho \le 2\int_T^\infty f(s,\, c\rho(s+\alpha(s)))\ \underset{\alpha(s)}{\overset{\beta(s)}{V}}\ d\eta(s,\sigma)\, ds$$

We apply the Lebesgue dominated convergence theorem to conclude from (4.7.35) that $\lim_{n\to\infty} \|\Phi y_n - \Phi y\|_\rho = 0$. This proves the continuity of Φ.

(iii) ΦY is precompact: It suffices to show that the family of functions $\{\rho^{-1}\Phi y : y \in Y\}$ is uniformly bounded and equicontinuous on $[T,\infty)$. It is easy to see that it is uniformly bounded; we need only to demonstrate the equicontinuity. This will be accomplished if we show that for any given $\epsilon > 0$, the interval $[T, \infty)$ can be decomposed into a finite number of subintervals in such a way that on each subinterval all functions of the family have oscillations less than ϵ.

If $y \in Y$ then we have, for $t_2 > t_1 \geq T$,

$$(\rho^{-1}\Phi y)(t_2) - (\rho^{-1}\Phi y)(t_1)$$

$$= \int_{t_1}^{t_2} \int_{\alpha(s)}^{\beta(s)} f(s, y(s + \sigma)) \, d\eta(s,\sigma) \, ds$$

$$+ \rho(t_2)^{-1} \int_{t_2}^{\infty} \rho(s) \int_{\alpha(s)}^{\beta(s)} f(s, y(s + \sigma)) \, d\eta(s, \sigma) \, ds$$

$$- \rho(t_1)^{-1} \int_{t_1}^{\infty} \rho(s) \int_{\alpha(s)}^{\beta(s)} f(s, y(s + \sigma)) \, d\eta(s, \sigma) \, ds$$

$$= \int_{t_1}^{t_2} \int_{\alpha(s)}^{\beta(s)} f(s, y(s + \sigma)) \, d\eta(s, \sigma) \, ds$$

$$+ [\rho(t_2)^{-1} - \rho(t_1)^{-1}] \int_{t_1}^{\infty} \rho(s) \int_{\alpha(s)}^{\beta(s)} f(s, y(s + \sigma)) \, d\eta(s, \sigma) \, ds$$

$$- \rho(t_2)^{-1} \int_{t_1}^{t_2} \rho(s) \int_{\alpha(s)}^{\beta(s)} f(s, y(s + \sigma)) \, d\eta(s, \sigma) \, ds$$

$$\leq 3 \int_{t_1}^{\infty} \int_{\alpha(s)}^{\beta(s)} f(s, y(s + \sigma)) \, d\eta(s, \sigma) \, ds$$

$$\leq 3 \int_{t_1}^{\infty} f(s, c\rho(s + \alpha(s))) \int_{\alpha(s)}^{\beta(s)} d\eta(s, \sigma) \, ds \to 0$$

as $t_1 \to \infty$. Hence given an $\epsilon > 0$, there exists a $T^* > T$ such that

$$|(\rho^{-1}\Phi y)(t_2) - (\rho^{-1}\Phi y)(t_1)| < \epsilon \qquad \text{if} \ \ t_2 > t_1 \geq T^*$$

This shows that the oscillations of all $\rho^{-1}\Phi y$, $y \in Y$, on $[T^*, \infty)$ are less than ϵ. Now, let $T \leq t_1 < t_2 \leq T^*$. Then, we see that

$$
|(\rho^{-1}\Phi y)(t_2) - (\rho^{-1}\Phi y)(t_1)| \leq \int_{t_1}^{t_2} f(s, c\rho(s + \alpha(s))) \int_{\alpha(s)}^{\beta(s)} d\eta(s, \sigma)\, ds
$$

$$
+ \frac{\rho(t_1)}{\rho(t_2)} \int_{t_2}^{t_1} f(s, c\rho(s + \alpha(s))) \int_{\alpha(s)}^{\beta(s)} d\eta(s, \sigma)\, ds
$$

$$
+ \rho(t_2)^{-1} |\rho(t_2) - \rho(t_1)| \int_{t_1}^{\infty} f(s, c\rho(s + \alpha(s))) \int_{\alpha(s)}^{\beta(s)} d\eta(s,\sigma)\, ds
$$

$$
\leq \frac{c}{4} \rho(T^*)^{-1} |\rho(t_2) - \rho(t_1)| + \left| 1 + \frac{\rho(t_1)}{\rho(t_2)} \right| \int_{t_1}^{t_2} f(s, c\rho(s + \alpha(s))) \int_{\alpha(s)}^{\beta(s)} d\eta(s,\sigma)\, ds
$$

This ensures the existence of a $\delta > 0$ such that for all $y \in Y$

$$
|(\rho^{-1}\Phi y)(t_1) - (\rho^{-1}\Phi y)(t_1)| < \epsilon \qquad \text{if } t_2 - t_1 < \delta
$$

Consequently, we can divide the interval $[T, T^*]$, and hence the whole interval $[T, \infty)$, into a finite number of subintervals on each of which every $\rho^{-1}\Phi y$, $y \in Y$, has oscillation less than ϵ. It follows that $\overline{\Phi Y}$ is a compact subset of Y.

From the preceding discussion, we see that Schauder's fixed point theorem can be applied to the operator Φ. Let $y \in Y$ be a fixed point of Φ. Then by the definition of Φ, $y(t)$ is a solution of the integral equation (4.7.34) for $t \geq T$. Moreover,

$$
\lim_{t \to \infty} \frac{y(t)}{\rho(t)} \text{ exists}
$$

We conclude that $y(t)$ is a solution of equation (4.7.18) with the required asymptotic property. The proof is complete.

REMARK 4.7.1 For the sublinear case, we can establish a similar theorem.

REMARK 4.7.2 Consider a simple advanced type equation

$$
(r(t)y'(t))' + f(t, y(g(t))) = 0 \tag{4.7.36}
$$

From Theorem 4.7.4, we obtain a necessary and sufficient condition for (4.7.36) to have a nonoscillatory solution $y(t)$ satisfying (4.7.28), namely,

$$
\int^{\infty} f(t, c\rho(g(t)))\, dt < \infty \qquad \text{for some } c \neq 0 \tag{4.7.37}
$$

As mentioned earlier, when (4.7.36) is sublinear, condition (4.7.37) remains valid also.

THEOREM 4.7.5 In addition to (4.7.19), assume that (4.7.18) satisfies the conditions of Theorem 4.7.3. Then a necessary and sufficient condition for (4.7.18) to have a nonoscillatory solution $y(t)$ with the property $\lim_{t \to \infty} y(t) = \text{const} \neq 0$, is that

$$\int_{\alpha(t)}^{\infty} \rho(t)\,|f(t,c)|\ \overset{\beta(t)}{\underset{\alpha(t)}{V}}\ \eta(t,s)\,dt < \infty \qquad \text{for some } c \neq 0 \tag{4.7.38}$$

The proof is similar to the proof of Theorem 4.6.1.

THEOREM 4.7.6 Assume that (4.7.18) satisfies the conditions of Theorem 4.7.3 and f is strongly superlinear. Then a necessary and sufficient condition for (4.7.18) to be oscillatory is that

$$\int^{\infty} |f(t,\ c\rho(t+\alpha(t)))|\ \overset{\beta(t)}{\underset{\alpha(t)}{V}}\ \eta(t,s)\,dt = \infty \qquad \text{for all } c \neq 0 \tag{4.7.39}$$

Proof: The necessity of condition (4.7.39) for (4.7.18) to be oscillatory follows from Theorem 4.7.4. To prove the sufficiency part suppose there exists a nonoscillatory solution $y(t)$ of (4.7.18). Without loss of generality, we may suppose that $y(t) > 0$ for $t \geq t_0$. From (4.7.8), $(r(t)y'(t))' \leq 0$ for $t \geq t_0$, so that $y'(t)$ is eventually of constant sign. If $y'(t) \geq 0$, integrating (4.7.18) we get

$$r(\infty)y'(\infty) - r(t_1)y'(t_1) + \int_{t_1}^{\infty} \int_{\alpha(t)}^{\beta(t)} f(t, y(t+s))\,d\eta(t,s)\,dt = 0$$

Hence

$$\int_{t_1}^{\infty} \int_{\alpha(t)}^{\beta(t)} f(t,\ y(t+s))\,d\eta(t,s)\,dt < \infty$$

But $y(t+s) \geq y(t+\alpha(t))$, $s \in [\alpha(t),\ \beta(t)]$, so

$$\int_{t_1}^{\infty} \int_{\alpha(t)}^{\beta(t)} f(t,\ y(t+\alpha(t)))\,d\eta(t,s)\,dt < \infty \tag{4.7.40}$$

By using (4.7.21), (4.7.40) leads to a contradiction with the assumption (4.7.39). Therefore, we have

$$y'(t) < 0 \qquad \text{for } t \geq t_1$$

From (4.7.20),

$$y(t) \geq -r(t)y'(t)\rho(t) \geq k\rho(t) \qquad \text{for } t \geq t_2$$

where $k = -r(t_2)y'(t_2) > 0$. Using the above inequality and strong super-linearity of f, we obtain

$$[y(g(t))]^{-\sigma} f(t, y(g(t)) \geq [k\rho(g(t))]^{-\sigma} f(t, k\rho(g(t)))$$

where σ is a constant bigger than one.

Now, we have

$$\{-[-r(t)y'(t)]^{1-\sigma}\} = (\sigma-1)[-r(t)y'(t)]^{-\sigma} \int_{\alpha(t)}^{\beta(t)} f(t, y(t+s))\, d\eta(t,s)$$

$$= (\sigma-1)[-r(t)y'(t)]^{-\sigma} \int_{\alpha(t)}^{\beta(t)} \frac{f(t,\ y(t+s))}{[y(t+s)]^{\sigma}} [y(t+s)]^{\sigma}\, d\eta(t,s)$$

$$\geq (\sigma-1)[-r(t)y'(t)]^{-\sigma} \int_{\alpha(t)}^{\beta(t)} \frac{f(t,\ k\rho(t+s))}{[k\rho(t+s)]^{\sigma}} [y(t+s)]^{\sigma}\, d\eta(t,s)$$

$$\geq (\sigma-1)[-r(t)y'(t)]^{-\sigma} [k\rho(t+\alpha(t))]^{-\sigma} f(t, k\rho(t+\alpha(t)))y(t+\alpha(t))^{\sigma} \overset{\beta(t)}{\underset{\alpha(t)}{V}} \eta(t,s)$$

$$\geq (\sigma-1)[-r(t)y'(t)]^{-\sigma} [-r(t+\alpha(t))y'(t+\alpha(t))\rho(t+\alpha(t))]^{\sigma}$$

$$[k\rho(t+\alpha(t))]^{-\sigma} f(t, k\rho(t+\alpha(t))) \overset{\beta(t)}{\underset{\alpha(t)}{V}} \eta(t,s)$$

$$\geq (\sigma-1)k^{-\sigma} f(t, k\rho(t+\alpha(t))) \overset{\beta(t)}{\underset{\alpha(t)}{V}} \eta(t,s)$$

Integrating the above inequality, we obtain

$$(\sigma-1)k^{-\sigma} \int_{t_2}^{t} f(s_1, k\rho(s_1+\alpha(s_1))) \overset{\beta(s_1)}{\underset{\alpha(s_1)}{V}} \eta(s_1, s)\, ds_1$$

$$\leq [-r(t_2)y'(t_2)]^{1-\sigma} - [-r(t)y'(t)]^{1-\sigma}$$

which implies that

$$\int_{t_1} f(s_1, k\rho(s_1+\alpha(s_1))) \overset{\beta(s_1)}{\underset{\alpha(s_1)}{V}} \eta(s_1, s)\, ds_1 < \infty$$

This is a contradiction. The proof is complete.

Next we shall present a series of results relative to (4.7.18) with $\int^{\infty} dt/r(t) = \infty$.

LEMMA 4.7.1 Assume that $r(t) > 0$ is continuous, $\int^{\infty} dt/r(t) = \infty$, $\eta(t,s)$ satisfies the conditions of Section 3.8 and f is either superlinear or sublinear. If $y(t)$ is an eventually positive solution of (4.7.18), then there are positive numbers t_1, a_1, a_2 such that

$$y'(t) > 0 \quad \text{for} \quad t \geq t_1 \tag{4.7.41}$$

and

$$a \leq y(t) \leq a_2 R(t) \quad \text{for} \quad t \geq t_1 \tag{4.7.42}$$

Proof: From (4.7.18), $(r(t)y'(t))' < 0$. If $r(t)y'(t) < 0$, as $t \geq t_1$, then $y'(t) < r(t_1)y'(t_1)/r(t)$. This implies that $y(t)$ will become negative for sufficiently large t. This contradicts the hypothesis $y(t) > 0$. Therefore $y'(t) > 0$ for $t \geq t_1$, and (4.7.38) holds.

We shall merely state the following results; proofs are left to the reader.

THEOREM 4.7.7 Assume that the conditions of Lemma 4.7.1 hold. Then a necessary condition for (4.7.18) to have a nonoscillatory solution $y(t)$ with the property $\lim_{t \to \infty} y(t)/R(t) = \text{const} \neq 0$ is that

$$\int^{\infty} |f(t,\ cR(t + \alpha(t)))| \overset{\beta(t)}{\underset{\alpha(t)}{V}} \eta(t,s)\, dt < \infty \quad \text{for some} \quad c \neq 0 \tag{4.7.43}$$

and a sufficient condition is that

$$\int^{\infty} |f(t,\ cR(t + \beta(t)))| \overset{\beta(t)}{\underset{\alpha(t)}{V}} \eta(t,s)\, dt < \infty \quad \text{for some} \quad c \neq 0 \tag{4.7.44}$$

whenever f is superlinear. Similarly, a necessary condition for (4.7.18) to have a nonoscillatory solution $y(t)$ with the property $\lim_{t \to \infty} y(t)/R(t) = \text{const} \neq 0$ is that

$$\int^{\infty} |f(t,\ cR(t + \beta(t)))| \overset{\beta(t)}{\underset{\alpha(t)}{V}} \eta(t,s)\, dt < \infty \quad \text{for some} \quad c \neq 0 \tag{4.7.45}$$

and a sufficient condition

$$\int^{\infty} \frac{R(t + \beta(t))}{R(t + \alpha(t))} |f(t,\ cR(t + \alpha(t)))| \overset{\beta(t)}{\underset{\alpha(t)}{V}} \eta(t,s)\, dt < \infty \tag{4.7.46}$$

whenever f is sublinear.

THEOREM 4.7.8 Assume that the conditions of Lemma 4.7.1 hold; then a necessary and sufficient condition for (4.7.18) to have a bounded nonoscillatory solution is that

$$\int_{\alpha(t)}^{\infty} R(t)\,|f(t,c)| \overset{\beta(t)}{\underset{}{V}} \eta(t,s)\,dt < \infty \quad \text{for some } c \neq 0 \qquad (4.7.47)$$

THEOREM 4.7.9 Assume that the conditions of Lemma 4.7.1 hold, but f is strongly superlinear. Then a necessary and sufficient condition for (4.7.18) to be oscillatory is that

$$\int_{\alpha(t)}^{\infty} R(t)|f(t,c)| \overset{\beta(t)}{\underset{}{V}} \eta(t,s)\,dt = \infty \quad \text{for all } c \neq 0 \qquad (4.7.48)$$

THEOREM 4.7.10 Assume that the conditions of Lemma 4.7.1 hold, but f is strongly sublinear and satisfies

$$\int^{\infty} \left[\frac{R(t)}{R(t+\beta(t))}\right] |f(t,\,cR(t+\beta(t)))| \overset{\beta(t)}{\underset{\alpha(t)}{V}} \eta(t,s)\,dt = \infty \qquad (4.7.49)$$

for all $c \neq 0$. Then (4.7.18) is oscillatory.

REMARK 4.7.3 If we assume that $f(t,y)$ is nondecreasing in y for each t, then we can replace (4.7.49) by the following weaker condition:

$$\int_{\alpha(t)}^{\infty} |f(t,\,cR(t))| \overset{\beta(t)}{\underset{}{V}} \eta(t,s)\,dt = \infty \qquad (4.7.50)$$

for all $c \neq 0$.

EXAMPLE 4.7.1 Consider the advanced equation

$$[t^{-2}y'(t)]' + 2t^{-1}[y(t^2)]^{1/3} = 0 \qquad (4.7.51)$$

which has a nonoscillatory solution $y(t) = t$. It is obvious that (4.7.51) does not satisfy (4.7.49) and it is easy to see that there is a gap between (4.7.46) and (4.7.49).

4.8 NOTES

The classification of solutions of second order linear differential equations with deviating arguments is discussed by Kaminskii [105] (see also Norkin's book [197]). In Section 4.1 we extend the results of [197] to equations (4.1.1). The material in Section 4.2 is taken from Koplatadze [124] (for related work see [189] and [197]). The existence of bounded oscillatory solutions of equation (4.3.3) is investigated by Ladas and Lakshmikantham [148]. Theorem 4.3.1 is adapted from Gustafson [94]. For Theorem 4.3.2 see [124]. Condition (4.3.12) is also obtained by Ladas and Lakshmikantham [148]. Theorem 4.3.3 is due to Ladas, Ladde, and Papadakis [147]. The results of Section 4.4 are due to Ladde [158], [159], whereas the results of Section 4.5 follow the development of Zhang [300]. Theorems 4.5.1 to 4.5.8 are new. The results of Section 4.6 are adapted from Kusano and Onose [134, 137]. For related work, see also Angelova and Bainov [3, 4], Bobisud [15], Bradley [17], Brands [18], Burkowski [22, 23], Burton [25], Bykov [29], Chiou [40], Dahiya [46, 48], Erbe [60], Garner [67], Gollwitzer [68], Hino [100], Koplatadze [121], Kung [128], Lalli [164], Liossatos [171], Macki and Wong [177], McCalla [186], Nababan [190], Myskis [189], Norkin [197], Ohriska [202], Philos [223], Shevelo [234], Singh [249], Teufel [270–271], Travis [279], Waltman [283], Yan [291, 293], Yeh [294], and Zhang [298, 303]. For the special case of (4.7.1) see Myskis [189] and McCalla [186]. Theorems 4.7.1 to 4.7.10 are generalizations of Kusano and Onose [131, 132, 135]. For related work see also Burton and Haddock [26, 27], Chen [31], Koplatadze [124], Ladde [158], Lovelady [175], Norkin [197], Sficas [231], Werbowski [285], and Zhang [298, 300].

5

Higher Order Differential Equations

5.0 INTRODUCTION

This chapter is devoted to the study of the oscillation of high order differential equations and inequalities with deviating arguments. We begin with the study of third and fourth order differential equations with deviating arguments. We then discuss even order differential equations in Section 5.2 and present several results which are important in studying differential equations and inequalities with oscillation of high order. Section 5.3 investigates higher order linear differential equations and inequalities. In Section 5.4, we consider a class of arbitrary order delay equations. Differential equations with deviating arguments of mixed type are presented in Section 5.5. In Section 5.6, we discuss a class of nonlinear differential inequalities with deviating arguments. Section 5.7 deals with oscillatory behavior of differential equations with forcing terms.

5.1 THIRD AND FOURTH ORDER DIFFERENTIAL EQUATIONS WITH DEVIATING ARGUMENTS

We consider a third order differential equation with deviating argument

$$y'''(t) + f(t, y(t), y(g(t))) = 0 \qquad t \geq 0 \tag{5.1.1}$$

THEOREM 5.1.1 Assume that

    (i) $g \in C[R_+, R_+]$ and $\lim_{t \to \infty} g(t) = \infty$

    (ii) $f \in C[R_+ \times R^2, R]$ and $f(t, x, y)x > 0$ for $xy > 0$

    (iii) $|f(t, x_1, y_1)| \leq |f(t, x_2, y_2)|$ for $|x_1| \leq |x_2|$, $|y_1| \leq |y_2|$,

        $x_1 x_2 > 0$, $y_1 y_2 > 0$

    (iv) $\int^{\infty} t^2 |f(t, a, a)| \, dt < \infty$ for some $a \neq 0$. \qquad\qquad (5.1.2)

Then (5.1.1) has a bounded nonoscillatory solution.

Proof: From (iv), one can find a sufficiently large T such that

$$\int_{}^{\infty} t^2 |f(t, a, a)| \, dt < |a| \tag{5.1.3}$$

Set $T_0 = \inf_{t \geq T} g(t)$ and let C be the space of bounded continuous functions on $[T_0, \infty)$. Let $\bar{Y} \subset C$ be defined by $\bar{Y} = \{y \in C : |a|/2 \leq y \operatorname{sgn} a \leq |a|\}$. Then $\bar{Y}$ is a bounded convex closed subset of C. Define an operator ψ on $\bar{Y}$ by

$$(\psi y)(t) = \begin{cases} \dfrac{a}{2} + \dfrac{1}{2!} \displaystyle\int_t^{\infty} (s - t)^2 f(s, y(s), y(g(s))) \, ds & t \geq T \\[2em] \dfrac{a}{2} + \dfrac{1}{2!} \displaystyle\int_T^{\infty} (s - T)^2 f(s, y(s), y(g(s))) \, ds & t \in [T_0, T] \end{cases}$$

(a) ψ maps $\bar{Y}$ into itself: In fact,

$$\frac{|a|}{2} \leq (\psi y)(t) \operatorname{sgn} a \leq \frac{|a|}{2} + \frac{1}{2!} \int_t^{\infty} s^2 |f(s, a, a)| \, ds \leq |a|$$

because of (iii) and (5.1.3).

(b) ψ is continuous   To prove this, let $\{y_n\}$ be a Cauchy sequence in $\bar{Y}$, and let $\lim_{n \to \infty} \|y_n - y\| = 0$. Because $\bar{Y}$ is closed, $y \in \bar{Y}$. To prove the continuity of ψ, we see that

$$|(\psi y_n)(t) - (\psi y)(t)| \leq \frac{1}{2!} \int_t^{\infty} (s - t)^2 \, |f(s, y_n(s), y_n(g(s))) - f(s, y(s), y(g(s)))| \, ds$$

Set

$$G_n(s) = s^2 |f(s, y_n(s), y_n(g(s))) - f(s, y(s), y(g(s)))|$$

Then the above inequality reduces to

$$\|\psi y_n - \psi y\| \leq \frac{1}{2!} \int_T^{\infty} G_n(s) \, ds \tag{5.1.4}$$

noting the fact that $(s - t)^2 \leq s^2$ for $s \geq t \geq 0$. It is obvious that $\lim_{n \to \infty} G_n(s) = 0$. From the definition of G_n and (iii) we obtain $G_n(s) \leq 2s^2 |f(s, a, a)|$. The above relations, (5.1.3), (5.1.4), and the Lebesgue convergence theorem give us

$$\lim_{n \to \infty} \| \psi y_n - \psi y \| = 0$$

which means that ψ is continuous.

(c)   To show $\psi \bar{Y}$ is precompact, we see that $(\psi y)(t)$, $y \in \bar{Y}$, is uniformly bounded. Now we will prove that $\psi \bar{Y}$ is an equicontinuous family of functions on $[T_0, \infty)$.

For $y \in \bar{Y}$ and $t_2 > t_1$, we have

$$| (\psi y)(t_2) - (\psi y)(t_1) | \leq \int_{t_1}^{\infty} s^2 | f(s, a, a) | \, ds$$

For any given $\epsilon > 0$, there exists $T^* > T$ such that $\int_{T^*}^{\infty} s^2 | f(s, a, a) | \, ds < \epsilon$. Hence, for any $t_2 > t_1 \geq T^*$, we have

$$| (\psi y)(t_2) - (\psi y)(t_1) | < \epsilon \qquad \text{for all } y \in \bar{Y}$$

For $T \leq t_1 < t_2 \leq T^*$,

$$| (\psi y)(t_2) - (\psi y)(t_1) | \leq \frac{1}{2!} \left| \int_{t_2}^{\infty} (s - t_2)^2 - (s - t_1)^2 \, f(s, y(s), y(g(s))) \, ds \right|$$

$$+ \frac{1}{2!} \int_{t_1}^{t_2} (s - t_1)^2 \, | f(s, y(s), y(g(s))) | \, ds$$

According to condition (iii), we have

$$| (\psi y)(t_2) - (\psi y)(t_1) | \leq M | t_2 - t_1 | + \int_{t_1}^{t_2} s^2 | f(s, a, a) | \, ds$$

Hence, for any given $\epsilon > 0$, there exists a $\delta > 0$ such that

$$| (\psi y)(t_2) - (\psi y)(t_1) | < \epsilon , \qquad | t_2 - t_1 | < \delta , \quad \text{for all } y \in \bar{Y}$$

That is, the interval $[T_0, +\infty)$ can be divided into a finite number of sub-intervals on which every $(\psi y)(t)$, $y \in \bar{Y}$, has oscillation less than ϵ.

Therefore $\psi \bar{Y}$ is an equicontinuous family on $[T_0, +\infty)$. Hence $\overline{\psi \bar{Y}}$ is a compact subset of $\bar{Y}$. According to the Schauder fixed point theorem there exists a $y \in \bar{Y}$ such that $y = \psi y$. This y is a bounded nonoscillatory solution of (5.1.1). The proof is complete.

THEOREM 5.1.2 Assume that conditions (i) to (iii) of Theorem 5.1.1 hold. In addition, assume that

$$\int^{\infty} t^2 | f(t, a, a) | \, dt = \infty \qquad \text{for any } a \neq 0 \tag{5.1.5}$$

Then every bounded solution y of (5.1.1) is either oscillatory or y, y', and y'' tend to zero as $t \to \infty$.

Proof: Assume that there exists a bounded nonoscillatory solution $y(t)$. Without loss of generality, suppose that $y(t) > 0$, $y(g(t)) > 0$ for $t \geq t_1$. From (5.1.1), $y'''(t) < 0$. Integrating (5.1.1) over $[s, t]$, we have

$$y''(t) - y''(s) + \int_s^t f(\tau, y(\tau), y(g(\tau))) \, d\tau = 0 \qquad (5.1.6)$$

Now we discuss two possible cases.

(1) The case $y''(t) > 0$ for $t \geq t_1$.

(a) If $y'(t) > 0$ for $t \geq t_2 \geq t_1$, then $y'(t) \geq y'(t_2) > 0$. From this, one can conclude that y is unbounded, which is impossible.

(b) If $y'(t) < 0$, for $t \geq t_2 \geq t_1$, $y(t)$ is decreasing and bounded, so there exists a limit $\lim_{t \to \infty} y(t) = c_1 \geq 0$. It is easy to see that $\lim_{t \to \infty} y'(t) = 0$, $\lim_{t \to \infty} y''(t) = 0$. We want to prove that $c_1 = 0$. Assume that $c_1 > 0$. From (5.1.1), we have

$$y''(t) \geq \int_t^\infty f(\tau, y(\tau), y(g(\tau))) \, d\tau$$

Integrating the above inequality, we have

$$-y'(t) \geq \int_t^\infty \left\{ \int_{t_1}^\infty f(\tau, y(\tau), y(g(\tau))) \, d\tau \right\} dt_1$$

$$= \int_t^\infty (\tau - t) f(\tau, y(\tau), y(g(\tau))) \, d\tau$$

Integrating it again from T to t, $T \geq t_2$, we get

$$y(T) - y(t) = \int_T^t \left[\int_{t_1}^\infty (\tau - t_1) f(\tau, y(\tau), y(g(\tau))) \, d\tau \right] dt_1$$

$$\geq \int_T^t \frac{(\tau - T)^2}{2} f(\tau, y(\tau), y(g(\tau))) \, d\tau$$

Letting $t \to \infty$ and noting that $y(t)$ is bounded, we obtain a contradiction to (5.1.5). Therefore $c_1 = 0$, i.e.,

$$\lim_{t \to \infty} y = 0, \quad \lim_{t \to \infty} y' = 0, \quad \lim_{t \to \infty} y'' = 0$$

(2) $y''(t) < 0$ for $t \geq t_3$.

It is easy to see that $y(t) < 0$ for sufficiently large t. This contradicts the assumption $y(t) > 0$. The proof is complete.

THEOREM 5.1.3 Assume that conditions (i)–(iii) of Theorem 5.1.1 hold. Then condition (5.1.2) is necessary and sufficient for the existence of a bounded nonoscillatory solution y such that $y(t) \to d \neq 0$ as $t \to \infty$.

This is a consequence of Theorems 5.1.1 and 5.1.2. The verification is left to the reader.

EXAMPLE 5.1.1 The equation

$$y'''(t) + \frac{15}{8} \frac{1}{t^{7/2}(1 + t^{-1/4})^3} y^3(\sqrt{t}) = 0 \tag{5.1.7}$$

satisfies conditions (i) to (iv) of Theorem 5.1.1. Therefore, it has a bounded nonoscillatory solution that tends to a nonzero limit as $t \to \infty$. In fact, $y(t) = 1 + t^{-\frac{1}{2}}$ is a nonoscillatory solution of (5.1.7) with $\lim_{t \to \infty} y(t) = \lim_{t \to \infty} (1 + t^{-\frac{1}{2}}) = 1$.

REMARK 5.1.1 For related results on third order functional differential equations see [229].

Now we consider a class of fourth order nonlinear functional differential equations

$$[r(t)y''(t)]'' + f(y(g(t)), t) = 0 \tag{5.1.8}$$

where $f(y,t)$ may be classified as superlinear, sublinear, strongly superlinear, or strongly sublinear. In the superlinear or sublinear cases, we shall present necessary and sufficient conditions for (5.1.8) to admit the existence of nonoscillatory solutions with special asymptotic properties. Similarly, in the case of strongly superlinear or strongly sublinear equalities we shall give sufficient conditions for all solutions to be oscillatory.

LEMMA 5.1.1 Assume that

(a) $r \in C[R_+, R_+]$, $r(t) > 0$, and $\int_0^\infty \frac{t\,dt}{r(t)} = +\infty$

(b) $g \in C[R_+, R_+]$ and $\lim_{t \to \infty} g(t) = \infty$

(c) $f \in C[R_+ \times R, R]$ and $yf(y,t) > 0$ for $y \neq 0$, $t \in R^+$.

Let $y(t)$ be an eventually positive solution of (5.1.8). Then

(i) One of the following statements holds:

(I) $y'(t) > 0$, $y''(t) > 0$, and $[r(t)y''(t)]' > 0$ for all sufficiently large t

(II) $y'(t) > 0$, $y''(t) < 0$, and $[r(t)y''(t)]' > 0$ for all sufficiently large t.

(ii) There are positive numbers T, a_1, a_2 such that

$$a_1 \leq y(t) \leq a_2 R(t) \quad \text{for } t \geq T \tag{5.1.9}$$

and

$$y(t) \geq R_T(t) [r(t)y''(t)] \quad \text{for } t \geq T \tag{5.1.10}$$

where

$$R(t) = \int_0^t \frac{(t-s)s}{r(s)} ds \qquad R_T(t) = \int_T^t \frac{s-T}{r(s)} ds \tag{5.1.11}$$

Proof: (i) Let $y(t)$ be an eventually positive solution of (5.1.8). Then there exists a $t_0 > 0$ such that $y(g(t)) > 0$ for $t \geq t_0$. It follows that $[r(t)y''(t)]'' < 0$ for $t \geq t_0$. Hence $[r(t)y''(t)]'$, $r(t)y''(t)$, and $y'(t)$ are eventually monotonic and of one sign. Suppose $[r(t_1)y''(t_1)]' = -c_1 < 0$, $t_1 > t_0$. Then $[r(t)y''(t)]' \leq -c_1$ for $t \geq t$. Integrating this inequality, we see that there are numbers $t_2 > t_1$ and $c_2 > 0$ such that $r(t)y''(t) \leq -c_2 t$ for $t \geq t_2$. From the last inequality it is easy to derive $\lim_{t\to\infty} y'(t) = -\infty$, which implies $\lim_{t\to\infty} y(t) = -\infty$. This contradicts the fact that $y(t) > 0$. Therefore we must have $[r(t)y''(t)]' > 0$ for $t \geq t_0$. Suppose $r(t)y''(t) < 0$ for $t \geq t_0$. Then $y'(t)$ must be eventually positive, otherwise we are led to $\lim_{t\to\infty} y(t) = -\infty$, which is a contradiction.

This verifies the case (II). Next suppose that there exists $t_3 > t_0$ such that $r(t)y''(t) > 0$ for $t \geq t_3$. Then we have $r(t)y''(t) \geq r(t_3)y''(t_3) = c_3 > 0$ for $t \geq t_3$. We multiply the above inequality by $t/r(t)$ and integrate it from t_3 to t. We have

$$\int_{t_3}^t sy''(s) \, ds = c_3 \int_{t_3}^t \frac{s \, ds}{r(s)}$$

or

$$ty'(t) + c = c_3 \int_{t_3}^t \frac{s \, ds}{r(s)} \to \infty \quad \text{as } t \to \infty$$

which shows that $y'(t)$ is eventually positive. This validates the case (I).

(ii) From (i), we have $y(t) > 0$, $y'(t) > 0$ for $t \geq T$, and so $y(t) \geq a_1$ for $t \geq T$.
    To prove the right-hand side of (5.1.9) we integrate $[r(t)y''(t)]'' < 0$ over $[t_0, t]$ and obtain

$$[r(t)y''(t)]' < k_0$$

Integrating it again over $[t_0, t]$, we get

$$r(t)y''(t) < k_0 t + k \qquad t \geq t_0$$

or

$$y''(t) < \frac{k_0 t + k_1}{r(t)}$$

Integrating it once again over $[t_0, t]$ we have

$$y'(t) < k_2 + k_0 \int_{t_0}^{t} \frac{s\,ds}{r(s)} + k_1 \int_{t_0}^{t} \frac{ds}{r(s)}$$

Integrating the above inequality further over $[t_0, t]$ yields

$$y(t) < k_3 + k_2 t + k_1 \int_{t_0}^{t} \frac{t-s}{r(s)}\,ds + k_0 \int_{t_0}^{t} \frac{(t-s)s}{r(s)}\,ds$$

We see that every term of the right-hand side of the above inequality is less than $R(t)$. Therefore, we obtain

$$y(t) \leq a_2 R(t) \qquad \text{for } t \geq T$$

To prove (5.1.10), let T be so large that $y(t)$ satisfies case (I) or case (II) of (i) for $t \geq T$. Assume that case (I) holds. Integrating (5.1.8) over $[T, t]$ gives

$$0 < [r(t)y''(t)]' \leq [r(T)y''(T)]'$$

Integrating the above inequality over $[T, t]$, we have

$$r(t)y''(t) \geq r(t)y''(t) - r(T)y''(T) \geq [r(t)y''(t)]'(t-T) \qquad t \geq T$$

Dividing the above inequality by $r(t)$ and integrating yields

$$y'(t) \geq \int_{T}^{t} \frac{(s-T)}{r(s)} [r(s)y''(s)]'\,ds \qquad t \geq T$$

or

$$y(t) \geq \int_{T}^{t} \int_{T}^{s} \frac{\sigma - T}{r(\sigma)} [r(\sigma)y''(\sigma)]'\,d\sigma\,ds$$

We note that $[r(t)y''(t)]'$ is nonincreasing and so

$$y(t) \geq \int_{T}^{t} [r(s)y''(s)]' \int_{T}^{s} \frac{\sigma - T}{r(\sigma)}\,d\sigma\,ds$$

$$= R_T(t)[r(t)y''(t)]' - \int_{T}^{t} R_T(s)[r(s)y''(s)]''\,ds$$

which implies (5.1.10). Assume now that case (II) holds. We multiply (5.1.8) by $R_T(t)$ and integrate it over $[T,t]$. By a repeated integration by parts, one obtains

$$R_T(t)[r(t)y''(t)]' - R_T(t)r(t)y''(t) + (t - T)y'(t)$$

$$- y(t) + y(T) + \int_T^t R_T(s)f(y(g(s)), s)\, ds = 0 \qquad (5.1.12)$$

According to (II), $y'(t) > 0$, $y''(t) < 0$. Therefore, from (5.1.12), it follows that

$$y(t) \geq R_T(t)[r(t)y''(t)]' \qquad \text{for } t \geq T$$

This completes the proof of Lemma 5.1.1.

REMARK 5.1.2 Obviously, a result similar to (5.1.9) and (5.1.10) holds for an eventually negative solution of (5.1.8).

THEOREM 5.1.4 Let (5.1.8) be either superlinear or sublinear. Assume that the conditions of Lemma 5.1.1 are satisfied. Then

(i) A necessary and sufficient condition for (5.1.8) to have a solution $y(t)$ such that $\lim_{t \to \infty} y(t)/R(t) = a \neq 0$ is that

$$\int^\infty |f(cR(g(t)), t)|\, dt < \infty \qquad \text{for some } c \neq 0 \qquad (5.1.13)$$

(ii) A necessary and sufficient condition for (5.1.8) to have a solution $y(t)$ such that $\lim_{t \to \infty} y(t) = b \neq 0$ is that

$$\int^\infty R(t)|f(c, t)|\, dt < \infty \qquad \text{for some } c \neq 0 \qquad (5.1.14)$$

Proof: Necessity: (i) Let $y(t)$ be a solution of (5.1.8) such that $\lim_{t \to \infty} y(t)/R(t) = a \neq 0$. Without loss of generality, we may suppose that $a > 0$. There are positive numbers t_0, a_1, a_2 such that

$$a_1 R(g(t)) \leq y(g(t)) \leq a_2 R(g(t)) \qquad \text{for } t \geq t_0$$

In view of the above, we see that

$$f(y(g(t)), t) \geq f(a_1 R(g(t)), t) \qquad (5.1.15)$$

or

$$f(y(g(t)), t) \geq \left(\frac{a_1}{a_2}\right) f(a_2 R(g(t)), t) \qquad (5.1.16)$$

depending on superlinearity or sublinearity of f. On the other hand, by

Lemma 5.1.1, we have $(r(t)y''(t))' > 0$. By integrating (5.1.8), we obtain

$$\int_{t_0}^{\infty} f(y(g(t)),\, t)\, dt < \infty \tag{5.1.17}$$

From (5.1.15) to (5.1.17) we conclude that $\int_0^{\infty} f(a_i R(g(t)),\, t)\, dt < \infty$, where $i = 1,\, 2$ corresponding to the superlinear or sublinear case.

(ii) Let $y(t)$ be a solution of (5.1.8) such that $\lim_{t \to \infty} y(t) = b > 0$. Then there are positive numbers t_0, b_1, b_2 for which $b_1 \leq y(g(t)) \leq b_2$ for $t \geq t_0$. Hence we have

$$f(y(g(t)),\, t) \geq f(b_1,\, t) \tag{5.1.18}$$

or

$$f(y(g(t)),\, t) \geq \frac{b_1}{b_2}\, f(b_2,\, t) \tag{5.1.19}$$

according as (5.1.8) is superlinear or sublinear. We now multiply (5.1.8) by $R(t)$ and integrate from t_0 to t to obtain

$$\int_{t_0}^{t} R(s)f(y(g(s)),\, s)\, ds = -\int_{t_0}^{t} R(s)[y(s)y''(s)]''\, ds$$

$$= -R(t)[r(t)y''(t)]' + R'(t)r(t)y''(t) - ty'(t) + y(t) + k \tag{5.1.20}$$

where k is a constant. Observing that $y(t)$ is subject to case (II) of Lemma 5.1.1 (i), from (5.1.20) we obtain

$$\int_{t_0}^{\infty} R(t)f(y(g(t)),\, t)\, dt < \infty \tag{5.1.21}$$

From (5.1.18), (5.1.19), and (5.1.21) it follows that $\int_{t_0}^{\infty} R(t)f(b_i,\, t)\, dt < \infty$ where $i = 1$ or 2 according as (5.1.8) is superlinear or sublinear.

Sufficiency: (i) Suppose (5.1.13) holds with $c > 0$. A similar argument holds if $c < 0$. Set $a = c/2$ or $a = c$ according as (5.1.8) is superlinear or sublinear. Take $T > 0$ so large that

$$\int_{T}^{\infty} f(cR(g(t)),\, t)\, dt < \frac{a}{4} \tag{5.1.22}$$

and $T_0 = \inf \{g(t) : t \geq T\} > 0$. Let $C_R[T_0, \infty)$ denote the linear space of all continuous functions $y(t)$ on $[T_0, \infty)$ such that

$$\|y\|_R = \sup \{R(t)^{-2}|y(t)| : t \geq T_0\} < \infty$$

It is clear that $C_R[T_0, \infty)$ is a Banach space with norm $\|\cdot\|_R$. Let $\bar{Y}$ be the set

$$\bar{Y} = \{y \in C_R[T_0, \infty) : aR(t) \le y(t) \le 2aR(t), \text{ for } t \ge T_0\}$$

which is a bounded, convex, and closed subset of $C_R[T_0, \infty)$. Let us now define the operator Φ as follows:

$$(\Phi y)(t) = aR(t) + R(t) \int_{t_0}^{\infty} f(y(g(s)), s) \, ds + \int_{T}^{t} R(s)f(y(g(s)), s) \, ds$$

$$+ \int_{T}^{t} \left(\int_{0}^{s} \frac{\sigma}{r(\sigma)} d\sigma \right) (t-s)f(y(g(s)), s) \, ds + \int_{T}^{t} \left(\int_{s}^{t} \frac{t-\sigma}{r(\sigma)} d\sigma \right) sf(y(g(s)), s) \, ds$$

$$\text{for } t \ge T$$

$$(\Phi y)(t) = aR(t) + R(t) \int_{T}^{\infty} f(y(g(s)), s) \, ds \quad \text{for } T_0 \le t \le T$$

It can be shown that the operator Φ is continuous and maps $\bar{Y}$ into a compact subset of $\bar{Y}$. Applying Schauder's fixed point theorem, we can conclude that Φ has a fixed point $y \in \bar{Y}$. Then by differentiation, we see that $y = y(t)$ is a solution of Eq. (5.1.8) for $t \ge T$. Moreover,

$$\lim_{t \to \infty} \frac{y(t)}{R(t)} = \lim_{t \to \infty} \frac{y'(t)}{R'(t)} = \lim_{t \to \infty} \frac{r(t) y''(t)}{r(t) R''(t)}$$

$$= \lim_{t \to \infty} \frac{r(t) y''(t)}{t} = \lim_{t \to \infty} [r(t) y''(t)]' = a$$

This shows that $y(t)$ is a solution of (5.1.8) with the above type of asymptotic behavior.

(ii) Suppose (5.1.14) holds with $c > 0$. Define a as in (i) and let T be so large that

$$\int_{T}^{\infty} R(t)f(c, t) \, dt < \frac{a}{4}$$

and $T_0 = \inf \{g(t) : t \ge T\} > 0$. The required solution is obtained as a fixed point of the operator defined by

$$(\Psi y)(t) = a + R(t) \int_{t}^{\infty} f(y(g(s)), s) \, ds + \int_{T}^{t} R(s)f(y(g(s)), s) \, ds$$

$$+ t\int_{t}^{\infty} \left(\int_{t}^{s} \frac{s-\sigma}{r(\sigma)} d\sigma \right) f(y(g(s)), s) \, ds + \int_{0}^{t} \frac{\sigma \, d\sigma}{r(\sigma)} \int_{t}^{\infty} (s-t)f(y(g(s)), s) \, ds$$

$$t \ge T$$

$$(\Psi y)(t) = (\Psi y)(T) \qquad \text{for} \quad T_0 \leq t \leq T$$

The underlying Banach space is $CB[T_0, \infty)$, the space of all bounded and continuous functions $y(t)$ on $[T_0, \infty)$ with norm $\|y\| = \sup\{|y(t)| : t \geq T_0\}$. It can be verified that Ψ is a continuous operator which maps the set

$$\bar{Y} = \{y \in CB[T_0, \infty) : a \leq y(t) \leq 2a \text{ for } t \geq T_0\}$$

into a compact subset of $\bar{Y}$. Therefore, by Schauder's fixed point theorem, Ψ has a fixed point $y \in \bar{Y}$, which is a solution of (5.1.8) on $[T, \infty)$. Since

$$y'(t) = \int_t^\infty \left(\int_t^s \frac{s - \sigma}{r(\sigma)} \, d\sigma \right) f(y(g(s)), s) \, ds > 0$$

it follows that $\lim_{t \to \infty} y(t) = b \in [a, 2a]$. Thus the proof of Theorem 5.1.4 is complete.

REMARK 5.1.3 It is easy to see that in Theorem 5.1.4 the superlinear or sublinear assumption can be replaced by the condition that $f(y, t)$ is monotone in y for each fixed $t \geq 0$. The following lemma will be useful in establishing an oscillation theorem for (5.1.8).

LEMMA 5.1.2 Let (i) (5.1.8) be strongly superlinear; (ii) $u(t)$, $v(t)$, $w(t)$, $\lambda(t)$ be positive continuous functions on $[T_0, \infty)$ satisfying the following inequalities: $\lambda(t) \geq 1$, $u(t) \geq kw(t)$,

$$u(t) \geq \lambda(t)w(t) \int_T^t v(s)f(u(s), s) \, ds$$

where k is a positive constant.
    Then

$$\int^\infty \lambda(t)v(t)f(kw(t), t) \, dt < \infty \tag{5.1.23}$$

Proof: Put $I(t) = \int_T^t v(s)f(u(s), s) \, ds$. Using the hypotheses, we obtain

$$-([I(t)]^{1-\sigma})' = (\sigma - 1)[I(t)]^{-\sigma}v(t)[u(t)]^{\sigma}[u(t)]^{-\sigma}f(u(t), t)$$

$$\geq (\sigma - 1)[I(t)]^{-\sigma}v(t)[\lambda(t)w(t)I(t)]^{\sigma}[kw(t)]^{-\sigma}f(kw(t), t)$$

$$= (\sigma - 1)k^{-\sigma}[\lambda(t)]^{\sigma}v(t)f(kw(t), t)$$

$$\geq (\sigma - 1)k^{-\sigma}\lambda(t)v(t)f(kw(t), t), \quad \text{where } \sigma > 1$$

Integrating the above from T' to t, $T' > T$, we get

$$(\sigma - 1)k^{-\sigma} \int_{T'}^{t} \lambda(s)v(s)f(kw(s),\ s)\ ds \leq [I(T')]^{1-\sigma} - [I(t)]^{1-\sigma}$$

which implies (5.1.23).

LEMMA 5.1.3 Let (i) (5.1.8) be strongly sublinear; (ii) $u(t)$, $v(t)$, $w(t)$, $\mu(t)$ be positive continuous functions on $[T,\infty)$ such that $\mu(t) \leq 1$, $u(t) \leq kw(t)$, and

$$u(t) \geq \mu(t)w(t) \int_{t}^{\infty} v(s)f(u(s),\ s)\ ds$$

where k is a positive constant.

Then

$$\int^{\infty} \mu(t)v(t)f(kw(t),\ t)\ dt < \infty \qquad (5.1.24)$$

Proof: Put

$$J(t) = \int_{t}^{\infty} v(s)f(u(s),\ s)\ ds$$

Then we compute to get

$$-([J(t)]^{1-\tau})' = (1 - \tau)[J(t)]^{-\tau}v(t)[u(t)]^{\tau}[u(t)]^{-\tau}f(u(t),\ t)$$

$$\geq (1 - \tau)[J(t)]^{-\tau}v(t)[\mu(t)w(t)J(t)]^{\tau}[kw(t)]^{-\tau}f(kw(t),\ t)$$

$$\geq (1 - \tau)k^{-\tau}\mu(t)v(t)f(kw(t),\ t)$$

where $0 < \tau < 1$.

An integration of the above shows that (5.1.24) is true.

THEOREM 5.1.5 Assume that the conditions of Lemma 5.1.1 are satisfied. In addition, let (5.1.8) be strongly sublinear, $g_*(t) = \min(g(t),\ t)$, and

$$\int^{\infty} \frac{R_T(g_*(t))}{R(g(t))}\ f(cR(g(t)),\ t)\ dt = \infty \qquad \text{for all } c \neq 0 \qquad (5.1.25)$$

Then all solutions of (5.1.25) are oscillatory.

Proof: Let there exist a nonoscillatory solution $y(t)$ of (5.1.8). Without loss of generality, we may suppose $y(t) > 0$. We observe that, by Lemma 5.1.1, $[r(t)y''(t)]'$ is positive and decreasing. Moreover, $y(t)$ is positive and increasing for $t \geq T$. Let $T' \geq T$ be so large that $g_*(t) \geq T$ for $t \geq T'$. Integrating (5.1.8), we obtain

$$[r(t)y''(t)]' \geq \int_t^\infty f(y(g(s)), s)\, ds \qquad t \geq T'$$

Because of (5.1.10), we have

$$y(g(t)) \geq y(g_*(t)) \geq R_T(g_*(t))[r(s)y''(s)]'_{s=g_*(t)}$$

$$\geq R_T(g_*(t))[r(t)y''(t)]' \qquad t \geq T'$$

Combining the above two inequalities, we get

$$y(g(t)) \geq R_T(g_*(t)) \int_t^\infty f(y(g(s)), s)\, ds \qquad t \geq T'$$

In view of (5.1.9) there is a constant $k > 0$ such that $y(g(t)) \leq kR(g(t))$, for $t \geq T'$. Hence we are able to apply Lemma 5.1.3 to the case t $u(t) \equiv y(g(t))$, $v(t) \equiv 1$, $w(t) \equiv R(g(t))$, $\mu(t) \equiv R_T(g_*(t))/R(g(t))$. From (5.1.24), we obtain

$$\int^\infty \frac{R_T(g_*(t))}{R(g(t))} f(kR(g(t)), t)\, dt < \infty$$

But this contradicts (5.1.25), and the proof is complete.

REMARK 5.1.4 If in addition $f(y,t)$ is assumed to be nondecreasing in y, then the assertion of Theorem 5.1.5 remains valid, provided

$$\int^\infty f(cR(g_*(t)), t)\, dt = \infty \qquad \text{for all } c \neq 0$$

This relation is weaker than (5.1.25).

Consider the delay case where $g(t) \leq t$, that is, (5.1.8) is a strongly sublinear retarded equation. In this case Theorem 5.1.5 says that all solutions of (5.1.8) are oscillatory if

$$\int^\infty |f(cR(g(t)), t)|\, dt = \infty \qquad \text{for all } c \neq 0 \tag{5.1.26}$$

On the other hand, since strong sublinearity implies sublinearity, from Theorem 5.1.4 (i) it follows that (5.1.26) holds if all solutions of (5.1.8) are oscillatory. Combining these results we obtain the following theorem.

THEOREM 5.1.6 Let (5.1.8) be a strongly sublinear retarded equation and assume that the conditions of Lemma 5.1.1 are satisfied. Then (5.1.26) is a necessary and sufficient condition for all solutions to be oscillatory.

The situation becomes different if (5.1.8) is not a retarded equation. The following example illustrates this.

EXAMPLE 5.1.1 Consider the advanced equation

$$[t^{1/2}y''(t)]'' + \frac{1}{2} t^{-7/2} |y(t^2)|^{1/2} \operatorname{sgn} y(t^2) = 0 \qquad t \geq 1$$

This equation satisfies condition (5.1.26), but it has nonoscillatory solution $y(t) = t^2$. Of course, (5.1.25) is violated.

A similar discussion is valid for the strongly superlinear case.

THEOREM 5.1.7 Let (5.1.8) be strongly superlinear. Assume that there is a differentiable function $h(t)$ on R^+ such that

$$h(t) \leq g_*(t) , \qquad h'(t) \geq 0 , \qquad \lim_{t \to \infty} h(t) = \infty$$

and

$$\int^{\infty} R(h(t)) \, |f(c,t)| \, dt = \infty \qquad \text{for all } c \neq 0 \tag{5.1.27}$$

Then all solutions of (5.1.8) are oscillatory.

EXAMPLE 5.1.2 Consider the equation

$$t^{1/2}y''(t)]'' + \frac{1}{2} t^{-7/2} [y(t^{1/2})]^2 \operatorname{sgn} y(t^{1/2}) = 0 \qquad t \geq 1$$

which has the nonoscillatory solution $y(t) = t^2$, even though the condition

$$\int^{\infty} R(t)|f(c,t)| \, dt = \infty \qquad \text{for all } c \neq 0 \tag{5.1.28}$$

is satisfied.

From Theorems 5.1.4 and 5.1.7 we have the following result for advanced type equations.

THEOREM 5.1.8 Let (5.1.8) be a strongly superlinear advanced equation. Assume that the conditions of Lemma 5.1.1 hold. Then (5.1.28) is a necessary and sufficient condition for all solutions of (5.1.8) to be oscillatory.

5.2 EVEN ORDER EQUATIONS WITH DEVIATING ARGUMENTS

We shall first prove some lemmas that will be useful for our discussion.

LEMMA 5.2.1 Let $y(t)$ be a n times differentiable function on R^+ of constant sign, $y^{(n)}(t)$ be of constant sign and not identically zero in any interval $[t_1, \infty)$ and

$$y^{(n)}(t)y(t) \le 0 \qquad (5.2.1)$$

Then

(i) There exists a number $t_2 \ge t_1$ such that the functions $y^{(j)}(t)$, $j = 1, 2, \ldots, n - 1$, are of a constant sign on $[t_2, +\infty)$

(ii) There exists a number $k \in \{1, 3, 5, \ldots, n - 1\}$ when n is even, or $k \in \{0, 2, 4, \ldots, n - 1\}$ when n is odd, such that

$$y(t)y^{(j)}(t) > 0 \quad \text{for} \quad j = 0, 1, \ldots, k, \quad t \ge t_2 \qquad (5.2.2)$$

$$(-1)^{n+j-1}y(t)y^{(j)}(t) > 0 \quad \text{for} \quad j = k + 1, \ldots, n, \quad t \ge t_2$$

(iii)

$$|y(t)| \ge \frac{(t - t_0)^{n-1}}{(n - 1) \cdots (n - k)} |y^{(n-1)}(2^{n-k-1}t)| \qquad (5.2.3)$$

(iv) Either

$$\begin{cases} \text{sign } y(s) = \text{sign } \lim_{t \to \infty} y^{(j)}(t) \quad \text{for} \quad j = 0, 1, 2, \ldots, q, \ s \ge t_2 \\[2mm] \lim_{t \to \infty} y^{(j)}(t) = 0 \quad \text{for} \quad j = q + 1, \ldots, n - 1 \\[2mm] q = k \quad \text{if } y(s) \cdot \lim_{t \to \infty} y^{(k)}(t) > 0 \\[2mm] q = k - 1 \quad \text{if } k > 0 \text{ and } \lim_{t \to \infty} y^{(k)}(t) = 0 \end{cases} \qquad (5.2.4)$$

or

$$\begin{cases} \lim_{t \to \omega} y^{(j)}(t) = 0 \text{ for } j = 0, 1, 2, \ldots, n - 1 \\[2mm] \text{if } k = 0 \text{ and } \lim_{t \to \infty} y^{(k)}(t) = \lim_{t \to \infty} y(t) = 0 \end{cases}$$

Proof: From (5.2.1), without loss of generality we assume that $y(t) > 0$, $y^{(n)}(t) \le 0$ for $t \ge t_1 \ge T$. Then $y^{(n-1)}(t)$ is a nonincreasing function for $t \ge t_1$ and is not constant on any (T, ∞) for large T. This implies that exactly one of the following is true:

(a_1) $y^{(n-1)}(t) > 0 \quad \text{for} \quad t \ge t_1$

(b_1) $y^{(n-1)}(t) < 0 \quad \text{for} \quad t \ge T^{(1)}_{n-1} \ge t_1$

From (b_1) together with $y(t) > 0$, $y^{(n)}(t) \leq 0$, it follows that there exists a number $T_{n-2}^{(1)} \geq T_{n-1}^{(1)}$ such that $y^{(n-2)}(t) \leq 0$ for $t > T_{n-2}^{(1)}$. Likewise, we have $y^{(n-3)}(t) < 0$ for $t \geq T_{n-3}^{(1)} \geq T_{n-2}^{(1)} \geq \cdots$, and hence $y(t) < 0$ for $t \geq T_0^{(1)} \geq T_1^{(1)}$, which is a contradiction. Since $y(t) > 0$ for $t \geq t_1$, then (a_1) holds. Now we know that $y^{(n-2)}$ is increasing and concave for $t \geq t_1$. Therefore exactly one of the following possibilities holds true:

(a_2) $y^{(n-2)}(t) \geq 0$ for $t \geq T_{n-2}^{(2)} \geq t_1$

(b_2) $y^{(n-2)}(t) < 0$ for $t \geq t_1$

From (a_2) and (a_1), we obtain $y^{(n-3)}(t) > 0$ for $t \geq T_{n-3}^{(2)} \geq T_{n-2}^{(2)}$. Analogously, we get $y^{(n-4)}(t) > 0$ for $t \geq T_{n-4}^{(2)} \geq T_{n-3}^{(2)} \geq \cdots$, and hence $y(t) > 0$ for $t \geq T_0^{(2)} \geq T_1^{(2)}$. Thus the functions $y^{(j)}(t)$ $(j = 1, \ldots, n-1)$ are of constant sign for t sufficiently large.

If (b_2) holds, then $y^{(n-3)}$ is decreasing and convex for $t \geq t_1$. Then exactly one of the following is true:

(a_3) $y^{(n-3)}(t) > 0$ for $t \geq t_1$

(b_3) $y^{(n-3)}(t) < 0$ for $t \geq T_{n-3}^{(3)} \geq t_1$

Thus we can repeat the above argument and show that the functions $y^{(j)}(t)$ $(j = 1, 2, \ldots, n-1)$ are of constant sign for t sufficiently large. This proves (i) and (ii) of Lemma 5.2.1.

Now we shall prove that (5.2.3) holds. Without loss of generality, we can assume that

$$y(t) \geq 0$$

From (5.2.2) and $y(t) \geq 0$, it follows that

$$-y^{(n-2)} = -y^{(n-2)}(\infty) + \int_t^\infty y^{(n-1)}(\tau) \, d\tau$$

$$\geq \int_t^{2t} y^{(n-1)}(\tau) \, d\tau \geq t y^{(n-1)}(2t)$$

By integrating this inequality, we have

$$y^{(n-3)}(t) \geq -\int_t^\infty y^{(n-2)}(\tau)\,d\tau \geq \int_t^{2t} \tau y^{(n-1)}(2\tau)\,d\tau \geq t^2 y^{(n-1)}(4t)$$

Hence

$$y^{(k)}(t) \geq t^{n-k-1} y^{(n-1)}(2^{n-k-1} t)$$

Since, $y^{(k-1)}(t) \geq 0$, from the last inequality we obtain

$$y^{(k-1)}(t) = y^{(k-1)}(t_0) + \int_{t_0}^t y^{(k)}(\tau)\,d\tau$$

$$\geq \int_{t_0}^t (\tau - t_0)^{n-k-1} y^{(n-1)}(2^{n-k-1}\tau)\,d\tau$$

$$\geq \frac{(t - t_0)^{n-k}}{n - k} y^{(n-1)}(2^{n-k-1} t)$$

Hence, after $(k - 1)$-fold integration, we arrive at the required inequality (5.2.3).

Now we shall prove the last part of Lemma 5.2.1. By (ii) we have $\lim_{t\to\infty} y^{(k)}(t) \geq 0$, and if $k > 0$, then $\lim_{t\to\infty} y^{(j)}(t) > 0$ for $j = 0, 1, 2, \ldots,$ $k - 1$. Moreover, (ii) implies that $\lim_{t\to\infty} y^{(k+1)}(t) \leq 0$. Suppose $\lim_{t\to\infty} y^{(k+1)}(t) = -a^2$ $(a \neq 0)$. Then $y^{(k)}(t) < 0$ for $t \geq t_2$ sufficiently large, which is a contradiction, since $y^{(k)}(t) > 0$ if $t \geq t_2$. Therefore $\lim_{t\to\infty} y^{(k+1)}(t) = 0$. Further, (ii) implies $\lim_{t\to\infty} y^{(k+2)}(t) \geq 0$. If we assume $\lim_{t\to\infty} y^{(k+2)}(t) = a^2$ $(a \neq 0)$, then by integrating, we obtain $y^{(k+1)}(t) > 0$ for sufficiently large $t \geq t_2$, which is a contradiction to $y^{(k+1)}(t) < 0$ if $t \geq t_2$. Thus $\lim_{t\to\infty} y^{(k+2)}(t) = 0$. Analogously, we get

$$\lim_{t\to\infty} y^{(k+3)}(t) = \lim_{t\to\infty} y^{(k+4)}(t) = \cdots = \lim_{t\to\infty} y^{(n-1)}(t) = 0$$

This completes the proof.

LEMMA 5.2.2 Assume that a function y together with its derivatives of order up to $n - 1$ is absolutely continuous and of constant sign on the interval $(t_0, +\infty)$. Moreover,

$$y^{(n)}(t)y(t) \geq 0 \tag{5.2.5}$$

Then either

$$y^{(k)}(t)y(t) \geq 0 \qquad k = 0, 1, \ldots, n \tag{5.2.6}$$

or one can find a number ℓ, $0 \leq \ell \leq n-2$, which is even when n is even and odd when n is odd, such that

$$y^{(k)}(t)y(t) \geq 0 \qquad k = 0, 1, \ldots, \ell$$

$$(-1)^{n+k}y^{(k)}(t)y(t) \geq 0 \qquad k = \ell+1, \ldots, n \tag{5.2.7}$$

and inequality (5.2.3) is satisfied.

Proof: There are two possible cases, namely, either $y^{(n-1)}(t)y(t) \geq 0$ or $y^{(n-1)}(t)y(t) \leq 0$. If the first of these holds, then by (5.2.5) we easily conclude that (5.2.6) holds. If $y^{(n-1)}(t)y(t) \leq 0$, then by Lemma 5.2.1 one can find a number ℓ, $0 \leq \ell \leq n-2$, which is odd when $n-1$ is even, and even when $n-1$ is odd, so that (5.2.7) holds and

$$|y(t)| \geq \frac{(t - t_0)^{n-2}}{(n-2) \cdots (n-\ell-1)} |y^{(n-2)}(2^{n-\ell-2}t)|$$

Hence, taking into account the fact that $|y^{(n-2)}(t)| \geq t|y^{(n-1)}(2t)|$, we immediately obtain inequality (5.2.3).

LEMMA 5.2.3 Assume that the hypotheses of Lemma 5.2.1 hold. Assume further that y satisfies the following relation:

$$y^{(n-1)}(t)y^{(n)}(t) \leq 0 \qquad \text{for every } t \geq t_0$$

Then for every λ, $0 < \lambda < 1$, there exists an $M > 0$ such that

$$y(\lambda t) \geq Mt^{n-1}|y^{(n-1)}(t)| \qquad \text{for all large t} \tag{5.2.8}$$

Proof: By Taylor's formula and Lemma 5.2.1, we have

$$y(s) = y(t_n) + \frac{y'(t_n)}{1!}(s-t_n) + \cdots + \frac{y^{(\ell)}(s^*)}{\ell!}(s-t_n)^\ell \geq \frac{y^{(\ell)}(s^*)}{\ell!}(s-t_n)^\ell$$

where $t_n \leq s^* \leq s$. But $y^{(n-1)}y^{(n)} \leq 0$ on $[t_n,\infty)$. So

$$y(s) \geq \frac{y^{(\ell)}(s)}{\ell!}(s-t_n)^\ell \qquad \text{for every } s \geq t_n$$

and consequently for $s = t\lambda$, $0 < \lambda < 1$, we have

$$y(\lambda t) \geq \frac{1}{\ell!}\left(\lambda - \frac{t_n}{t}\right)^\ell t^\ell y^{(\ell)}(\lambda t) \qquad \text{for every } t \geq t_n$$

Obviously, there exists a $T \geq t_n$ such that

$$y(\lambda t) \geq \frac{\lambda^\ell}{\ell! \, 2^\ell} t^\ell y^{(\ell)}(\lambda t) \qquad \text{for every } t \geq T \tag{5.2.9}$$

Set $\ell = n - 1$; then

$$y(\lambda t) \geq M t^{n-1} y^{(n-1)}(\lambda t) \geq M t^{n-1} y^{(n-1)}(t)$$

where $M = \lambda^{n-1}/(n-1)! \, 2^{n-1}$. That is, (5.2.8) is proved whenever $\ell = n - 1$. Therefore, it remains to prove (5.2.8) when $\ell < n - 1$. In this case, again using Taylor's formula, we obtain

$$y^{(\ell)}(\lambda t) = y^{(\ell)}(t) + (-1) \frac{y^{(\ell+1)}(t)}{1!}(t - \lambda t) + \cdots + (-1)^{n-1-\ell} \frac{y^{(n-1)}(t^*)}{(n-1-\ell)!}(t - \lambda t)^{n-1-\ell}$$

where $\lambda t \leq t^* \leq t$. Hence, by Lemma 5.2.1 we get

$$y^{(\ell)}(\lambda t) \geq (-1)^{n-\ell-1} \frac{y^{(n-1)}(t^*)}{(n-1-\ell)!}(t - \lambda t)^{n-1-\ell} = \frac{|y^{(n-1)}(t^*)|}{(n-1-\ell)!}(t - \lambda t)^{n-1-\ell}$$

Since

$$\frac{d}{dt}|y^{(n-1)}(t)|^2 = 2 y^{(n-1)}(t) y^{(n)}(t) \leq 0 \qquad \text{for every } t \geq T$$

the function $|y^{(n-1)}|$ is nonincreasing on $[T, \infty)$ and therefore

$$y^{(\ell)}(\lambda t) \geq \frac{(1-\lambda)^{n-1-\ell}}{(n-1-\ell)!} t^{n-1-\ell} |y^{(n-1)}(t)| \qquad \text{for every } t \geq T \tag{5.2.10}$$

From (5.2.9) and (5.2.10), we obtain

$$y(\lambda t) \geq \frac{\lambda (1-\lambda)^{n-1-\ell}}{2^\ell \, \ell! (n-1-\ell)!} t^{n-1} |y^{(n-1)}(t)|$$

This proves (5.2.8) with

$$M = \frac{\lambda(1-\lambda)^{n-1-\ell}}{2^\ell \, \ell! (n-1-\ell)!}$$

REMARK 5.2.1 If, in addition, $\lim_{t \to \infty} y(t) \neq 0$, then, because of the monotonicity of y, it is easy to see that for all large t,

$$y(t) \geq \tfrac{1}{2} y(\tfrac{1}{2} t)$$

Thus, applying (5.2.8) for $\lambda = \tfrac{1}{2}$, we derive

$$y(t) \geq Mt^{n-1}|y^{(n-1)}(t)| \qquad \text{for all large } t$$

where $M > 0$. This means that in the particular case of $\lim_{t \to \infty} y(t) \neq 0$, inequality (5.2.8) holds also for $\lambda = 1$.

LEMMA 5.2.4 If y is as in Lemma 5.2.1, and for some $k = 0, 1, \ldots, n - 2$,

$$\lim_{t \to \infty} y^{(k)}(t) = c \qquad c \in R$$

then

$$\lim_{t \to \infty} y^{(k+1)}(t) = 0$$

The proof is very simple and left as an exercise.

Now we consider even order differential equations with deviating arguments of the form

$$y^{(n)}(t) + q(t)f(y(t), y(g_1(t)), \ldots, y(g_m(t))) = 0 \tag{5.2.11}$$

THEOREM 5.2.1 Assume that

(i) $q, g_i \in C[R_+, R_+]$, $i = 1, 2, \ldots, m$, $f \in C[R^{m+1}, R]$
and

$$y_i f(y_1, \ldots, y_{m+1}) > 0$$

for $y_i \neq 0$ with same sign, and $i = 1, 2, \ldots, m + 1$

(ii) There exists a continuously differentiable function $\sigma \in C[R_+, R_+]$ such that $0 < \sigma(t) \leq g_i(t)$, $1 \geq \sigma'(t) > 0$ for $t \geq T$, and $\sigma(t) \to \infty$ as $t \to \infty$

(iii) There exists a positive number $c > 0$ such that

$$\liminf_{|y| \to \infty} \left| \frac{f(y_1, \ldots, y_{m+1})}{y} \right| \geq c > 0 \tag{5.2.12}$$

where $|y_i| \geq |y|$, $i = 1, 2, \ldots, m + 1$

(iv) $q(t) \geq 0$ and $q(t)$ is not eventually identically equal to zero on any subinterval $[t_1, +\infty)$

(v) There exists a positive continuously differentiable function $\rho(t)$ on R_+ such that

$$\lim_{t\to\infty} \frac{1}{t^{m-1}} \int_{t_0}^{t} \frac{(t-s)^{m-3}(\rho'(s)(t-s) - (m-1)\rho(s))^2}{\sigma'(s)\sigma^{n-2}(s)\rho(s)}\, ds < \infty \qquad (5.2.13)$$

and

$$\lim_{t\to\infty} \frac{1}{t^{m-1}} \int_{t_0}^{t} (t-s)^{m-1}\rho(s)q(s)\, ds = +\infty \qquad (5.2.14)$$

where $m > 2$ is some integer.

Then every solution of equation (5.2.11) oscillates.

Proof: Without loss of generality, let $y(t) > 0$ be a nonoscillatory solution of equation (5.2.11) for $t \geq t_0$. We choose a t_1 so that $t_1 \geq t_0$ and $g_i(t) \geq t_0$ for $t \geq t_1$, $i = 1, 2, \ldots, m$. By Lemma 5.2.1, there exists a $t_2 \geq t_1$ such that $y^{(n-1)}(t) > 0$ and $y'(t) > 0$ for $t \geq t_2$. Choose a $t_3 \geq t_2$ so that $\sigma(t) \geq 2t_2$ for $t \geq t_3$. We apply Lemma 5.2.3 for $y'(t)$ with $\lambda = \frac{1}{2}$ and conclude that there exist numbers M and t_4 such that $M > 0$, $t_4 \geq t_3$, $y[\frac{1}{2}\sigma(t)] \geq M$, $\sigma^{n-2}(t)y^{(n-1)}(\sigma(t)) \geq M_1\sigma^{n-2}(t)y^{(n-1)}(t)$, for $t \geq t_4$. Set $w(t) = y^{(n-1)}(t)/y[\frac{1}{2}\sigma(t)]$. Thus $w(t)$ satisfies

$$w'(t) = -q(t)\frac{f(y(t), y(g_1(t)), \ldots, y(g_m(t)))}{y[\frac{1}{2}\sigma(t)]} - \tfrac{1}{2}\sigma(t)w(t)\frac{\dot{y}[\frac{1}{2}\sigma(t)]}{y[\frac{1}{2}\sigma(t)]}$$

Since $y'(t) > 0$ for $t \geq t_4$, $\lim_{t\to\infty} y(t)$ exists in the sense of extended real numbers. Assume that $\lim_{t\to\infty} y(t) = b$ and b is finite. This implies

$$\lim_{t\to\infty} \frac{f(y(t), y(g_1(t)), \ldots, y(g_m(t)))}{y(\frac{1}{2}\sigma(t))} = \frac{f(b, \ldots, b)}{b} > 0$$

In case $\lim_{t\to\infty} y(t) = +\infty$, we have

$$\frac{f(y(t), y(g_1(t)), \ldots, y(g_m(t)))}{y(\frac{1}{2}\sigma(t))} \geq \frac{c}{2} > 0 \qquad \text{for large } t$$

On the other hand

$$\tfrac{1}{2}\sigma'(t)w(t)\frac{y'(\frac{1}{2}\sigma(t))}{y(\frac{1}{2}\sigma(t))} \geq \tfrac{1}{2}\sigma'(t)w(t)\left[M_1\sigma^{n-2}(t)\frac{y^{(n-1)}(t)}{y[\frac{1}{2}\sigma(t)]}\right]$$

$$= \frac{M_1}{2}\sigma'(t)\sigma^{n-2}(t)w^2(t)$$

So we obtain

$$w'(t) \leq -\frac{c_0}{2}q(t) - \frac{M_1}{2}\sigma'(t)\sigma^{n-2}(t)w^2(t)$$

where $c_0 = \min(c, f(b, \ldots, b)/b)$.

Therefore

$$\frac{c_0}{2}\rho(t)q(t) \leq -\rho(t)w'(t) - \frac{M_1}{2}\sigma'(t)\sigma^{n-2}(t)\rho(t)w^2(t) \qquad t \geq t_4$$

Multiplying the above inequality by $(t-s)^{m-1}$ and integrating, we see that

$$\frac{c_0}{2}\int_{t_4}^{t}(t-s)^{m-1}\rho(s)q(s)\,ds$$

$$\leq -\int_{t_4}^{t}(t-s)^{m-1}\rho(s)w'(s)\,ds - \frac{M_1}{2}\int_{t_4}^{t}(t-s)^{m-1}\sigma'(s)\sigma^{n-2}(s)\rho(s)w^2(s)\,ds$$

$$= (t-t_4)^{m-1}\rho(t_4)w(t_4) + \int_{t_4}^{t}(t-s)^{m-2}w(s)[\rho'(s)(t-s) - (m-1)\rho(s)]\,ds$$

$$- \frac{M_1}{2}\int_{t_4}^{t}(t-s)^{m-1}\sigma'(s)\sigma^{n-2}(s)\rho(s)w^2(s)\,ds$$

$$= (t-t_4)^{m-1}\rho(t_4)w(t_4) - \int_{t_4}^{t}\left[\sqrt{\frac{M_1}{2}}(t-s)^{(m-1)/2}\sqrt{\sigma'(s)\sigma^{n-2}(s)\rho(s)}\,w(s)\right.$$

$$\left. - \frac{1}{2}\sqrt{\frac{2}{M_1}}\frac{(t-s)^{(m-3)/2}}{\sqrt{\sigma'(s)\sigma^{n-2}(s)\rho(s)}}(\rho'(s)(t-s) - (m-1)\rho(s))\right]^2\,ds$$

$$+ \frac{1}{2M_1}\int_{t_4}^{t}\frac{(t-s)^{m-3}}{\sigma'(s)\sigma^{n-2}(s)\rho(s)}(\rho'(s)(t-s) - (m-1)\rho(s))^2\,ds$$

$$\leq (t-t_4)^{m-1}\rho(t_4)w(t_4) + \frac{1}{2M_1}\int_{t_4}^{t}\frac{(t-s)^{m-3}}{\sigma'(s)\sigma^{n-2}(s)\rho(s)}(\rho'(s)(t-s) - (m-1)\rho(s))^2\,ds$$

But, for every $t \geq t_4$,

$$\int_{t_0}^{t} (t-s)^{m-1}\rho(s)q(s)\,ds - \int_{t_4}^{t} (t-s)^{m-1}\rho(s)q(s)\,ds$$

$$= \int_{t_0}^{t_4} (t-s)^{m-1}\rho(s)q(s)\,ds \le (t-t_0)^{m-1}\int_{t_0}^{t_4}\rho(s)q(s)\,ds$$

and so

$$\frac{c_0}{2t^{m-1}}\int_{t_0}^{t}(t-s)^{m-1}\rho(s)q(s)\,ds \le \frac{c_0}{2}\left(1-\frac{t_0}{t}\right)^{m-1}\int_{t_0}^{t_4}\rho(s)q(s)\,ds$$

$$+\left(1-\frac{t_4}{t}\right)^{m-1}\rho(t_4)w(t_4) + \frac{1}{2M_1 t^{m-1}}\int_{t_0}^{t}\frac{(t-s)^{m-3}(\rho'(s)(t-s)-(m-1)\rho(s))^2}{\sigma'(s)\sigma^{n-2}(s)\rho(s)}\,ds$$

for all $t \ge t_4$.

This gives

$$\lim_{t\to\infty}\frac{c_0}{2t^{m-1}}\int_{t_0}^{t}(t-s)^{m-1}\rho(s)q(s)\,ds \le \rho(t_4)w(t_4)$$

$$+\frac{c_0}{2}\int_{t_0}^{t_4}\rho(s)q(s)\,ds + \lim_{t\to\infty}\frac{1}{2M_1 t^{m-1}}\int_{t_0}^{t}\frac{(t-s)^{m-3}(\rho'(s)(t-s)-(m-1)\rho(s))^2}{\sigma'(s)\sigma^{n-2}(s)\rho(s)}\,ds$$

which contradicts condition (5.2.14).

A similar proof holds if $y(t) < 0$, for $t \ge t_0$.

COROLLARY 5.2.1 Assume that the conditions of Theorem 5.2.1 are satisfied. Further, assume that $1 \ge \sigma'(t) > k > 0$ and there exists a number $\alpha \in [0, n-1)$ such that

$$\lim_{t\to\infty}\frac{1}{t^{m-1}}\int_{t_0}^{t}(t-s)^{m-1}s^{\alpha}q(s)\,ds = \infty \tag{5.2.15}$$

Then every solution of equation (5.2.11) oscillates.

Proof: We choose $\rho(s) = s^{\alpha}$. By direct computation one can show that for $\alpha \in [0, n-1)$, $\rho(t)$ satisfies condition (5.2.13). By application of Theorem 5.2.1 we obtain the conclusion of Corollary 5.2.1.

EXAMPLE 5.2.1 Consider the equation

$$y^{(n)}(t) + f(y(\ln t)) = 0, \quad n \text{ even}, \quad t > 1 \tag{5.2.16}$$

where

$$f(y) = \begin{cases} y \exp[y(1 + \sin y)], & \text{for } y \geq 0 \\ y, & \text{for } y \leq 0 \end{cases} \qquad (5.2.17)$$

$q(t) = 1$, $\sigma(t) = \ln t$, and $\sigma' = 1/t$.

Choose $\rho(t) = t^{-\alpha}$, $0 < \alpha < 1$. We note that $\rho(t)$ satisfies conditions (5.2.13) and (5.2.14). Therefore, all solutions of (5.2.16) are oscillatory.

Under a modification of the hypotheses of Theorem 5.2.1, we can obtain the following result.

THEOREM 5.2.2 Assume that all conditions of Theorem 5.2.1 hold except that condition (v) is replaced by the following condition.

(v') There exists a positive continuously differentiable function $\rho(t)$ on $[t_0, +\infty)$ such that

$$\int^{\infty} \rho(t)q(t)\, dt = \infty \qquad (5.2.18)$$

and

$$\int^{\infty} \frac{\rho'^2(s)}{\rho(s)\sigma'(s)\sigma^{n-2}(s)}\, ds < \infty \qquad (5.2.19)$$

Then the conclusion of Theorem 5.2.1 is true.

EXAMPLE 5.2.2 Consider

$$y^{(4)}(t) + t^{-4} \ln t\, y^{1/3}\left(\frac{t}{2}\right) y^{2/3}\left(\frac{t}{3}\right) = 0 \qquad (5.2.20)$$

We observe that (5.2.20) does not satisfy condition (5.2.15), but we can choose $\rho(t) = t^3/(\ln t)^2$ which satisfies conditions (5.2.18) and (5.2.19). Therefore all solutions of (5.2.20) are oscillatory.

REMARK 5.2.2 Let $n = 2$, $f(y_1, y_2, \ldots, y_{m+1}) = F(y_1)$, and $yF(y) > 0$ for $y \neq 0$, $F'(y) \geq k > 0$. Then $q(t)$ need not be a positive function to ensure the oscillation of (5.2.11).

5.3 LINEAR DIFFERENTIAL EQUATIONS AND INEQUALITIES WITH DEVIATING ARGUMENTS

We consider even order differential equation with deviating arguments,

$$y^{(n)}(t) = p(t)y(g(t)) \qquad (5.3.1)$$

where $p, g \in C[R_+, R_+]$, $p(t) > 0$, n is an even integer, and $\lim_{t \to \infty} g(t) = \infty$.

According to Lemma 5.2.2, the nonoscillatory solution $y(t)$ of (5.3.1) satisfies (5.2.7).

DEFINITION 5.3.1 A solution $y(t)$ of (5.3.1) is said to be <u>nonoscillatory of degree ℓ</u> if y satisfies (5.2.7). A set of all nonoscillatory solutions of (5.3.1) of degree ℓ is denoted by N_ℓ. We shall denote the set of all nonoscillatory solutions of (5.3.1) of degree $0 \leq \ell \leq n$ by N. In view of Lemma 5.2.2, we have

$$N = N_0 \cup N_2 \cup \cdots \cup N_n \tag{5.3.2}$$

We consider an even order delay differential inequality

$$y^{(n)}(t) - [p^n + q(t)]y(t - n\tau) \geq 0 \tag{5.3.3}$$

THEOREM 5.3.1 Assume that n is a positive even integer, $p > 0$, $\tau > 0$, $p\tau e > 1$, and $q \in C[R_+, R_+]$. Then (5.3.3) has no eventually positive bounded solutions.

Proof: Otherwise, there exists an eventually positive bounded solution $y(t)$ of (5.3.3) for $t \geq t_0$. Then $y(t - n\tau) > 0$ for $t > t_0 + n\tau$ and $y^{(n)}(t) > 0$ for $t > t_0 + n\tau$. Since $y(t)$ is bounded, by Lemma 5.2.2, $\ell = 0$ and

$$(-1)^k y^{(k)}(t) > 0 \qquad k = 0, 1, 2, \ldots, n$$

Set

$$x(t) = y^{(n-1)}(t) - py^{(n-2)}(t - \tau) + p^2 y^{(n-3)}(t - 2\tau) - \cdots - p^{n-1}y(t - (n-1)\tau) \tag{5.3.4}$$

For sufficiently large t, $x(t)$ is negative. Differentiating both sides of (5.3.4), we obtain

$$x'(t) = y^{(n)}(t) - py^{(n-1)}(t - \tau) + p^2 y^{(n-2)}(t - 2\tau) - \cdots - p^{n-1}y'(t - (n-1)\tau)$$

Observe that

$$x'(t) + px(t - \tau) = y^{(n)}(t) - p^n y(t - n\tau) \geq q(t)y(t - n\tau) \geq 0$$

That is,

$$x'(t) + px(t - \tau) \geq 0$$

which has an eventually negative solution. This contradicts the conclusion of Theorem 2.1.1.

Similarly, we can prove the following theorem.

THEOREM 5.3.2 Consider the delay differential inequality

$$y^{(n)}(t) - [p^n + q(t)]y(t - n\tau) \leq 0 \qquad \text{for even } n \qquad (5.3.5)$$

Assume that conditions of Theorem 5.3.1 are satisfied. Then (5.3.5) has no eventually negative bounded solutions.

Combining Theorems 5.3.1 and 5.3.2, we obtain the following result.

COROLLARY 5.3.1 Consider the delay differential equation

$$y^{(n)}(t) - [p^n + q(t)]y(t - n\tau) = 0 \qquad n : \text{even} \qquad (5.3.6)$$

Assume that conditions of Theorem 5.3.1 are satisfied. Then every bounded solution of (5.3.6) oscillates.

Consider an odd order delay differential inequality

$$y^{(n)}(t) + [p^n + q(t)]y(t - n\tau) \leq 0 \qquad (5.3.7)$$

THEOREM 5.3.3 Assume that n is an odd positive integer, $p > 0$, $\tau > 0$ are constants, $q \in C[R_+, R_+]$. Further assume

$$p\tau e > 1 \qquad (5.3.8)$$

Then (5.3.7) has no eventually positive solutions.

Proof: Otherwise there exists a solution $y(t)$ of (5.3.7) such that for t_0 sufficiently large

$$y(t) > 0 \qquad t > t_0$$

Then $y(t - n\tau) > 0$ for $t > t_0 + n\tau$, and from (5.3.7), $y^{(n)}(t) < 0$ for $t > t_0 + n\tau$. From Lemma 5.2.1 there exists an even integer ℓ, such that $0 \leq \ell < n$, and for $t > t_0 + n\tau$,

$$y^{(k)}(t) > 0 \qquad \text{for } k = 0, 1, 2, \ldots, \ell$$

$$(-1)^k y^{(k)}(t) > 0 \qquad \text{for } k = \ell+1, \ell+2, \ldots, n \qquad (5.3.9)$$

We claim that $\ell = 0$, i.e., for $t > t_0 + n\tau$,

$$(-1)^k y^{(k)}(t) > 0 \qquad \text{for } k = 0, 1, 2, \ldots, n \qquad (5.3.10)$$

To prove this, assume that $\ell > 0$. Then integrating (5.3.7) $n - \ell$ times from t_1 to t, for t_1 sufficiently large, we obtain

$$y^{(\ell)}(t) \leq \sum_{k=0}^{n-\ell-1} \frac{(t-t_1)^k}{k!} y^{(\ell+k)}(t_1) - \int_{t_1}^{t} \frac{(t-s)^{n-\ell-1}}{(n-\ell-1)!} [p^n + q(s)]y(s - n\tau) \, ds$$

$$\leq \sum_{k=0}^{n-\ell-1} \frac{(t - t_1)^k}{k!} y^{(\ell+k)}(t_1) - \frac{y(t_1 - n\tau)}{(n - \ell - 1)!} p^n \int_{t_1}^{t} (t - s)^{n-\ell-1} \, ds$$

$$= \sum_{k=0}^{n-\ell-1} \frac{(t - t_1)^k}{k!} y^{(\ell+k)}(t_1) - c(t - t_1)^{n-\ell}$$

where

$$c = \frac{y(t_1 - n\tau)}{(n - \ell)!} p^n$$

is a positive constant. This implies that

$$y^{(\ell)}(t) \to \infty \quad \text{as } t \to \infty$$

which contradicts (5.3.9) and proves (5.3.10).

Set

$$x(t) = y^{(n-1)}(t) - py^{(n-2)}(t - \tau) + p^2 y^{(n-3)}(t - 2\tau) - \cdots + p^{n-1} y(t - (n - 1)\tau)$$

Then, in view of (5.3.10),

$$x(t) > 0 \qquad\qquad (5.3.11)$$

Observe that

$$x'(t) = y^{(n)}(t) - py^{(n-1)}(t - \tau) + p^2 y^{(n-2)}(t - 2\tau) - \cdots + p^{n-1} y'(t - (n - 1)\tau)$$

and therefore

$$x'(t) + px(t - \tau) = y^{(n)}(t) + p^n y(t - n\tau) \leq -q(t)y(t - n\tau) \leq 0$$

That is,

$$x'(t) + px(t - \tau) \leq 0 \qquad\qquad (5.3.12)$$

According to Theorem 2.1.1, (5.3.8) implies that (5.3.12) has no positive solutions. The proof is complete.

THEOREM 5.3.4 Consider the delay differential inequality

$$y^{(n)}(t) + [p^n + q(t)]y(t - n\tau) \geq 0 \qquad n \text{ odd} \qquad (5.3.13)$$

subject to the hypotheses of Theorem 5.3.3. Then (5.3.13) has no eventually negative solutions.

Proof: The result follows immediately from the observation that if $y(t)$ is a solution of (5.3.13) then $-y(t)$ is a solution of (5.3.7).

Combining Theorems 5.3.3 and 5.3.4 we obtain the following result.

COROLLARY 5.3.2 Consider the delay differential equation

$$y^{(n)}(t) + [p^n + q(t)]y(t - n\tau) = 0 \qquad n \text{ odd} \tag{5.3.14}$$

subject to hypotheses of Theorem 5.3.3. Then every solution of (5.3.14) oscillates.

THEOREM 5.3.5 Let $p > 0$ and $\tau > 0$ be constants. Then (5.3.8) is a necessary and sufficient condition for all nonoscillatory solutions of the delay differential inequality

$$\left[y^{(n)}(t) - p^n y(t - n\tau)\right] \operatorname{sgn} y(t - n\tau) \geq 0 \tag{5.3.15}$$

to be of degree n, where n is an even integer.

Proof: From Theorems 5.3.1 and 5.3.2, $N_0 = \emptyset$ for (5.3.15) ($\emptyset$ denotes empty set) if and only if (5.3.8) holds. Suppose that there exists a nonoscillatory solution $y(t)$ of (5.3.15); then $y(t) \in N_\ell$, $\ell > 0$. Without loss of generality, we may assume that $y(t) > 0$ for $t \geq t_0$. Putting $t_2 = t_1 + n\tau$ and applying Lemma 5.2.2, we get

$$y^{(\ell)}(t) - y^{(\ell)}(t_2) = \sum_{j=\ell+1}^{n-1} \frac{y^{(j)}(t_2)}{(j - \ell)!} (t - t_2)^{j-\ell} + \int_{t_2}^{t} \frac{(t - s)^{n-\ell-1}}{(n - \ell - 1)!} y^{(n)}(s)\, ds$$

$$\geq \sum_{j=\ell+1}^{n-1} \frac{y^{(j)}(t_2)}{(j - \ell)!} (t - t_2)^{j-\ell} + p^n y(t_1) \int_{t_2}^{t} \frac{(t - s)^{n-\ell-1}}{(n - \ell - 1)!}\, ds$$

$$= \sum_{j=\ell+1}^{n-1} \frac{(y^{(j)}(t_2)}{(j - \ell)!} (t - t_2)^{j-\ell} + p^n y(t_1) \frac{(t - t_2)^{n-\ell}}{(n - \ell)!}$$

for $\ell < n$ and $t \geq t_2$. It follows that $y^{(\ell)}(t) \to \infty$ as $t \to \infty$. But this is impossible, since $y^{(\ell)}(t)$ is bounded above. Hence we must have $\ell = n$. The proof is complete. A result similar to Theorem 5.3.4 holds for an advanced differential inequality.

THEOREM 5.3.6 Let $p > 0$ and $\tau > 0$ be constants. Then (5.3.8) is a necessary and sufficient condition for all nonoscillatory solutions of the inequality

$$\left[y^{(n)}(t) - p^n y(t + n\tau)\right] \operatorname{sgn} y(t + n\tau) \geq 0 \tag{5.3.16}$$

to be of degree 0.

Proof: Suppose (5.3.8) holds. Let $y(t)$ be a positive solution of (5.3.16) of degree $\ell > 0$. Proceeding exactly as in the proof of Theorem 5.3.5 we see that $\ell = n$, i.e., $y(t) > 0$, $y'(t) > 0$, ..., $y^{(n-1)}(t) > 0$ for all large t, say $t \geq t_1 > t_0$. Define

$$z(t) = y^{(n-1)}(t) + py^{(n-2)}(t + \tau) + \cdots + p^{n-1}y(t + (n-1)\tau)$$

Then

$$z'(t) - pz(t + \tau) = y^{(n)}(t) - p^n y(t + n\tau) \geq 0 \qquad t \geq t_1$$

has a positive solution. This contradicts Theorem 2.4.1. Therefore every eventually positive solution of (5.3.16) is necessarily of degree 0. The same is true of all eventually negative solutions of (5.3.16).

Suppose $pe\tau \leq 1$. Then it is easy to see that (5.3.16) has a solution $y(t) = \exp(1/\tau)t$ which is of degree n. This completes the proof.

Now we consider a class of mixed type equations

$$y^{(n)}(t) = p^n y(t - n\sigma) + q^n y(t + n\tau) \tag{5.3.17}$$

THEOREM 5.3.7 Let n be even and p, q, σ, and τ be positive constants. Furthermore, $p\sigma e > 1$ and $q\tau e > 1$. Then all solutions of equation (5.3.17) are oscillatory.

Proof: Otherwise, without loss of generality, there exists an eventually positive solution y(t) of (5.3.17) which satisfies the relation

$$y^{(n)}(t) \geq p^n y(t - n\sigma)$$

From Theorem 5.3.1, $y(t) \in N_\ell$, $\ell > 0$. It is obvious that y(t) also satisfies

$$y^{(n)}(t) \geq q^n y(t + n\tau)$$

According to Theorem 5.3.6, y(t) is of degree 0. From this together with the fact that $y \in N_\ell$, $\ell > 0$, we have a contradiction. The proof is complete.

EXAMPLE 5.3.1 Consider

$$y^{(4)}(t) = 8y(t - \pi) + 8y(t + \pi) \tag{5.3.18}$$

This is a special case of (5.3.17) in which $n = 4$, $p = q = 2^{3/4}$, and $\sigma = \tau = \pi/4$, and $p\sigma e = q\tau e > 1$. Thus, by Theorem 5.3.7, all solutions of (5.3.18) are oscillatory. In fact, $y = \sin 2t$ is an oscillatory solution.

Now we consider an even order differential inequality with several delays, of the form

$$y^{(n)}(t) - \sum_{i=1}^{m} [p_i^n + q_i(t)]y(t - n\tau_i) \geq 0 \tag{5.3.19}$$

THEOREM 5.3.8 Assume that p_i, τ_i are positive constants and n is an even positive integer, $q_i \in C[R_+, R_+]$, i = 1, 2, $\ldots$, m. Let $\tau_i = \min(\tau_1, \cdots, \tau_m)$. Further assume that one of the following conditions holds:

(i) $p_i \tau_i e > 1$ for some $i \in \{1, 2, \cdots, m\}$

(ii) $\left(\Sigma_{i=1}^{m} p_i^n\right)^{1/n} \tau_1 e > 1$

Then (5.3.19) has no eventually positive bounded solutions.

Proof: Assume that (i) holds. Otherwise, there exists an eventually positive bounded solution y(t) which satisfies

$$y^{(n)}(t) - [p_i^n + q_i(t)]y(t - n\tau_i) \geq 0$$

From Theorem 5.3.1, this is impossible.

Assume that (ii) holds and y(t) is an eventually positive bounded solution. By Lemma 5.2.2, we have

$$(-1)^k y^{(k)}(t) > 0 \qquad k = 0, 1, 2, \cdots, n$$

Set

$$x(t) = y^{(n-1)}(t) - \left(\sum_{i=1}^{m} p_i^n\right)^{1/n} y^{(n-2)}(t - \tau_1) + \cdots - \left(\left(\sum_{i=1}^{m} p_i^n\right)^{1/n}\right)^{n-1} y(t - (n-1)\tau_1)$$

It is obvious that x(t) < 0, and

$$x'(t) = y^{(n)}(t) - \left(\sum_{i=1}^{m} p_i^n\right)^{1/n} y^{(n-1)}(t - \tau_1) + \cdots - \left(\left(\sum_{i=1}^{m} p_i^n\right)^{1/n}\right)^{n-1} y'(t - (n-1)\tau_1)$$

Therefore

$$x'(t) + \left(\sum_{i=1}^{m} p_i^n\right)^{1/n} x(t - \tau_1) = y^{(n)}(t) - \sum_{i=1}^{m} p_i^n y(t - n\tau_1)$$

$$\geq \sum_{i=1}^{m} p_i^n y(t - n\tau_i) - \sum_{i=1}^{m} p_i^n y(t - n\tau_1)$$

$$\geq \left[\sum_{i=1}^{m} p_i^n [y(t - n\tau_i) - y(t - n\tau_1)]\right] \geq 0$$

This contradicts the conclusion of Theorem 2.1.1. The proof is complete.

Similar reasoning, together with Theorem 5.3.4, leads to the following result.

THEOREM 5.3.9 Assume that the conditions of Theorem 5.3.8 hold. Then

$$y^{(n)}(t) - \sum_{i=1}^{m} [p_i^n + q_i(t)] y(t - n\tau_i) \leq 0 \tag{5.3.20}$$

has no eventually negative bounded solutions.

Combining Theorems 5.3.8 and 5.3.9 we obtain the following result.

COROLLARY 5.3.3 Assume that the conditions of Theorem 5.3.8 hold. Then every bounded solution of

$$y^{(n)}(t) - \sum_{i=1}^{m} [p_i^n + q_i(t)] y(t - n\tau_i) = 0 \tag{5.3.21}$$

is oscillatory.

Consider the odd order differential inequalities

$$y^{(n)}(t) + \sum_{i=1}^{m} [p_i^n + q_i(t)] y(t - n\tau_i) \leq 0 \tag{5.3.22}$$

$$y^{(n)}(t) + \sum_{i=1}^{m} [p_i^n + q_i(t)] y(t - n\tau_i) \geq 0 \tag{5.3.23}$$

with differential equation

$$y^{(n)}(t) + \sum_{i=1}^{m} [p_i^n + q_i(t)] y(t - n\tau_i) = 0 \tag{5.3.24}$$

THEOREM 5.3.10 Assume that the conditions of Theorem 5.3.8 hold with n being an odd integer. Then

(i) (5.3.22) has no eventually positive solutions
(ii) (5.3.23) has no eventually negative solutions
(iii) Every solution of (5.3.24) oscillates.

Proof: To prove (i), we assume that there exists a solution $y(t)$ of (5.3.22) such that $y(t) > 0$ for $t > t_0$, t_0 being sufficiently large. Then $y(t - n\tau_i) > 0$, $i = 1, 2, \ldots, m$, for $t_0 + n\tau_{max}$. Moreover, from (5.3.22), $y^{(n)}(t) < 0$ for $t > t_0 + n\tau_{max}$.

From the above facts and by Lemma 5.2.1, we conclude that there exists an even integer ℓ such that

$$y^{(k)}(t) > 0 \qquad \text{for } k = 0, 1, 2, \ldots, \ell$$

$$(-1)^k y^{(k)}(t) > 0 \quad \text{for } k = \ell + 1, \ldots, n$$

for $t > t_0 + n\tau_{max}$.

We claim that $\ell = 0$, i.e., for $t > t_0 + n\tau_{max}$,

$$(-1)^k y^{(k)}(t) > 0 \quad \text{for } k = 0, 1, 2, \ldots, n \qquad (5.3.25)$$

To prove this, assume that $\ell > 0$. Then integrating (5.3.22) $n - \ell$ times from t_1 to t for t_1 sufficiently large, we obtain

$$y^{(\ell)}(t) \le \sum_{k=0}^{n-\ell-1} \frac{(t-t_1)^k}{k!} y^{(\ell+k)}(t_1) - \int_{t_1}^{t} \frac{(t-s)^{n-\ell-1}}{(n-\ell-1)!} \left[\sum_{i=1}^{m} [p_i^n + q_i(s)] y(s - n\tau_i) \right] ds$$

$$\le \sum_{k=0}^{n-\ell-1} \frac{(t-t_1)^k}{k!} y^{(\ell+k)}(t_1) - \sum_{i=1}^{m} \frac{y(t_1 - n\tau_i)}{(n-\ell-1)!} p_i^n \int_{t_1}^{t} (t-s)^{n-\ell-1} ds$$

$$= \sum_{k=0}^{n-\ell-1} \frac{(t-t_1)^k}{k!} y^{(\ell+k)}(t_1) - c(t-t_1)^{n-\ell}$$

where

$$c = \frac{\Sigma_{i=1}^{m} y(t_1 - n\tau_i) p_i^n}{(n-\ell)!}$$

is a positive number. This implies that

$$y^{(\ell)}(t) \to -\infty \quad \text{as } t \to \infty$$

which contradicts Lemma 5.2.1. This proves (5.3.25).

Set

$$x(t) = y^{(n-1)}(t) - \left(\sum_{i=1}^{m} p_i^n \right)^{1/n} y^{(n-2)}(t - \tau_1) + \cdots + \left(\left(\sum_{i=1}^{m} p_i^n \right)^{1/n} \right)^{n-1} y(t - (n-1)\tau_1)$$

Then, in view of (5.3.25),

$$x(t) > 0$$

Observe that

$$x'(t) = y^{(n)}(t) - \left(\sum_{i=1}^{m} p_i^n \right)^{1/n} y^{(n-1)}(t - \tau_1) + \cdots + \left(\left(\sum_{i=1}^{m} p_i^n \right)^{1/n} \right)^{n-1} y'(t - (n-1)\tau_1)$$

Hence

$$x'(t) + \left(\sum_{i=1}^{m} p_i^n \right)^{1/n} x(t - \tau_1) = y^{(n)}(t) + \sum_{i=1}^{m} p_i^n y(t - n\tau_1)$$

$$\leq - \sum_{i=1}^{m} [p_i^n + q_i(t)] y(t - n\tau_i) + \sum_{i=1}^{m} p_i^n y(t - n\tau_1)$$

$$\leq - \sum_{i=1}^{m} p_i^n (y(t - n\tau_i) - y(t - n\tau_1)) \leq 0$$

This together with condition (ii) leads to a contradiction because of Theorem 2.1.1.

In the case of condition (i), we set

$$x(t) = y^{(n)}(t) - p_i y^{(n-1)}(t - \tau_i) + \cdots + p_i^{n-1} y(t - (n-1)\tau_i)$$

Using the above arguments we are led to a similar contradiction.

The rest of the proof is similar to the proof of Theorem 5.3.9 and Corollary 5.3.3.

Consider

$$y^{(n)}(t) + (-1)^{n+1} \sum_{i=1}^{m} p_i^n y(t - n\tau_i) = 0 \tag{5.3.26}$$

THEOREM 5.3.11 Assume that $p_i > 0$ and $\tau_i > 0$, $i = 1, 2, \ldots, m$, and there exist $N_i > 0$ such that $\sum_{1 \leq i \leq m} N_i = 1$ and

$$\ln \frac{\sqrt[n]{N_i}}{p_i \tau_i} < 1 \qquad i = 1, 2, \ldots, m \tag{5.3.27}$$

Then (i) all solutions of (5.3.26) oscillate if n is odd; (ii) all bounded solutions of (5.3.26) oscillate if n is even.

Proof: To prove (ii), we can write the characteristic equation of (5.3.26) in the form

$$F(\lambda) = \sum_{i=1}^{m} \left(N_i \lambda^n - p_i^n e^{-n\tau_i \lambda} \right) = \sum_{i=1}^{m} f_i(\lambda)$$

where $f_i(\lambda) = N_i \lambda^n - p_i^n e^{-n\tau_i \lambda}$. Assume that

$$F(\lambda) = 0$$

has a negative real root. Then at least for one i, $f_i(\lambda) = 0$ has a real root. Now we try to find a sufficient condition for every $f_i(\lambda) = 0$ to have no negative real root.

Set

$$\bar{f}_i(\lambda) = \sqrt[n]{N_i}\, \lambda - p_i e^{\tau_i \lambda}$$

Then

$$\bar{f}'_i(\lambda) = \sqrt[n]{N_i} - p_i \tau_i e^{\tau_i \lambda}$$

It is easy to see that

$$\lambda_{max} = \frac{1}{\tau_i} \ln \frac{\sqrt[n]{N_i}}{p_i \tau_i}$$

and

$$\max \bar{f}_i(\lambda) = \frac{\sqrt[n]{N_i}}{\tau_i}\left(\ln \frac{\sqrt[n]{N_i}}{p_i \tau_i} - 1 \right)$$

Condition (5.3.27) implies that

$$\max_i \bar{f}_i(\lambda) < 0 \tag{5.3.28}$$

Assume that, for some i, $f_i(\lambda) = 0$ has a negative real root λ_0, i.e., $f_i(\lambda_0) = 0$ and $\lambda_0 < 0$. It is easy to see that $\bar{f}_i(-\lambda_0) = 0$. This contradicts the relation (5.3.28). Thus, we have proved that $f_i(\lambda) = 0$ has no negative real roots for every i. The proof is complete in the case of (ii). The proof of the case (i) can be formulated analogous to the proof of (ii).

THEOREM 5.3.12 Assume that $p_i > 0$, $\tau_i > 0$, for $i = 1, 2, \ldots, m$. Further assume that

$$\sum_{i=1}^{m} (p_i \tau_i e)^n > 1 \tag{5.3.29}$$

Then the conclusions of Theorem 5.3.11 remain valid. Moreover, if

$$(\tau_{max})^n \sum_{i=1}^{k} (p_i e^{\tau_i/\tau_{max}})^n \leq 1 \tag{5.3.30}$$

then Eq. (5.3.26) has a nonoscillatory solution y(t) with negative exponential.

Proof: We choose

$$N_i = \frac{(p_i \tau_i)^n}{\sum_{i=1}^m (p_i \tau_i)^n}$$

It is easy to see that $N_i > 0$ and $\sum_{1 \leq i \leq m} N_i = 1$. From (5.3.29),

$$\sum_{i=1}^m (p_i \tau_i)^n > e^{-n}$$

Hence

$$-\frac{1}{n} \ln \left(\sum_{i=1}^m (p_i \tau_i)^n \right) - 1 < 0$$

that is,

$$\ln \frac{\sqrt[n]{N_i}}{p_i \tau_i} < 1$$

From Theorem 5.3.11, the first part of the proof follows immediately. To prove (5.3.30), if n is even, we note that $F(0) < 0$,

$$F\left(-\frac{1}{\tau_{max}}\right) = \left(-\frac{1}{\tau_{max}}\right)^n - \sum_{i=1}^m p_i^n e^{n\tau_i / \tau_{max}}$$

$$= \left(-\frac{1}{\tau_{max}}\right)^n \left[1 - \sum_{i=1}^m (p_i e^{\tau_i / \tau_{max}} \tau_{max})^n \right] \geq 0$$

Therefore (5.3.30) guarantees that $F(\lambda) = 0$ has a negative real root λ_0. Thus $e^{\lambda_0 t}$ is a nonoscillatory solution of (5.3.26). The same argument can be used in case n is odd.

EXAMPLE 5.3.1 Consider

$$y^{(4)}(t) - \frac{2}{3} y\left(t - \frac{4}{e}\right) - \frac{1}{2e^8} y(t - 4e) = 0 \tag{5.3.31}$$

It is easy to check that this equation satisfies condition (5.3.29), where

$$p_1^4 = \frac{2}{3}, \quad p_2^4 = \frac{1}{2e^8}, \quad \tau_1 = \frac{1}{e}, \quad \tau_2 = e \tag{5.3.32}$$

Therefore every bounded solution of (5.3.31) is oscillatory.

REMARK 5.3.1 From Theorem 5.3.11, we can obtain several sufficient conditions for the validity of conclusions of Theorem 5.3.11 depending
on the choice of N_i. For example, we may choose either

$$N_i = \frac{p_i^n}{\Sigma_{i=1}^m p_i^n} \qquad i = 1, 2, \ldots, m \tag{5.3.33}$$

or

$$N_i = \frac{\tau_i^n}{\Sigma_{i=1}^m \tau_i^n} \qquad i = 1, 2, \ldots, m \tag{5.3.34}$$

Let $m = 1$. Then $p\tau e > 1$ is a necessary and sufficient condition for the
conclusions of Theorem 5.3.11 to be true.

REMARK 5.3.2 For $m > 1$, there is a gap between (5.3.29) and
(5.3.30). This is an open problem.

The above argument can be applied to advanced type equations

$$y^{(n)}(t) - \sum_{j=1}^{\ell} q_j^n \, y(t + n\sigma_j) = 0 \tag{5.3.35}$$

The formulation and verification of these results will be left to the reader.

THEOREM 5.3.13 Let $g(t)$ be of mixed type argument in (5.3.1) and n
be an even integer. Further, suppose that there is a constant $\epsilon > 0$ such that

$$\int^{\infty} [g_*(t)]^{n-1-\epsilon} \, p(t) \, dt = \infty \tag{5.3.36}$$

where $g_*(t) = \min\{g(t), t\}$. Suppose further that there exist two sequences
$\{t_k\}$, $\{\tau_k\}$ such that $t_k \in A_g$, $t_k \to \infty$, $\tau_k \in R_g$, $\tau_k \to \infty$ as $k \to \infty$,

$$\int_{t_k}^{g(t_k)} [g(s) - g(t_k)]^{n-1} \, p(s) \, ds \geq (n-1)! \tag{5.3.37}$$

and

$$\int_{g(\tau_k)}^{\tau_k} [g(\tau_k) - g(s)]^{n-1} \, p(s) \, ds \geq (n-1)! \tag{5.3.38}$$

for all $k = 1, 2, \ldots$, where $A_g = \{t \in [0,\infty) : g(t) > t\}$ and $R_g = \{t \in [0,\infty) : g(t) < t\}$. Then all solutions of (5.3.1) are oscillatory.

Proof: We first show that condition (5.3.36) guarantees that $N_\ell = \emptyset$ for $\ell \in \{2, 4, \ldots, n-2\}$. Assume the contrary. This would imply that there exists a positive solution $y \in N_\ell$ for some $\ell \in \{2, 4, \ldots, n-2\}$. Note that Lemma 5.2.2 holds for $t \geq t_0$. Choose a $t_1 > t_0$ so that $g_*(t) \geq t_0$ for $t \geq t_1$. We observe that

$$y^{(\ell)}(t) = \sum_{i=\ell}^{n-1} \frac{(t-T)^{i-1}}{(i-1)!} y^{(i)}(T) + (-1)^{n-\ell-1} \int_T^t \frac{(s-t)^{n-\ell-1}}{(n-\ell-1)!} y^{(n)}(s)\, ds \quad (5.3.39)$$

for any t, $T \geq t_1$. Using Lemma 5.2.2 and the fact that $n - \ell$ is even, relation (5.3.39) reduces to

$$y^{(\ell)}(t) \geq \int_t^T \frac{(s-t)^{n-\ell-1}}{(n-\ell-1)!} p(s) y(g(s))\, ds \qquad T \geq t \geq t_1 \quad (5.3.40)$$

On the other hand, from equation

$$y'(t) = \sum_{i=1}^{\ell-1} \frac{(t-t_1)^{i-1}}{(i-1)!} y^{(i)}(t_1) + \int_{t_1}^t \frac{(t-s)^{\ell-2}}{(\ell-2)!} y^{(\ell)}(s)\, ds$$

and Lemma 5.2.2, we note that $y^\ell(t)$ is decreasing and hence

$$y'(t) \geq \int_{t_1}^t \frac{(t-s)^{\ell-2}}{(\ell-2)!} y^{(\ell)}(s)\, ds \geq \frac{(t-t_1)^{\ell-1}}{(\ell-1)!} y^{(\ell)}(t) \qquad t \geq t_1 \quad (5.3.41)$$

Combining (5.3.40) and (5.3.41), we have

$$y'(t) \geq \frac{(t-t_1)^{\ell-1}}{(\ell-1)!} \int_t^\infty \frac{(s-t)^{n-\ell-1}}{(n-\ell-1)!} p(s) y(g(s))\, ds \quad (5.3.42)$$

Let us take a $t_2 > t_1$ so that $g_*(t) \geq t_1$ for $t \geq t_2$. Since $\ell \leq n-2$, $y^{(n-1)}(t) < 0$ for $t \geq t_1$. Further there is a constant $a \geq 1$ such that $y(t) \leq at^{n-2}$ for $t \geq t_1$, and hence when $s \geq t_1$, we have

$$y(t) \leq a[g_*(s)]^{n-2} \qquad \text{for } t_1 \leq t \leq g_*(s) \quad (5.3.43)$$

Dividing both sides of (5.3.42) by $(y(t))^{1+\delta}$, $\delta = \epsilon/(n-2)$, and integrating it over $[t_1, t_3]$ with $t_3 > t_2$, we obtain

$$\int_{t_1}^{t_3} \frac{y'(t)\,dt}{(y(t))^{1+\delta}} \geq \int_{t_1}^{t_3} \frac{(t-t_1)^{\ell-1}}{(\ell-1)!(y(t))^{1+\delta}} \int_{t}^{t_3} \frac{(s-t)^{n-\ell-1}}{(n-\ell-1)!} p(s)y(g(s))\,ds\,dt$$

$$= \int_{t_1}^{t_3}\int_{t_1}^{s} \frac{(s-t)^{n-\ell-1}(t-t_1)^{\ell-1}p(s)y(g(s))}{(n-\ell-1)!(\ell-1)!(y(t))^{1+\delta}}\,dt\,ds$$

$$\geq \int_{t_2}^{t_3}\int_{t_1}^{g_*(s)} \frac{(g_*(s)-t)^{n-\ell-1}(t-t_1)^{\ell-1}p(s)y(g(s))}{(n-\ell-1)!(\ell-1)!(y(t))^{1+\delta}}\,dt\,ds \quad (5.3.44)$$

Noting that $y(t)$ is increasing, and taking (5.3.43) into account, we get

$$\frac{p(s)y(g(s))}{y(t)^{1+\delta}} \geq \frac{y(g_*(s))p(s)}{y(t)y(t)^{\delta}} \geq \frac{p(s)}{a^{\delta}(g_*(s))^{(n-2)\delta}} \quad (5.3.45)$$

for $t_1 \leq t \leq g_*(s)$ and $t_2 \leq s \leq t_3$. From (5.3.44) and (5.3.45) we obtain

$$\int_{t_2}^{t_3} \frac{[g_*(s)-t_1]^{n-1}}{[g_*(s)]^{(n-2)\delta}} p(s)\,ds \leq a^{\delta}(n-1)! \int_{t_1}^{t_3} \frac{y'(t)\,dt}{y(t)^{1+\delta}} < \frac{a^{\delta}(n-1)!}{\delta(y(t_1))^{\delta}}$$

Letting $t_3 \to \infty$ in the above, we conclude that

$$\int_{t_2}^{\infty} [g_*(s)]^{n-1-\epsilon} p(s)\,ds < \infty$$

This contradicts (5.3.36). A parallel argument holds if we assume that (5.3.1) has a negative solution of degree $\ell \in \{2,\, 4,\, \ldots,\, n-2\}$.

Next we shall show that condition (5.3.37) does not allow (5.3.1) to have nonoscillatory solutions of degree n. Suppose there is a solution $y \in N_n$. Without loss of generality, we may assume that $y(t)$ is positive. From Lemma 5.2.2 we have

$$y^{(i)}(t) > 0 \quad \text{on } t \geq t_0 \text{ for } 0 \leq i \leq n \quad (5.3.46)$$

Let $t_1 > t_0$ be as before. From the equation

$$y(u) = \sum_{i=0}^{n-1} \frac{(u-v)^i}{i!} y^{(i)}(v) + \frac{1}{(n-1)!} \int_{v}^{u} (u-t)^{n-1} y^{(n)}(t)\,dt \quad (5.3.47)$$

and (5.3.46), it follows that

$$y(u) \geq \frac{(u-v)^{n-1}}{(n-1)!} y^{(n-1)}(v) \quad \text{for } u \geq v \geq t_1$$

Let $t_k \in A_g$ and set $u = g(s)$ and $v = g(t_k)$ in the above, where $t_k \le s \le g(t_k)$. Then

$$y(g(s)) \ge \frac{[g(s) - g(t_k)]^{n-1}}{(n-1)!} y^{(n-1)}(g(t_k)) \qquad t_k \le s \le g(t_k) \qquad (5.3.48)$$

Integrating (5.3.1) over $[t_k, g(t_k)]$ and using (5.3.48), we get

$$y^{(n-1)}(g(t_k)) - y^{(n-1)}(t_k) \ge y^{(n-1)}(g(t_k)) \int_{t_k}^{g(t_k)} \frac{[g(s) - g(t_k)]^{n-1}}{(n-1)!} p(s)\, ds$$

or

$$y^{(n-1)}(g(t_k)) \left[1 - \int_{t_k}^{g(t_k)} \frac{[g(s) - g(t_k)]^{n-1}}{(n-1)!} p(s)\, ds \right] \ge y^{(n-1)}(t_k)$$

This, together with (5.3.37) shows that $y^{(n-1)}$ must take on nonpositive values for arbitrarily large t. This contradicts (5.3.46).

Finally, we shall show that condition (5.3.38) precludes nonoscillatory solutions of degree 0. Suppose that (5.3.1) has a positive solution $y \in N_0$. Then from Lemma 5.2.2,

$$(-1)^i y^{(i)}(t) > 0, \qquad t \ge t_0, \quad 0 \le i \le n \qquad (5.3.49)$$

Let t_1 be as before. From (5.3.47) and (5.3.49) we obtain the inequality

$$y(u) \ge \frac{(u-v)^{n-1}}{(n-1)!} y^{(n-1)}(v) \qquad \text{for} \ \ v \ge u \ge t_1$$

Let $\tau_k \in R_g$ and set $u = g(s)$ and $v = g(\tau_k)$, where $g(\tau_k) \le s \le \tau_k$. From this we have

$$y(g(s)) \ge \frac{(g(\tau_k) - g(s))^{n-1}}{(n-1)!} (-1)^{n-1} (g(\tau_k)) \qquad g(\tau_k) \le s \le \tau_k \qquad (5.3.50)$$

Integrating (5.3.1) and using (5.3.50), we get

$$y^{(n-1)}(\tau_k) - y^{(n-1)}(g(\tau_k)) \ge (-1)^{n-1} y^{(n-1)}(g(\tau_k)) \int_{g(\tau_k)}^{\tau_k} \frac{(g(\tau_k) - g(s))^{n-1}}{(n-1)!} p(s)\, ds$$

or

$$y^{(n-1)}(\tau_k) \geq (-1)^{n-1} y^{(n-1)}(g(\tau_k)) \left[\int_{g(\tau_k)}^{\tau_k} \frac{[g(\tau_k) - g(s)]^{n-1}}{(n-1)!} p(s) \, ds - 1 \right]$$

According to condition (5.3.38) this contradicts the fact that $y^{(n-1)}(t)$ is negative. Similarly, (5.3.1) has no negative solutions of degree 0. The proof is complete.

A result similar to Theorem 5.3.13 can be obtained for an odd order differential equations.

THEOREM 5.3.14   Let $n \geq 3$ be odd. Assume that (5.3.36), (5.3.37), and (5.3.38) hold. Then all solutions of (5.3.1) are oscillatory.

EXAMPLE 5.3.2   Consider

$$y^{(n)}(t) = py(t + \sin t) \tag{5.3.51}$$

where $p > 0$ is constant and $n \geq 2$ is an integer. The deviating argument $g(t) = t + \sin t$ is of mixed type. It is easy to see that (5.3.36) is satisfied. Put

$$\phi_{n-1}(t) = \int_{t}^{t+\sin t} [(s + \sin s) - (t + \sin t)]^{n-1} \, ds$$

Taking $t_k = \pi/2 + 2k\pi$, $k = 1, 2, \ldots$, $t_k \in A_g$, and $\phi_1(t_k) = \sin 1 - \frac{1}{2} > 0$, by Jensen's inequality this yields $\phi_{n-1}(t_k) \geq (\phi_1(t_k))^{n-1} = (\sin 1 - \frac{1}{2})^{n-1}$. For even n, we have

$$\phi_{n-1}(t) = \int_{t+\sin t}^{t} [t + \sin t - (s + \sin s)]^{n-1} \, ds$$

Setting $\tau_k = -\pi/2 + 2k\pi$, $k = 1, 2, \ldots$, it follows that $\tau_k \in R_g$ and $\phi_{n-1}(\tau_k) \geq (\phi_1(\tau_k))^{n-1} = (\sin 1 - \frac{1}{2})^{n-1}$. Thus if p is so large that $(\sin 1 - \frac{1}{2})^{n-1} p \geq (n-1)!$ then from Theorems 5.3.13 and 5.3.14 it follows that all solutions of (5.3.51) are oscillatory.

5.4 A CLASS OF ARBITRARY ORDER DELAY EQUATIONS

We consider the equation

$$y^{(n)}(t) + p(t) |y(g(t))|^{\alpha} \, \text{sign} \, y(g(t)) = 0 \tag{5.4.1}$$

THEOREM 5.4.1 Assume that

(i) $p \in C[R_+, R_+]$ and $p(t)$ is not identically zero in any neighborhood of ∞.

(ii) $g \in C^1[R_+, R_+]$, $g(t) \leq t$, $g'(t) \geq 0$, and $\lim_{t \to \infty} g(t) = \infty$

(iii) $n \geq 2$, $\alpha > 1$

(iv) (5.4.1) has an unbounded nonoscillatory solution

Then

$$\limsup_{t \to \infty} \left[t^{n-1} \int_{\nu^*(t)}^{\infty} p(x)\, dx \right] = 0, \quad \nu^*(t) = \sup\{s \geq t_0 : g(s) \leq t\} \text{ for } t \geq t_0 \quad (5.4.2)$$

Proof: Let $y(t)$ be a nonoscillatory unbounded solution of (5.4.1); without loss of generality, we may assume that $y(t) > 0$ for $t \geq T \geq t_0$. Then there would exist $t_1 \geq T$ such that $y(g(t)) > 0$ for $t \geq t_1$. Using conclusion (iv) of Lemma 5.2.1 and integrating (5.4.1) successively $n - q - 1$ times from $t \ (\geq t_2)$ to ∞, we have

$$y^{(n-1)}(t) \geq \int_t^{\infty} p(s) y^{\alpha}(g(s))\, ds$$

$$-y^{(n-2)}(t) \geq \int_t^{\infty} (s - t) p(s) y^{\alpha}(g(s))\, ds \tag{5.4.3}$$

$$(-1)^{n-q} y^{(q+1)}(t) \geq \frac{1}{(n-q-2)!} \int_t^{\infty} (s-t)^{n-q-2} p(s) y^{\alpha}(g(s))\, ds$$

Integrating (5.4.3) from v_1 to v_2 $(t_2 < v_1 < v_2)$ we obtain

$$(-1)^{n-q} y^{(q)}(v_2) - (-1)^{n-q} y^{(q)}(v_1) \geq \frac{1}{(n-q-1)!} \int_{v_1}^{v_2} (s-v_1)^{n-q-1} p(s) y^{\alpha}(g(s))\, ds$$

$$+ \frac{1}{(n-q-1)!} \int_{v_2}^{\infty} [(s-v_1)^{n-q-1} - (s-v_2)^{n-q-1}] p(s) y^{\alpha}(g(s))\, ds \tag{5.4.4}$$

It is easy to verify by induction that

$$(v_2 - v_1)^{n-q-1} \leq (t - v_1)^{n-q-1} - (t - v_2)^{n-q-1}$$

for $v_1 < v_2 \leq t$. Therefore, from (5.4.4), we have

$$(-1)^{n-q} y^{(q)}(v_2) - (-1)^{n-q} y^{(q)}(v_1) \geq \frac{1}{(n-q-1)!} \int_{v_1}^{v_2} (t - v_1)^{n-q-1} p(t) y^{\alpha}(g(t))\, dt$$

$$+ \frac{1}{(n-q-1)!} (v_2 - v_1)^{n-q-1} \int_{v_2}^{\infty} p(t) y^{\alpha}(g(t))\, dt \qquad (5.4.5)$$

Let $n-q$ be even. Then conclusion (ii) of Lemma 5.2.1 implies that k is odd if n is even and k is even if n is odd. This implies that $n-k$ is always odd. Thus $q \neq k$, by Lemma 5.2.1, so $q = k-1$, and we have

$$y^{(q+1)}(t) = y^{(k)}(t) > 0 \quad \text{for } t \geq t_2$$

and

$$y^{(q)}(t) = y^{(k-1)}(t) > 0 \quad \text{for } t \geq t_2$$

Hence $y^{(q)}(t)$ is an increasing and positive function for $t \geq t_2$. It is clear that the first integral on the right-hand side of (5.4.5) is nonnegative. Therefore, from (5.4.5) we have

$$y^{(q)}(v_2) \geq \frac{1}{(n-q-1)!} (v_2 - v_1)^{n-q-1} \int_{v_2}^{\infty} p(t) y^{\alpha}(g(t))\, dt \qquad (5.4.6)$$

Let $q > 0$. Integrating (5.4.6) from v_1 to t ($t_2 \leq v_1 < t$), we get

$$y^{(q-1)}(t) - y^{(q-1)}(v_1) \geq \frac{1}{(n-q)!} (t-v_1)^{n-q} \int_{t}^{\infty} p(s) y^{\alpha}(g(s))\, ds$$

$$+ \int_{v_1}^{t} (v_2 - v_1)^{n-q} p(v_2) y^{\alpha}(g(v_2))\, dv_2$$

The second integral on the right-hand side of this inequality is nonnegative and $y^{(q-1)}(v_1) > 0$ (by (ii) of Lemma 5.2.1). Thus we can write

$$y^{(q-1)}(t) \geq \frac{1}{(n-q)!} (t - v_1)^{n-q} \int_{t}^{\infty} p(s) y^{\alpha}(g(s))\, ds \qquad (5.4.7)$$

Analogously, integrating (5.4.7) $q-1$ times from v_1 to t we obtain

$$y^{(q-2)}(t) \geq \frac{1}{(n-q-1)!} (t-v_1)^{n-q+1} \int_{t}^{\infty} p(s) y^{\alpha}(g(s))\, ds \qquad (5.4.8)$$

$$y(t) \geq \frac{1}{(n-1)!} (t-v_1)^{n-1} \int_{t}^{\infty} p(s) y^{\alpha}(g(s))\, ds \qquad (5.4.9)$$

Note that for $q = 0$, the inequality (5.4.9) reduces to the inequality (5.4.6) (just replace v_2 by t). Therefore, for $n - q$ even, we can consider the case $q > 0$ together with the case $q = 0$.

Let us denote $\nu^*(t) = \sup\{s \geq t_0 \,|\, g(s) \leq t\}$ for $t \geq t_0$. From condition (ii), $t \leq \nu^*(t)$ and $g(\nu^*(t)) = t$. In fact, assume that $g(\nu^*(t)) > t$. Then for s sufficiently close to $\nu^*(t)$, $g(s) > t$. This contradicts the definition of $\nu^*(t)$. So $g(\nu^*(t)) \leq t$. Assume that $g(\nu^*(t)) < t$. Then for s sufficiently close to $\nu^*(t)$, $g(s) < t$. This implies $\nu^*(t) > s$, which is impossible. Hence $g(\nu^*(t)) = t$.

Since $y'(t) > 0$ for $t \geq t_2$, $y(g(s)) \geq y(t)$ for $s \geq \nu^*(t)$, and from (5.4.9), we have

$$y(t) \geq \frac{y^\alpha(t)}{(n-1)!} (t - v_1)^{n-1} \int_{\nu^*(t)}^\infty p(s)\, ds$$

and also

$$\frac{y(t)}{y^\alpha(t)} \geq \frac{(t - v_1)^{n-1}}{(n-1)!} \int_{\nu^*(t)}^\infty p(s)\, ds \tag{5.4.10}$$

Since $y(t)$ is an unbounded positive solution, the last inequality directly implies that

$$\limsup_{t \to \infty} t^{n-1} \int_{\nu^*(t)}^\infty p(s)\, ds = 0$$

Thus in the case when $n - q$ is an even number and the conclusion (iv) of Lemma 5.2.1 holds, the theorem is proved.

Next, we consider the case that conclusion (iv) holds and $n - q$ is odd. Then, by Lemma 5.2.1, we obtain $q = k$. Thus from conclusion (ii) of Lemma 5.2.1, we have

$$y^{(q+1)}(t) = y^{(k+1)}(t) < 0 \qquad \text{for } t \geq t_2$$

and

$$y^{(q)}(t) = y^{(k)}(t) > 0 \qquad \text{for } t \geq t_2$$

Hence $y^{(q)}(t)$ is a positive and decreasing function for $t \geq t_2$. It is easy to see that the second integral on the right-hand side of the inequality (5.4.5) is positive. Therefore, (5.4.5) yields

$$y^{(q)}(v_1) - y^{(q)}(v_2) \geq \frac{1}{(n-q-1)!} \int_{v_1}^{v_2} (s - v_1)^{n-q-1} p(s) y^\alpha(g(s))\, ds$$

and for $v_2 \to \infty$,

$$y^{(q)}(v_1) \geq \frac{1}{(n-q-1)!} \int_{v_1}^{\infty} (s-v_1)^{n-q-1} p(s) y^{\alpha}(g(s))\, ds \tag{5.4.11}$$

Let us consider the case $q > 0$. Integrating (5.4.11) from v_3 to t ($t_2 < v_3 < t$) we obtain

$$y^{(q-1)}(t) - y^{(q-1)}(v_3) \geq \frac{1}{(n-q)!} \int_{v_3}^{t} (s-v_3)^{n-q} p(s) y^{\alpha}(g(s))\, ds$$

$$+ \frac{1}{(n-q)!} \int_{t}^{\infty} [(s-v_3)^{n-q} - (s-t)^{n-q}] p(s) y^{\alpha}(g(s))\, ds \tag{5.4.12}$$

It is evident that $(t-v_3)^{n-q} \leq (s-v_3)^{n-q} - (s-t)^{n-q}$ for $v_3 < t \leq s$. The first integral on the right-hand side of (5 4.12) is nonnegative, and by conclusion (ii) of Lemma 5.2.1, $y^{(q-1)}(v_3) = y^{(k-1)}(v_3) > 0$. Thus replacing v_3 by v_1 in (5.4.12) we obtain

$$y^{(q-1)}(t) \geq \frac{1}{(n-q)!} (t-v_1)^{n-q} \int_{t}^{\infty} p(s) y^{\alpha}(g(s))\, ds \tag{5.4.13}$$

The inequality (5.4.13) coincides with the inequality (5.4.7). Consequently, continuing as above we successively get the inequalities (5.4.9) and (5.4.10) and the assertion of the theorem follows.

Now let $q = 0$. Then $n - q = n$, which is odd. Since $q = k$ in Lemma 5.2.1, conclusion (ii) of Lemma 5.2.1 gives $y'(t) < 0$ for each $t \geq t_2$. But this is a contradiction to the assumption that $y(t)$ is a positive and unbounded solution of (5.4.1).

Now consider the case when condition (5.2.4) is not satisfied. In this case, by Lemma 5.2.1, we have $y(t) \to 0$ as $t \to \infty$. This is again a contradiction to the assumption that $y(t)$ is a positive and unbounded nonoscillatory solution of (5.4.1). The proof is complete.

REMARK 5.4.1 From the proof of Theorem 5.4.1, we obtain the following consequence. Assume $\alpha = 1$ and equation (5.4.1) has an unbounded nonoscillatory solution. Then

$$\limsup_{t \to \infty} t^{n-1} \int_{\nu^*(t)}^{\infty} p(s)\, ds \leq (n-1)! \tag{5.4.14}$$

THEOREM 5.4.2 Let $\alpha > 0$, assumptions (i)–(iii) be satisfied, and

$$\limsup_{t \to \infty} t^{n-1} \int_{t}^{\infty} p(s)\, ds = \infty \tag{5.4.15}$$

Then, for n even, all nonoscillatory solutions of (5.4.1) are unbounded, while for n odd, every nonoscillatory solution of (5.4.1) is either unbounded or strongly monotone.

Proof: Yet $y(t)$ be a nonoscillatory solution of (5.4.1). We may assume that $y(t) > 0$ for $t \geq T \geq t_0$. Then $y(g(t)) > 0$ for $t \geq t_1 = \nu^*(T)$. Now by (5.4.1), we have $y^{(n)}(t) \leq 0$ for $t \geq t_1$. By Lemma 5.2.1 and by following the argument that is used in the proof of Theorem 5.4.1, we successively obtain the inequalities (5.4.3)–(5.4.5).

Let $n - q$ be even. Then, as in the corresponding part of the proof of Theorem 5.4.1, $q = k - 1$ and we get (5.4.6).

In the case $q > 0$, from (5.4.6) we obtain (5.4.8) and (5.4.9). In the case $q = 0$, the inequality (5.4.9) reduces to the inequality (5.4.6) (just replace v_2 by t). Therefore, for $n - q$ even, we can consider the case $q > 0$ together with the case $q = 0$. Note that for $q \geq 0$ we have $k \geq 1$. Thus, by Lemma 5.2.1, assuming that (5.2.4) holds, $y'(t) > 0$ for $t \geq t_2$. Since $s \geq t \geq t_2 \geq t_1 = \nu^*(T)$ in (5.4.9) and $y(t)$ is an increasing function for $t \geq t_2$, we get

$$y(t) \geq \frac{y^{\alpha}(t_2)}{(n-1)!} (t - v_1)^{n-1} \int_t^{\infty} p(s) \, ds \qquad (5.4.16)$$

for $t \geq \nu^*(t_2)$.

From the last inequality, by (5.4.15), we have

$$\lim_{t \to \infty} y(t) = \infty \qquad (5.4.17)$$

and the theorem is proved in the case when (5.2.4) holds and $n - q$ is an even number.

Next, we consider the case, (5.2.4) holds and $n - q$ is odd. Then $q = k$. Continuing as in the corresponding part of the proof of Theorem 5.4.1, we get (5.4.11). Further, for $q > 0$, we get (5.4.12) and (5.4.13), which coincides with the inequality (5.4.7). From (5.4.7), we obtain (5.4.8), (5.4.9), and (5.4.16) as above. Finally, from (5.4.16) we obtain (5.4.17).

Now let $q = 0$. Then $n - q = n$ is odd. Since $q = k$, by (ii) of Lemma 5.2.1, we get $y'(t) < 0$ for $t \geq t_2$. This means that $y(t)$ is a positive and decreasing function for $t \geq t_2$ and, by (5.2.4), $\lim_{t \to \infty} y(t) = c > 0$. In this case we shall show that the condition (5.4.15) is not satisfied.

We know that

$$0 < y(t) \leq y(t_2) \qquad \text{for} \quad t \geq t_2$$

and for $t \geq \nu^*(t_2)$. Hence

$$c < y(t) \leq y(g(t)) \leq y(t_2)$$

These inequalities imply that

$$c^{\alpha} p(t) < p(t) y^{\alpha}(g(t)) = -y^{(n)}(t)$$

or

$$c^{\alpha} p(t) < -y^{(n)}(t)$$

Integrating the last inequality n times from t to ∞ for $t \geq \nu^*(t_2)$ and using (5.4.2), we obtain

$$\frac{c^{\alpha}}{(n-1)!} \int_t^{\infty} (s-t)^{n-1} p(s)\, ds \leq y(t) - c$$

whence

$$\int_t^{\infty} (s-t)^{n-1} p(s)\, ds \leq \frac{(n-1)!\, y(t_2)}{c^{\alpha}}$$

because $c > 0$ and $y(t) \leq y(t_2)$.

Without loss of generality, we can assume that $t > 0$. Then

$$(2t)^{n-1} \int_{2t}^{\infty} p(s)\, ds \leq 2^{n-1} \int_{2t}^{\infty} (s-t)^{n-1} p(s)\, ds \leq 2^{n-1} \int_t^{\infty} (s-t)^{n-1} p(s)\, ds$$

$$\leq 2^{n-1} \frac{(n-1)!\, y(t_2)}{c^{\alpha}}$$

where

$$S^{n-1} \int_S^{\infty} p(\sigma)\, d\sigma \leq 2^{n-1} \frac{(n-1)!\, y(t_2)}{c^{\alpha}}$$

and $S = 2t$. From this we see that

$$\limsup_{S \to \infty} S^{n-1} \int_S^{\infty} p(\sigma)\, d\sigma < \infty$$

This completes the proof of the theorem whenever (5.2.4) holds and $n - q$ is an odd number.

Finally, assume that (5.2.4) is not satisfied. But by Lemma 5.2.1, we note that in this case the nonoscillatory solution $y(t)$ is strongly monotone. On the other hand, we know, by Lemma 5.2.1, (5.2.4) is false iff $k = 0$. Moreover, $k = 0$ iff n is odd. This completes the proof of Theorem 5.4.2.

Combining Theorems 5.4.1 and 5.4.2 we obtain the following results.

THEOREM 5.4.3 Assume that conditions (i)-(iii) of Theorem 5.4.1 are satisfied. Suppose further that

$$\limsup_{t \to \infty} t^{n-1} \int_{\nu^*(t)}^{\infty} p(s)\, ds > 0 \qquad (5.4.18)$$

and

$$\limsup_{t \to \infty} t^{n-1} \int_{t}^{\infty} p(s)\, ds = \infty \qquad (5.4.19)$$

Then every solution of (5.4.1) is oscillatory if n is even, and is either oscillatory or strongly monotone if n is odd.

Proof: Let y(t) be a nonoscillatory solution of (5.4.1). Moreover, let y(t) be not strongly monotone if n is odd. Then, by (5.4.2) and Theorem 5.4.1, y(t) must be bounded. On the other hand, by (5.4.19) and Theorem 5.4.2, y(t) must be unbounded if n is even. Hence y(t) must be either unbounded or strongly monotone if n is odd. This is a contradiction and the theorem is proved.

By Remark 5.4.1, we obtain the following theorem.

THEOREM 5.4.4 Assume that the conditions (i) and (ii) of Theorem 5.4.1 are satisfied. Further assume that $n \geq 2$, $\alpha = 1$,

$$\limsup_{t \to \infty} t^{n-1} \int_{\nu^*(t)}^{\infty} p(s)\, ds > (n-1)! \qquad (5.4.20)$$

and

$$\limsup_{t \to \infty} t^{n-1} \int_{t}^{\infty} p(s)\, ds = \infty \qquad (5.4.21)$$

Then every solution of (5.4.1) is oscillatory if n is even, and is either oscillatory or strongly monotone if n is odd.

REMARK 5.4.2 Conditions (5.4.19) and (5.4.21) can be replaced by condition

$$\int^{\infty} t^{n-1} p(t)\, dt = \infty \qquad (5.4.22)$$

5.5 EQUATIONS WITH DEVIATING ARGUMENTS OF MIXED TYPE

We consider the higher order functional differential equations of the form

$$y^{(n)}(t) - \sum_{i=1}^{N} p_i(t) f_i(y(\tau_1(t)), \cdots, y(\tau_m(t))) = 0 \qquad (5.5.1)$$

and

$$y^{(n)}(t) + \sum_{i=1}^{N} p_i(t) f_i(y(\tau_1(t)), \cdots, y(\tau_m(t))) = 0 \qquad (5.5.2)$$

We define the subsets A_j and R_j of $[a, \infty)$ as follows:

$$A_j = \{t \in [a, \infty) : \tau_j(t) > t\}$$

$$R_j = \{t \in [a, \infty) : \tau_j(t) < t\}$$

A_j and R_j are the sets of which the deviating argument $\tau_j(t)$ is advanced and retarded, respectively.

Set

$$A = A_1 \cap A_2 \cap \cdots \cap A_m$$

$$R = R_1 \cap R_2 \cap \cdots \cap R_m$$

THEOREM 5.5.1 Assume that

(i) $p_i, \tau_j \in C[R_+, R_+]$, $p_i(t) \geq 0$, $\lim_{t \to \infty} \tau_j(t) = \infty$ for $i = 1, 2, \ldots, N$, $j = 1, 2, \ldots, m$

(ii) $f_i \in C[R^m, R]$ and $f_i(u, \cdots, u_m)u_1 > 0$ for $u_1 u_j > 0$, $j = 2, \ldots, m$, $i = 1, 2, \ldots, N$

(iii) $n \geq 3$ is an odd integer

(iv) There are integers i, j, and k, $1 \leq i, j, k \leq N$, and a positive number M such that the following conditions are satisfied:

$$\int_A [\min_{1 \leq j \leq m} \tau_j(t) - t]^{n-1} p_i(t)\, dt = \infty \qquad (5.5.3)$$

$$\int_A t^{n-2} [\min_{1 \leq j \leq m} \tau_j(t) - t] p_j(t)\, dt = \infty \qquad (5.5.4)$$

$$\int_A t^{n-1} p_k(t)\, dt = \infty \qquad (5.5.5)$$

(v) $f_h(u_1, \ldots, u_m)$ is monotonically increasing for every $|u_i| \geq M$ and

$$\int_M^\infty \frac{du}{f_h(u, \ldots, u)} < \infty \qquad \int_{-M}^{-\infty} \frac{du}{f_h(u, \ldots, u)} < \infty \qquad (5.5.6)$$

where $h = i, j, k$.

Then all solutions of (5.5.1) are oscillatory.

Proof: Let $y(t)$ be a nonoscillatory solution of (5.5.1). Without loss of generality we may suppose that $y(t)$ is eventually positive. From (5.5.1), $y^{(n)}(t)$ is eventually positive. By Lemma 5.2.2, there is an integer $\ell \in [0, 1, 2, \ldots, n]$, such that

$$y^{(i)}(t) \geq 0 \qquad 0 \leq i \leq \ell$$

$$(-1)^{i-\ell} y^{(i)}(t) \geq 0 \qquad \ell \leq i \leq n \qquad (5.5.7)$$

for all sufficiently large t, say $t \geq t_0 \geq a$. Let $T \geq t_0$ be so large that $\tau_j(t) \geq t_0$ for $t \geq T$, $1 \leq j \leq m$.

First suppose that $\ell = n$. Then integrating (5.5.1) and using (5.5.7), we have

$$y'(t) = \sum_{i=1}^{n-1} \frac{(t-T)^{i-1}}{(i-1)!} y^{(i)}(T) + \int_T^t \frac{(t-s)^{n-2}}{(n-2)!} y^{(n)}(s)\, ds$$

$$\geq \int_T^t \frac{(t-s)^{n-2}}{(n-2)!} p_i(s) f_i(y(\tau_1(s)), \ldots, y(\tau_m(s)))\, ds \qquad t \geq T \qquad (5.5.8)$$

Take any $T' > T$ and let $\bar{T} = \sup_{T \leq t \leq T'} \{\min(\tau_1(t), \ldots, \tau_m(t)), t\}$. Dividing (5.5.8) by $f_i(y(t), \ldots, y(t))$ and integrating over $[T, \bar{T}]$, we obtain

$$\int_T^{\bar{T}} \frac{y'(t)\, dt}{f_i(y(t), \ldots, y(t))}$$

$$\geq \int_T^{\bar{T}} \frac{1}{f_i(y(t), \ldots, y(t))} \int_T^t \frac{(t-s)^{n-2}}{(n-2)!} p_i(s) f_i(y(\tau_1(s)), \ldots, y(\tau_m(s)))\, ds\, dt$$

$$= \int_T^{\bar{T}} p_i(s) \int_s^{\bar{T}} \frac{(t-s)^{n-2}}{(n-2)!} \frac{f_i(y(\tau_1(s)), \ldots, y(\tau_m(s)))}{f_i(y(t), \ldots, y(t))}\, dt\, ds$$

$$\geq \int_T^{\bar{T}} p_i(s) \int_s^{\bar{T}} \frac{(t-s)^{n-2}}{(n-2)!} \frac{f_i(y(\tau_1(s)), \ldots, y(\tau_m(s)))}{f_i(y(t), \ldots, y(t))}\, dt\, ds$$

$$\geq \int_{A\cap[T,T']} p_i(s) \int_s^{\min_{1\leq j\leq m} \tau_j(s)} \frac{(t-s)^{n-2}}{(n-2)!} \frac{f_i(y(\tau_1(s)), \ldots, y(\tau_m(s)))}{f_i(y(t), \ldots, y(t))} \, dt \, ds$$

Since $\ell = n \geq 2$, $\lim_{t\to\infty} y(t) = \infty$ and the number T may be chosen so large that $y(s) \geq M$ for $s \geq T$. Since $f_i(y(t), \ldots, y(t))$ is increasing for $t \geq T$,

$$\frac{f_i(y(\tau_1(s)), \ldots, y(\tau_m(s)))}{f_i(y(t), \ldots, y(t))} \geq 1 \quad \text{for } s \leq t < \min_{\substack{1\leq j\leq m \\ s\in A\cap[T,T']}} \tau_j(s)$$

Hence, we have

$$\int_T^{\bar{T}} \frac{y'(t)\, dt}{f_i(y(t), \ldots, y(t))} \geq \int_{A\cap[T,T']} p_i(s) \int_s^{\min_{1\leq j\leq m} \tau_j(s)} \frac{(t-s)^{n-2}}{(n-2)!} \, dt \, ds$$

$$= \frac{1}{(n-1)!} \int_{A\cap[T,T']} p_i(s) [\min_{1\leq j\leq m} \tau_j(s) - s]^{n-1} \, ds$$

Letting $T' \to \infty$ in the above and using $(5.5.6)$ we see that

$$\int_{A\cap[T,\infty)} p_i(s) [\min_{1\leq j\leq m} \tau_j(s) - s]^{n-1} \, ds \leq (n-1)! \int_{y(T)}^{\infty} \frac{du}{f_i(u, \ldots, u)} < \infty$$

which contradicts $(5.5.3)$.

In the case of $2 \leq \ell \leq n$, we have

$$y'(t) = \sum_{i=1}^{\ell-1} \frac{(t-T)^{i-1}}{(i-1)!} y^{(i)}(T) + \int_T^t \frac{(t-s)^{\ell-2}}{(\ell-2)!} y^{(\ell)}(s) \, ds$$

$$\geq \int_T^t \frac{(t-s)^{\ell-2}}{(i-1)!} y^{(\ell)}(s) \, ds \qquad t \geq T$$

and

$$y^{(\ell)}(t) = \sum_{i=\ell}^{n-1} (-1)^{i-\ell} \frac{(s-t)^{i-\ell}}{(i-\ell)!} y^{(i)}(s) + (-1)^{n-\ell} \int_t^s \frac{(r-t)^{n-\ell-1}}{(n-\ell-1)!} y^{(n)}(r) \, dr$$

$$\geq \int_t^s \frac{(r-t)^{n-\ell-1}}{(n-\ell-1)!} p_j(r) f_j(y(\tau_1(r)), \ldots, y(\tau_m(r))) \, dr \qquad t \geq T$$

Letting $s \to \infty$, we obtain

$$y^{(\ell)}(t) \geq \int_t^\infty \frac{(r-t)^{n-\ell-1}}{(n-\ell-1)!} p_j(r) f_j(y(\tau_1(r)), \ldots, y(\tau_m(r)))\, dr \quad t \geq T$$

$$(5.5.9)$$

Combining the above two inequalities, we have for $t \geq T$

$$y'(t) \geq \int_T^t \frac{(t-s)^{\ell-2}}{(\ell-2)!} \int_s^\infty \frac{(r-s)^{n-\ell-1}}{(n-\ell-1)!} p_j(r) f_j(y(\tau_1(r)), \ldots, y(\tau_m(r)))\, dr\, ds$$

$$\geq \int_T^t \left(\int_T^r \frac{(t-s)^{\ell-2}(r-s)^{n-\ell-1}}{(\ell-2)!(n-\ell-1)!}\, ds \right) p_j(r) f_j(y(\tau_1(r)), \ldots, y(\tau_m(r)))\, dr$$

$$\geq \int_T^t \left(\int_T^r \frac{(r-s)^{n-3}}{(\ell-2)!(n-\ell-1)!}\, ds \right) p_j(r) f_j(y(\tau_1(r)), \ldots, y(\tau_m(r)))\, dr$$

$$= c(n, \ell) \int_T^t (r-T)^{n-2} p_j(r) f_j(y(\tau_1(r)), \ldots, y(\tau_m(r)))\, dr \qquad (5.5.10)$$

where $c(n, \ell)$ is a positive constant depending only on n and ℓ. Take any $T' > T$ and let $\bar{T} = \sup_{T \leq t \leq T'} \max \{\min (\tau_1(t), \ldots, \tau_m(t)), t\}$. We divide (5.5.10) by $f_j(y(t), \ldots, y(t))$ and integrate over $[T, \bar{T}]$ to get

$$\int_T^{\bar{T}} \frac{y'(t)\, dt}{f_j(y(t), \ldots, y(t))}$$

$$\geq c(n, \ell) \int_T^{\bar{T}} \frac{1}{f_j(y(t), \ldots, y(t))} \int_T^t (s-T)^{n-2} p_j(s) f_j(y(\tau_1(s)), \ldots, y(\tau_m(s)))\, ds\, dt$$

$$= c(n, \ell) \int_T^{\bar{T}} (s-T)^{n-2} p_j(s) \int_s^{\bar{T}} \frac{f_j(y(\tau_1(s)), \ldots, y(\tau_m(s)))}{f_j(y(t), \ldots, y(t))}\, dt\, ds$$

$$\geq c(n, \ell) \int_T^{T'} (s-T)^{n-2} p_j(s) \int_s^{\bar{T}} \frac{f_j(y(\tau_1(s)), \ldots, y(\tau_m(s)))}{f_j(y(t), \ldots, y(t))}\, dt\, ds$$

$$\geq c(n, \ell) \int_{A \cap [T, T']} (s-T)^{n-2} p_j(s) \int_s^{\min \tau_j(s)} \frac{f_j(y(\tau_1(s)), \ldots, y(\tau_m(s)))}{f_j(y(t), \ldots, y(t))}\, dt\, ds$$

Note that $\lim_{t \to \infty} y(t) = \infty$, since $\ell \geq 2$. We can take T so large that $y(s) \geq M$ for $s \geq M$. Since

$$\frac{f_j(y(\tau_1(s)), \; \cdots, \; y(\tau_m(s)))}{f_j(y(t), \; \cdots, \; y(t))} \geq 1$$

for $s \leq t \leq \min_j \tau_j(s)$ and $s \in A \cap [T, T']$. Hence, we have

$$\int_{A \cap [T, T']} (s - T)^{n-2} p_j(s) [\min_j \tau_j(s) - s] \, ds \leq c(n, \ell)^{-1} \int_{y(T)}^{\infty} \frac{du}{f_j(u, \; \cdots, \; u)} < \infty$$

which contradicts (5.5.4).

In the case of $\ell = 1$, it follows from (5.5.9) that

$$y'(t) \geq \int_t^{\infty} \frac{(r-t)^{n-2}}{(n-2)!} p_k(r) f_k(y(\tau_1(r)), \; \cdots, \; y(\tau_m(r))) \, dr \quad t \geq T \quad (5.5.11)$$

Take any $T' > T$ and let $\bar{T} = \sup_{T \leq t \leq T'} \{\min(\tau_1(t), \; \cdots, \; \tau_m(t)), t\}$. Dividing (5.5.9) by $f_k(y(t), \; \cdots, \; y(t))$ and integrating over $[T, \bar{T}]$, we obtain

$$\int_T^{\bar{T}} \frac{y'(t) \, dt}{f_k(y(t), \; \cdots, \; y(t))}$$

$$\geq \int_T^{\bar{T}} \frac{1}{f_k(y(t), \; \cdots, \; y(t))} \int_t^{\infty} \frac{(r-t)^{n-2}}{(n-2)!} p_k(r) f_k(y(\tau_1(r)), \; \cdots, \; y(\tau_m(r))) \, dr \, dt$$

$$\geq \int_T^{\bar{T}} p_k(r) \int_T^r \frac{(r-t)^{n-2}}{(n-2)!} \frac{f_k(y(\tau_1(r)), \; \cdots, \; y(\tau_m(r)))}{f_k(y(t), \; \cdots, \; y(t))} \, dt \, dr$$

$$\geq \int_{A \cap [T, T']} p_k(r) \int_T^r \frac{(r-t)^{n-2}}{(n-2)!} \frac{f_k(y(\tau_1(r)), \; \cdots, \; y(\tau_m(r)))}{f_k(y(t), \; \cdots, \; y(t))} \, dt \, dr \quad (5.5.12)$$

If we suppose that $\lim_{t \to \infty} y(t) = y_0 < \infty$, then it follows from the above inequality that

$$\int_{A \cap [T, \infty)} p_k(r)(r - T)^{n-1} \, dr \leq c(n-2)! \int_{y(T)}^{y_0} \frac{du}{f_k(u, \; \cdots, \; u)} < \infty$$

where c is a positive constant. But this contradicts (5.5.5), and so we must have $\lim_{t \to \infty} y(t) = \infty$. We choose T so large that $y(t) \geq M$ for $t \geq T$. Then, from (5.5.12), we get

$$\int_{T}^{\bar{T}} \frac{y'(t)\ dt}{f_k(y(t),\ \cdots,\ y(t))}$$

$$\geq \int_{A \cap [T,\, T']} p_k(r) \int_{T}^{r} \frac{(r-t)^{n-2}}{(n-2)!}\ \frac{f_k(y(\tau_1(r)),\ \cdots,\ y(\tau_m(r)))}{f_k(y(t),\ \cdots,\ y(t))}\ dt\ dr$$

$$\geq \int_{A \cap [T,\, T']} p_k(r) \int_{T}^{r} \frac{(r-t)^{n-2}}{(n-2)!}\ dt\ dr$$

$$= \frac{1}{(n-1)!} \int_{A \cap [T,\, T']} p_k(r)(r-T)^{n-1}\ dr$$

and letting $T' \to \infty$ and using (5.5.6), we are again led to a contradiction to (5.5.5). The proof is complete.

THEOREM 5.5.2 Assume that conditions (i) and (ii) of Theorem 5.5.1 are satisfied. Let n be even. Suppose that there are integers i and j, $1 \leq i \leq j \leq N$, and a positive number M such that conditions (5.5.3), (5.5.4), and (5.5.6) for $h = i,\ j$ hold. In addition suppose that there is an integer k, $1 \leq k \leq N$, and a positive number D such that

$$\int_{R} [t - \max_{1 \leq k \leq m} \tau_k(t)]^{n-1} p_k(t)\ dt = \infty \tag{5.5.13}$$

If $f_k(u,\ \cdots,\ u_m)$ is monotonically increasing for every argument $|u_i| \leq D$ and

$$\int_{0^+}^{D} \frac{du}{f_k(u,\ \cdots,\ u)} < \infty, \qquad \int_{0^-}^{-D} \frac{du}{f_k(u,\ \cdots,\ u)} < \infty \tag{5.5.14}$$

then all solutions of (5.5.1) are oscillatory.

Proof: The argument of the proof of Theorem 5.5.1 can be applied here. In the case $2 \leq \ell \leq n$, the proof of this theorem follows from the proof of Theorem 5.5.1. It remains to prove the theorem in the case $\ell = 0$. In this case we have

$$-y'(t) = \sum_{i=1}^{n-1} (-1)^i y^{(i)}(s)\ \frac{(s-t)^{i-1}}{(i-1)!} + \int_{t}^{s} \frac{(r-t)^{n-2}}{(n-2)!}\ y^{(n)}(r)\ dr$$

for $s \geq t \geq T$. Choose $T' > T$ so that $\tau_i(t) \geq T$ for $t \geq T'$, $i = 1,\ 2,\ \cdots,\ m$, and let $T^* > T'$ be fixed. By applying Lemma 5.2.2 to (5.5.1), we get

$$-y'(t) \geq \int_t^{T^*} \frac{(s-t)^{n-2}}{(n-2)!} p_k(s) f_k(y(\tau_1(s)), \cdots, y(\tau_m(s))) \, ds \qquad T \leq t \leq T^*$$

Dividing the above inequality by $f_k(y(t), \cdots, y(t))$ and integrating over $[T, T^*]$, we have

$$-\int_T^{T^*} \frac{y'(t) \, dt}{f_k(y(t), \cdots, y(t))}$$

$$\geq \int_T^{T^*} \frac{1}{f_k(y(t), \cdots, y(t))} \int_t^{T^*} \frac{(s-t)^{n-2}}{(n-2)!} p_k(s) f_k(y(\tau_1(s)), \cdots, y(\tau_m(s))) \, ds \, dt$$

$$= \int_T^{T^*} p_k(s) \int_T^s \frac{(s-t)^{n-2}}{(n-2)!} \frac{f_k(y(\tau_1(s)), \cdots, y(\tau_m(s)))}{f_k(y(t), \cdots, y(t))} \, dt \, ds$$

$$\geq \int_{T'}^{T^*} p_k(s) \int_T^s \frac{(s-t)^{n-2}}{(n-2)!} \frac{f_k(y(\tau_1(s)), \cdots, y(\tau_m(s)))}{f_k(y(t), \cdots, y(t))} \, dt \, ds$$

$$\geq \int_{T'}^{T^*} p_k(s) \int_{\max \tau_i(s)}^s \frac{(s-t)^{n-2}}{(n-2)!} \frac{f_k(y(\tau_1(s)), \cdots, y(\tau_m(s)))}{f_k(y(t), \cdots, y(t))} \, dt \, ds \qquad (5.5.15)$$

If $\lim_{t\to\infty} y(t) = y_0 > 0$, then the above inequality implies that

$$\int_{T'}^{\infty} p_k(s)(s - \max_{1 \leq i \leq m} \tau_i(s))^{n-1} \, ds < \infty$$

which contradicts $(5.5.13)$. It follows that $\lim_{t\to\infty} y(t) = 0$.

In $(5.5.15)$, T may be taken so large that $y(s) \leq D$ and $y(\tau_k(s)) \leq D$, $k = 1, 2, \cdots, m$, for $s \geq T$. Noting that

$$\frac{f_k(y(\tau_1(s)), \cdots, y(\tau_m(s)))}{f_k(y(t), \cdots, y(t))} \geq 1$$

for $\max \tau_k(s) \leq t \leq s$ and $s \in R \cap [T', T^*]$, from $(5.5.15)$ we conclude that

$$\int_{R\cap[T',\infty)} [s - \max \tau_k(s)]^{n-1} p_k(s) \, ds \leq (n-1)! \int_{0^+}^{y(T)} \frac{du}{f_k(u, \cdots, u)} < \infty$$

which also contradicts $(5.5.13)$. Thus the proof of $5.5.2$ is complete.

THEOREM 5.5.3 Assume that conditions (i) and (ii) of Theorem 5.5.1 are satisfied. Suppose that n is even and there are integers i and j, $1 \leq i, j \leq N$, and a positive number M such that the following conditions are satisfied:

$$\int_A t^{n-2}[\min(\tau_k(t)) - t]p_i(t)\, dt = \infty \qquad (5.5.16)$$

$$\int_A t^{n-1} p_j(t)\, dt = \infty \qquad (5.5.17)$$

$f_h(u_1, u_2, \ldots, u_m)$ is monotonically increasing for every $|u_i| \geq M$, and

$$\int_M^\infty \frac{du}{f_h(u, \ldots, u)} < \infty, \qquad \int_{-M}^{-\infty} \frac{du}{f_h(u, \ldots, u)} < \infty \qquad (5.5.18)$$

for $h = i, j$.

Then all solutions of (5.5.2) are oscillatory.

Proof: Let $y(t)$ be a nonoscillatory solution of (5.5.2). We may suppose without loss of generality that $y(t)$ is eventually positive. From (5.5.2), $y^{(n)}(t)$ is eventually negative. By Lemma 5.2.1, there is an integer $\ell \in \{0, 1, \ldots, n-1\}$ such that $\ell \equiv n-1 \pmod 2$ and

$$y^{(i)}(t) \geq 0 \qquad 0 \leq i \leq \ell$$

$$(-1)^{j-\ell} y^{(i)}(t) \geq 0 \qquad \ell \leq i \leq n$$

for all sufficiently large t, say $t \geq t_0 \geq a$. Let $T \geq t_0$ be so large that $\tau_j(t) \geq t_0$ for $t \geq T$, $i = 1, 2, \ldots, m$.

First suppose that $2 \leq \ell \leq n-1$. Then we have

$$y'(t) = \sum_{i-1}^{\ell-1} \frac{(t-T)^{i-1}}{(i-1)!} y^{(i)}(T) + \int_T^t \frac{(t-s)^{\ell-2}}{(\ell-2)!} y^{(\ell)}(s)\, ds$$

$$\geq \int_T^t \frac{(t-s)^{\ell-2}}{(\ell-2)!} y^{(\ell)}(s)\, ds \qquad t \geq T \qquad (5.5.19)$$

and

$$y^{(\ell)}(t) = \sum_{i=\ell}^{n-1} (-1)^{i-\ell} \frac{(s-t)^{i-\ell}}{(i-\ell)!} y^{(i)}(s) + (-1)^{n-\ell} \int_t^s \frac{(r-t)^{n-\ell-1}}{(n-\ell-1)!} y^{(n)}(r)\, dr$$

$$\geq \int_t^s \frac{(r-t)^{n-\ell-1}}{(n-\ell-1)!} p_i(r) f_i(y(\tau_1(r)), \ldots, y(\tau_m(r)))\, dr \qquad s \geq t \geq T$$

$$(5.5.20)$$

Combining the two inequalities above yields inequality (5.5.10) and the subsequent proof proceeds exactly as in the corresponding part of the proof of Theorem 5.5.1.

Next suppose that $\ell = 1$. From (5.5.20) we then have

$$y'(t) \geq \int_t^\infty \frac{(r-t)^{n-2}}{(n-2)!} \, p_j(r) f_j(y(\tau_1(r)), \cdots, y(\tau_m(r))) \, dr \qquad t \geq T$$

This is exactly (5.5.11) and so the subsequent proof proceeds as in the corresponding part of the proof of Theorem 5.5.1. The proof is complete.

THEOREM 5.5.4   Assume that conditions (i) and (ii) of Theorem 5.5.1 are satisfied and $n \geq 3$ is odd. Further assume that there is an integer i, $1 \leq i \leq N$, and a positive number M such that the conditions (5.5.16) and (5.5.18) are satisfied for $h = i$. In addition suppose that there is an integer k, $1 \leq k \leq N$, and a positive number D such that

$$\int_R [t - \max_{1 \leq i \leq m} (\tau_i(t))]^{n-1} p_k(t) \, dt = \infty \tag{5.5.21}$$

$f_k(u_1, u_2, \cdots, u_m)$ is monotonically increasing for $|u_i| \leq D$, $1 \leq i \leq m$, and

$$\int_0^D \frac{du}{f_k(u, \cdots, u)} < \infty \,, \qquad \int_{0^-}^{-D} \frac{du}{f_k(u, \cdots, u)} < \infty \tag{5.5.22}$$

Then all solutions of (5.5.2) are oscillatory.

Proof: The argument of the proof of this theorem is similar to the argument of the proof of Theorem 5.5.3. Especially, the discussion with respect to $\ell \in [2, n-1]$ in Theorem 5.5.3 remains valid.

Now we consider the case where $\ell = 0$. Note that this is possible if n is odd. From the equation

$$-y'(t) = \sum_{i=1}^{n-1} (-1)^i y^{(i)}(s) \frac{(s-t)^{i-1}}{(i-1)!} - \int_t^s \frac{(r-t)^{n-2}}{(n-2)!} y^{(n)}(r) \, dr$$

and from Lemma 5.2.1, we have

$$-y'(t) \geq \int_t^{T^*} \frac{(s-t)^{n-2}}{(n-2)!} \, p_k(s) f_k(y(\tau_1(s)), \cdots, y(\tau_m(s))) \, ds \tag{5.5.23}$$

for $T \leq t \leq T^*$, where T^* is sufficiently large. Thus we obtain inequality (5.5.23) and from this point on the proof is entirely the same as the corresponding proof in Theorem 5.5.2. This completes the proof of Theorem 5.5.4.

Let us now consider the following equations which are particular cases of equation (5.5.1) and (5.5.2):

$$y^{(n)}(t) + p(t)f(y(g(t))) = 0 \tag{5.5.24}$$

The following corollaries are immediate consequences of Theorems 5.5.1 and 5.5.2.

COROLLARY 5.5.1   Assume that $p(t)$, $g(t)$, and f satisfy the conditions (i) and (ii) of Theorem 5.5.1. Suppose $n \geq 3$ is odd and there exists a positive number M such that $f(u)$ is monotonically increasing for $|u| \geq M$ and

$$\int_A [g(t) - t]^{n-1} p(t) \, dt = \infty$$

$$\int_A t^{n-2} [g(t) - t] p(t) \, dt = \infty$$

$$\int_A t^{n-1} p(t) \, dt = \infty$$

$$\int_M^{\infty} \frac{du}{f(u)} < \infty, \qquad \int_{-M}^{-\infty} \frac{du}{f(u)} < \infty, \qquad M > 0$$

Then all solutions of (5.5.24) are oscillatory.

COROLLARY 5.5.2   Assume that $p(t)$, $g(t)$, and f satisfy conditions (i) and (ii) of Theorem 5.5.1. Suppose n is even and there exist positive numbers M, m such that $f(u)$ is monotonically increasing for $|u| \geq M$, $|u| \leq m$, and

$$\int_A [g(t) - t]^{n-1} p(t) \, dt = \infty$$

$$\int_A t^{n-2} [g(t) - t] p(t) \, dt = \infty$$

$$\int_R [t - g(t)]^{n-1} p(t) \, dt = \infty$$

$$\int_M^{\infty} \frac{du}{f(u)} < \infty, \qquad \int_{-M}^{-\infty} \frac{du}{f(u)} < \infty$$

$$\int_{0+}^{m} \frac{du}{f(u)} < \infty, \qquad \int_{0-}^{-m} \frac{du}{f(u)} < \infty$$

Then all solutions of (5.5.24) are oscillatory.

The reader can easily formulate corresponding corollaries with respect to equation (5.5.25).

EXAMPLE 5.5.1 By Theorems 5.5.1–5.5.4 all solutions of the equations

$$y^{(n)}(t) - p|y(t + \sin t)|^{\alpha} \operatorname{sgn} y(t + \sin t) - q|y(t - \sin t)|^{\beta} \operatorname{sgn} y(t - \sin t) = 0$$

$$y^{(n)}(t) + p|y(t + \sin t)|^{\alpha} \operatorname{sgn} y(t + \sin t) + q|y(t - \sin t)|^{\beta} \operatorname{sgn} y(t - \sin t) = 0$$

are oscillatory provided $p > 0$, $q > 0$, $\alpha > 1$, and $0 < \beta < 1$.

EXAMPLE 5.5.2 Define the function $f(u)$ by

$$f(u) = u^{\alpha} \ (0 \le u < 1), \quad f(u) = u^{\beta} \ (u \ge 1), \quad f(-u) = -f(u)$$

where $0 < \alpha < 1$ and $\beta > 1$. Then from Corollaries 5.5.1, 5.5.2 it follows that all solutions of the equation

$$y^{(n)}(t) - f(y(t + \sin t)) = 0$$

are oscillatory.

5.6 NONLINEAR DIFFERENTIAL INEQUALITIES WITH DEVIATING ARGUMENTS

Consider a nonlinear differential inequality with deviating argument of the form

$$y^{(n)}(t) + p(t)f(y(g(t))) \le 0 \qquad n \ge 2 \tag{5.6.1}$$

LEMMA 5.6.1 Assume that

(i) $p \in C[R_+, R_+]$ and not eventually identically zero

(ii) $g \in C[R_+, R_+]$ and

$$0 < \varliminf_{t \to \infty} g'(t) \le \varlimsup_{t \to \infty} g'(t) < \infty$$

(iii) $f \in C[R, R]$, $yf(y) > 0$, $y \ne 0$

(iv) $y(t)$ is an eventually positive solution of (5.6.1) satisfying $\lim_{t \to \infty} y(t) > 0$.

Then there are positive numbers b_1 and b_2 such that

$$b_1 \le y(t) \le b_2 t^{n-1} \tag{5.6.2}$$

for all large t.

Proof: Let $y(t) > 0$ be a solution of (5.6.1). From Lemma 5.2.1, there exists an integer ℓ, $0 \leq \ell \leq n-1$, and ℓ is odd (even) if n is even (odd), such that

$$y^{(k)}(t) \geq 0 \qquad k = 0, 1, \ldots, \ell$$

$$(-1)^{n+k-1} y^{(k)}(t) \geq 0 \qquad k = \ell+1, \ldots, n$$

for all large t. However, $y^{(n)}(t) \leq 0$ and hence $y^{(n-1)}(t) \leq y^{(n-1)}(t_1)$, $t \geq t_1$. Integrating $n-1$ times repeatedly, we get $y(t) \leq b_2 t^{n-1}$.

On the other hand, since $y(t)$ is monotone positive and $\underline{\lim}_{t \to \infty} y(t) > 0$,

therefore $y(t) \geq b_1$. The proof is complete.

THEOREM 5.6.1 Assume that conditions (i), (ii), and (iii) of Lemma 5.6.1 are satisfied. Further assume that $p(t)$ can be decomposed in such a way that $p(t) = c(t)q(t)$, $c(t)$ is continuous on $[a, \infty)$, $0 < \underline{\lim}_{t \to \infty} c(t) \leq \overline{\lim}_{t \to \infty} c(t) < \infty$, and $t^\sigma q(t)$ is continuous nondecreasing on $[a, \infty)$, where σ is some real number. Then, a necessary and sufficient condition for (5.6.1) to have a solution $y(t)$ such that

$$\lim_{t \to \infty} [y(t)/t^{n-1}] \quad \text{and} \quad \lim_{t \to \infty} [y'(t)/t^{n-2}] \tag{5.6.3}$$

exist and are positive is that

$$\int^\infty p(s)f(cs^{n-1}) \, ds < \infty \qquad \text{for some } c > 0 \tag{5.6.4}$$

Proof: In view of the conditions on $c(t)$, we note that there are positive constants c_1 and c_2 such that

$$c_1 \leq c(t) \leq c_2 \tag{5.6.5}$$

for all large t.

Necessity: Let $y(t)$ be a solution of (5.6.1) satisfying (5.6.3). Then it satisfies the following inequalities:

$$\frac{b}{2}[g(t)]^{n-1} \leq y(g(t)) \leq 2b[g(t)]^{n-1} \tag{5.6.6}$$

$$\frac{b'}{2}[g(t)]^{n-2} \leq y'(g(t)) \leq 2b'[g(t)]^{n-2} \tag{5.6.7}$$

We note that condition (ii) implies that there exist positive constants ν_1, ν_2, β_1, β_2, and β_3 such that

$$\nu_1 \leq g'(t) \leq \nu_2 \tag{5.6.8}$$

$$\beta_1 t \leq g(t) \leq \beta_2 t \tag{5.6.9}$$

$$g_*(t) \equiv \min\{g(t), t\} \geq \beta_3 t \tag{5.6.10}$$

This, together with (5.6.6) and (5.6.7), yields

$$\frac{b}{2}\beta_1^{n-1}t^{n-1} \le y(g(t)) \le 2b\beta_2^{n-1}t^{n-1} \tag{5.6.11}$$

$$\frac{b'}{2}\beta_1^{n-2}\nu_1 t^{n-2} \le y'(g(t))g'(t) \le 2b'\beta_2^{n-2}\nu_2 t^{n-2} \tag{5.6.12}$$

It follows from (5.6.11) that

$$t \le \eta_1[y(g(t))]^{1/n-1} \quad \text{and} \quad \eta_2[y(g(t))]^{1/n-1} \le t \tag{5.6.13}$$

for $t \ge T$, where $\eta_1 = (b\beta_1^{n-1}/2)^{-1/(n-1)}$, $\eta_2 = (2b\beta_2^{n-1})^{-1/(n-1)}$ are positive constants, and $T > a$ is chosen so large that (5.6.11) and (5.6.12) hold, and $\eta_2[(b/2)\beta_1^{n-1}t^{n-1}]^{1/n-1} \ge a$ for $t \ge T$. Using the inequalities (5.6.13), (5.6.5), and the condition $p(t) = c(t)q(t)$, we compute the following:

$$p(t) = c(t)q(t) = c(t)t^{-\sigma}t^{\sigma}q(t)$$

$$\ge c_1 t^{-\sigma}(\eta_2[y(g(t))]^{1/(n-1)})^{\sigma}q(\eta_2[y(g(t))]^{1/(n-1)})$$

$$\ge c_1(\eta_1[y(g(t))]^{1/(n-1)})^{-\sigma}(\eta_2[y(g(t))]^{1/n-1})^{\sigma}q(\eta_2[y(g(t))]^{1/n-1})$$

$$= c_1\eta_1^{-\sigma}\eta_2^{\sigma}q(\eta_2[y(g(t))]^{1/(n-1)})$$

for $t \ge T$ and $\sigma \ge 0$. Furthermore,

$$p(t) \ge c_1(\eta_2[y(g(t))]^{1/(n-1)})^{-\sigma}(\eta_2[y(g(t))]^{1/(n-1)})^{\sigma}q(\eta_2[y(g(t))]^{1/(n-1)})$$

$$= c_1 q(\eta_2[y(g(t))]^{1/(n-1)})$$

for $t \ge T$ and $\sigma < 0$. In either case, we get

$$p(t) \ge dq(\eta_2[y(g(t))]^{1/(n-1)}) \tag{5.6.14}$$

for $t \ge T$, where $d = c_1\eta_1^{-\sigma}\eta_2^{\sigma}$ for $\sigma \ge 0$ and $d = c_1$ for $\sigma < 0$.

Now, integrating (5.6.1) from T to t, we obtain

$$y^{(n-1)}(t) - y^{(n-1)}(T) + \int_T^t p(s)f(y(g(s)))\, ds \le 0 \qquad t \ge T$$

Because of (5.6.11) and (5.6.12), we observe that there are positive constants c_1 and c_2 such that

$$c_1 \le \frac{d}{dt} \, \eta_2 [y(g(t))]^{1/(n-1)} \le c_2 \quad \text{for} \quad t \ge T$$

Hence, it follows that

$$\int_T^\infty q(\eta_2 [y(g(s))]^{1/(n-1)}) f(y(g(s))) \frac{d}{ds} \eta_2 [y(g(s))]^{1/(n-1)} \, ds < \infty$$

Letting $v = \eta_2 [y(g(s))]^{1/(n-1)}$, we obtain

$$\int_{T*}^\infty q(v) f(\eta_2^{-n+1} v^{n-1}) \, dv < \infty$$

where $T* = \eta_2 [y(g(T))]^{1/(n-1)}$. The desired inequality (5.6.4) is readily derived from $c_1 \le c(t) \le c_2$ and $p(t) = c(t)q(t)$.

Sufficiency: Assume that (5.6.4) holds. We suppose that (5.6.8), (5.6.9), and (5.6.5) hold for $t \ge T'$. Set $b = c\beta_1^{-n+1}$, $\eta_1 = (b\beta_1^{n-1})^{-1/(n-1)}$, $\eta_2 = (2b\beta_2^{n-1})^{-1/(n-1)}$, $d = c_2 \eta_2^{-\sigma} \eta_1^{\sigma}$ for $\sigma \ge 0$, $d = c_2$ for $\sigma < 0$, and $M = \eta_1 (2b\beta_2^{n-1})^{(-n+2)/(n-1)} b\beta_1^{n-2} v_2$. Choose a number $T > T'$ so large that $T_0 = \inf \{\min (g(t), t) : t \ge T\} > a$,

$$\int_T^\infty p(s)f(cs^{n-1}) \, ds \le b(n-1)! \, c_1 M d^{-1} \tag{5.6.15}$$

and $\eta_2 [b\beta_1^{n-1} t^{n-1}]^{1/(n-1)} \ge a$ for $t \ge T$. Let F denote the Frechet space of all continuously differentiable functions on $[T_0, \infty)$ with the family of semi-norms $\{\|\cdot\|_m, \ m = 1, 2, \ldots\}$ defined by

$$\|y\|_m = \sup \{|y(t)| + |y'(t)| : \ T_0 \le t \le T_0 + m\}$$

We have convergence $y_k \to y$ in the topology of F iff $y_k(t) \to y(t)$ and $y_k'(t) \to y'(t)$ uniformly on every compact subinterval of $[T_0, \infty)$. Let $\bar{y}$ be the subset of F such that

$$\bar{Y} = \{y \in F : bt^{n-1} \le y(t) \le 2bt^{n-1},$$

$$b(n-1)t^{n-2} \le y'(t) \le 2b(n-1)t^{n-2} \text{ for } t \ge T_0\}$$

which is a convex and closed subset of F. Define an operator Φ on $\bar{Y}$ by

$$(\Phi y)(t) = \begin{cases} 2bt^{n-1} - \dfrac{1}{(n-1)!} \displaystyle\int_T^t (t-s)^{n-1} p(s)f(y(g(s))) \, ds & t \ge T \\[1.5em] 2bt^{n-1} & T_0 \le t \le T \end{cases} \tag{5.6.16}$$

We seek a fixed point of the operator Φ in $\bar{Y}$ with the aid of the Schauder-Tychonov fixed point theorem.

We show that Φ defined by (5.6.16) satisfies the conditions of the Schauder-Tychonov fixed point theorem. In fact,

(i) Φ maps $\bar{Y}$ into $\bar{Y}$:

Let $y \in \bar{Y}$. It is obvious that $(\Phi y)(t)$ is continuously differentiable on $[T_0, \infty)$ and $(\Phi y)(t) \leq 2bt^{n-1}$, $(\Phi y)'(t) \leq 2b(n-1)t^{n-2}$ for $t \geq T_0$. To prove that $(\Phi y)(t) \geq bt^{n-1}$, $(\Phi y)'(t) \geq b(n-1)t^{n-2}$ for $t \geq T_0$, we consider the integral $\int_T^\infty p(s)f(y(g(s)))\,ds \leq b(n-1)!$. Hence we easily see that

$$(\Phi y)(t) \geq 2bt^{n-1} - \left(\frac{t^{n-1}}{(n-1)!}\right) b(n-1)! = bt^{n-1}$$

and

$$(\Phi y)'(t) \geq 2b(n-1)t^{n-2} - \left(\frac{(n-1)t^{n-2}}{(n-1)!}\right) b(n-1)! = b(n-1)t^{n-2}$$

for $t \geq T_0$.

(ii) Φ is continuous on $\bar{Y}$:

Let $\{y_k\}$, $k = 1, 2, \ldots$, and y be functions in $\bar{Y}$ such that $y_k(t) \to y(t)$,

$$y_k'(t) \to y'(t) \quad \text{as} \quad k \to \infty$$

uniformly on every compact subinterval of $[T_0, \infty)$. For $t \in [T, A]$, from (5.6.16),

$$| (\Phi y_k)(t) - (\Phi y)(t) | \leq \frac{(A-T)^{n-1}}{(n-1)!} \int_T^A p(s)\,|f(y_k(g(s))) - f(y(g(s)))|\,ds$$

$$| (\Phi y_k)'(t) - (\Phi y)'(t) | \leq \frac{(A-T)^{n-1}}{(n-1)!} \int_T^A p(s)\,|f(y_k(g(s))) - f(y(g(s)))|\,ds$$

Using the fact that $|f(y_k(g(s))) - f(y(g(s)))| \to 0$ as $k \to \infty$ for $s \geq T$, we conclude that $(\Phi y_k)(t) \to (\Phi y)(t)$, $(\Phi y_k)'(t) \to (\Phi y)'(t)$ as $k \to \infty$ uniformly on $[T, A]$. Thus $(\Phi y_k)(t)$ and $(\Phi y_k)'(t)$ converge to $(\Phi y)(t)$ and $(\Phi y)'(t)$, respectively, as $k \to \infty$ uniformly on every compact subinterval of $[T_0, \infty)$. This proves the continuity of Φ on $\bar{Y}$.

(iii) $\overline{\Phi \bar{Y}}$ is compact:

It is enough to show that, for any sequence $\{y_k\}$, $k = 1, 2, \ldots$, in $\bar{Y}$ there exist a subsequence $\{y_{k_i}\}$, $i = 1, 2, \ldots$, and a function y in F such that $(\Phi y_{k_i})(t) \to y(t)$, $(\Phi y_{k_i})'(t) \to y'(t)$ as $i \to \infty$ uniformly on compact subintervals of $[T_0, \infty)$.

Let $\{y_k\}$ be an arbitrary sequence in $\bar{Y}$. Differentiating $(\Phi y_k)(t)$ twice, we have, for the case of $n > 2$,

$$(\Phi y_k)''(t) = \begin{cases} 2b(n-1)(n-2)t^{n-3} - \dfrac{1}{(n-3)!}\displaystyle\int_T^t (t-s)^{n-3}p(s)f(y_k(g(s)))\,ds \\ \qquad\qquad\qquad\qquad\qquad\qquad\qquad\qquad\qquad\text{for } t \geq T \\[2ex] 2b(n-1)(n-2)t^{n-3} \qquad \text{for } T_0 \leq t \leq T \end{cases}$$

and in the case of $n = 2$,

$$(\Phi y_k)''(t) = \begin{cases} -p(t)f(y_k(g(t))) & \text{for } t \geq T \\[1ex] 0 & \text{for } T_0 \leq t \leq T \end{cases}$$

It is easy to see that, for $t \in [T,A]$, we have either

$$(\Phi y_k)''(t) \leq 2b(n-1)(n-2)A^{n-3} + \frac{1}{(n-3)!}(A-T)^{n-2}PL$$

or

$$|(\Phi y_k)''(t)| \leq PL$$

where $P = \max\{p(s) : T \leq S \leq A\}$ and $L = \max\{f(u) : b\beta_1^{n-1}T^{n-1} \leq u \leq 2b\beta_2^{n-1}A^{n-1}\}$ are constants independent of $k = 1, 2, \ldots$. Hence $\{(\Phi y_k)'(t)\}$ is equicontinuous on every subinterval of $[T_0, \infty)$. The boundedness of $\{(\Phi y_k)'(t)\}$ at every point of $[T_0, \infty)$ is evident. By applying Ascoli's theorem, we can choose a subsequence $\{(\Phi y_{k_i})'(t)\}$, $i = 1, 2, \ldots$, of $\{(\Phi y_k)'(t)\}$, $k = 1, 2, \ldots$, which is uniformly convergent on every compact subinterval of $[T_0, \infty)$. Let the limit function of $(\Phi y_{k_i})'(t)$ be denoted by $z(t)$; this is clearly continuous on $[T_0, \infty)$. Putting $y(t) = 2bT_0^{n-1} + \int_T^t z(s)\,ds$, $t \geq T_0$; we observe that $(\Phi y_{k_i})'(t) \to z(t) = y'(t)$ and $(\Phi y_{k_i})(t) = (\Phi y_{k_i})(T_0) + \int_T^t (\Phi y_{k_i})'(s)\,ds \to 2bT_0^{n-1} + \int_T^t z(s)\,ds = y(t)$ as $i \to \infty$, uniformly on compact subintervals of $[T_0, \infty)$. This proves that $\overline{\Phi\bar{Y}}$ is compact.

From the preceding considerations, we are able to apply the Schauder-Tychonov fixed point theorem to the operator Φ. Let $y(t) \in \bar{Y}$ be a fixed point of Φ. It is immediately clear that $y(t)$ is a solution of (5.6.1) for $t \geq T$ and has the property that $\lim_{t\to\infty}[y(t)/t^{n-1}]$ and $\lim_{t\to\infty}[y'(t)/t^{n-2}]$ exist and belong to $[b, 2b]$ and $[b(n-1), 2b(n-1)]$, respectively. Thus the proof of Theorem 5.6.1 is complete.

THEOREM 5.6.2 Assume that conditions (i), (ii), and (iii) of Lemma 5.6.1 are satisfied. Then, a necessary and sufficient condition for (5.6.1) to have a solution $y(t)$ such that $\lim_{t\to\infty} y(t) = b > 0$ exists is that

$$\int^{\infty} p(s)s^{n-1} \, ds < \infty \tag{5.6.17}$$

Proof: Necessity: Let $y(t)$ be a solution of (t.6.1) such that $\lim_{t\to\infty} y(t) = b$, $0 < b < \infty$. Hence we have

$$\frac{b}{2} \le y(t) \le 2b \quad \text{and} \quad \frac{b}{2} \le y(g(t)) \le 2b \tag{5.6.18}$$

for all large t. By Lemma 5.2.1 we can find an integer ℓ, $0 \le \ell \le n-1$. Since $y(t)$ is bounded, this integer ℓ must be equal to 0 or 1. Therefore, we obtain

$$(-1)^{n+k-1} y^{(k)}(t) \ge 0 \qquad k = 1, 2, \ldots, n \tag{5.6.19}$$

for all large t. Take a number $T > a$ so large that (5.6.18) and (5.6.19) hold for $t \ge T$. Multiplying (5.6.1) by t^{n-1} and integrating from T to t, we get

$$t^{n-1} y^{(n-1)}(t) - (n-1)t^{n-2} y^{(n-2)}(t) + \cdots$$

$$+ (-1)^{k+1}(n-1) \cdots (n-k+1)t^{n-k} y^{(n-k)}(t) + \cdots + (-1)^{n}(n-1) \cdots 2ty'(t)$$

$$+ (-1)^{n+1}(n-1)!y(t) + \int_{T}^{t} s^{n-1} p(s)f(y(g(s))) \, ds \le c$$

where c is a constant. In view of (5.6.18), (5.6.19), and the above inequality we get

$$\int_{T}^{\infty} s^{n-1} p(s)f(y(g(s))) \, ds < \infty$$

From this and (5.6.18) it follows that

$$m \int_{T}^{\infty} s^{n-1} p(s) \, ds < \infty$$

where $m = \min \{f(u) : b/2 \le u \le 2b\}$.

Sufficiency: Assume that (5.6.17) holds. Let β be an arbitrary positive number. We choose a number $T > a$ so large that $T_0 = \inf \{\min \{g(t), t\}, t \ge T\} > a$ and

$$\int_{T}^{\infty} p(s)s^{n-1} \, ds \le \beta(n-1)!M^{-1}$$

where $M = \max\{f(u) : \beta \le u \le 2\beta\}$. We consider the operator Φ defined
by

$$(\Phi y)(t) = \begin{cases} b + \dfrac{(-1)^{n-1}}{(n-1)!} \displaystyle\int_t^\infty (s-t)^{n-1} p(s)f(y(g(s)))\, ds & \text{for } t \ge T \\[2em] b + \dfrac{(-1)^{n-1}}{(n-1)!} \displaystyle\int_T^\infty (s-T)^{n-1} p(s)f(y(g(s)))\, ds & \text{for } T_0 \le t \le T \end{cases}$$

where $b = 2\beta$ for n even and $b = \beta$ for n odd. With the argument of the fixed
point theorem stated in the proof of Theorem 5.6.1, we seek a fixed point
of Φ. The underlying Frechet space F is the set of all continuous functions
on $[T_0, \infty)$ with the topology of uniform convergence on compact subintervals
of $[T_0, \infty)$. A convex and closed subset $\bar{Y}$ of F on which Φ is well defined is
$\bar{Y} = \{y \in F : \beta \le y \le 2\beta$ for $t \ge T_0 \}$. It can be shown that (i) the operator Φ
maps $\bar{Y}$ into $\bar{Y}$; (ii) Φ is continuous on $\bar{Y}$; and (iii) $\overline{\Phi \bar{Y}}$ is a compact subset
of $\bar{Y}$. Therefore, Φ has a fixed point $y \in \bar{Y}$. The function $y(t)$ is a solution
of (5.6.1) for $t \ge T$, and satisfies $\lim_{t \to \infty} y(t) = b > 0$. This sketches the

proof of the sufficiency part. The details are left to the reader.

5.7 EQUATIONS WITH FORCING TERMS

We consider equation of the form

$$(\tau_{n-1}(t)(\tau_{n-2}(t)(\cdots(\tau_1(t)y'(t))' \cdots)')')' + f(t, y(\Delta(t, y(t)))) = Q(t) \quad (5.7.1)$$

where $n \ge 2$ is a natural number. Under certain conditions imposed on the
coefficients of equation (5.7.1) sufficient conditions are found under which
all bounded nonoscillatory solutions tend to zero as $t \to \infty$.

The following lemmas are useful. The proofs can be found in [135].

LEMMA 5.7.1 Consider the equation

$$u' - \frac{\rho'(t)}{\rho(t)} u + \frac{\rho'(t)}{\rho(t)} \phi(t) = 0 \qquad\qquad (5.7.2)$$

with the following assumptions for the functions $\rho(t)$ and $\phi(t)$:

$$\phi \in C[R_+, R], \quad \rho \in C^1[R_+, R], \quad \text{for } t \ge T$$

$$\rho(t) > 0, \quad \rho'(t) < 0, \quad \text{for } t \ge T$$

$$\lim_{t \to \infty} \rho(t) = 0 \qquad\qquad (5.7.3)$$

Let $u(t)$ be a solution of equation (5.7.2) defined on the interval $[T, \infty)$ for which $u(T) = 0$. If $\lim_{t \to \infty} \phi(t) = \pm \infty$ then $\lim_{t \to \infty} u(t) = \pm \infty$.

LEMMA 5.7.2 Let $\sigma(t)$ and $\nu(t)$ be two functions for which $\sigma \in C[R_+, R]$, $\nu \in C^1[R_+, R]$ for $t \geq T$. If the boundary $\lim_{t \to \infty} (\sigma(t)\nu'(t) + \nu(t))$ exists on R^*, then the boundary $\lim_{t \to \infty} \nu(t)$ also exists on R^*, where R^* is the extended real line.

THEOREM 5.7.1 Assume that

(i) $\tau_i \in C[R_+, R_+ \setminus \{0\}]$, $i = 1, 2, \ldots, n-1$.

(ii) $\lim_{t \to \infty} \rho_i(t) = 0$, where $\rho_0(t) \equiv 1$, $\rho_i(t) = \int_t^\infty \rho_{i-1}(s)/\tau_i(s) \, ds$,
      $i = 1, 2, \ldots, n-1$.

(iii) $uf(t, u) > 0$ for $u \neq 0$, $f \in C[R_+ \times R, R]$.

(iv) If $|u_1| \leq |u_2|$ and $u_1 u_2 \geq 0$, then

        $|f(t, u_1)| \leq |f(t, u_2)|$ for $t \geq t_0$

(v) $\Delta(t, u) \in C[R_+ \times R, R]$, and

            $\Delta(t, u) = \infty$ for every fixed $u \in R$

      and for $t \in R^+$, the function $\Delta(t, u)$ is increasing for $u \geq 0$ and
      decreasing for $u \leq 0$.

(vi) $\displaystyle\int_{t_0}^\infty \rho_{n-1}(t) |f(t, u)| \, dt = \infty$ for $u \neq 0$ $\hspace{2cm}$ (5.7.4)

      and

      $\displaystyle\int_{t_0}^\infty \rho_{n-1}(t) |Q(t)| \, dt < \infty$ $\hspace{3cm}$ (5.7.5)

Then every bounded nonoscillatory solution of equation (5.7.1) tends to zero as $t \to \infty$.

Proof: Let $y(t)$ be a bounded nonoscillatory solution of equation (5.7.1). Without loss of generality suppose that

    $y(t) > 0$ for $t \geq t_1$

where $t_1 \geq t_0$. It follows by (v) that $\Delta(t, y(t)) \geq \Delta(t, 0)$ for $t \geq t_1$, and there exists a number $t_2 \geq t_1$ such that $\Delta(t, 0) \geq t_1$ for $t \geq t_2$. Then

    $y(\Delta(t, y(t))) > 0$ for $t \geq t_2$

We introduce the following notation:

$$g_0(t) = y(t), \qquad g_i(t) = \tau_i(t)g'_{i-1}(t), \qquad i = 1, 2, \ldots, n-1 \qquad (5.7.6)$$

$$u_k(t) = \int_{t_2}^{t} \rho_{n-k-1}(s)g'_{n-k-1}(s)\, ds, \qquad k = 0, 1, \ldots, n-1 \qquad (5.7.7)$$

Integrating equation (5.7.7) by parts, we get

$$u_{k-1}(t) = \int_{t_2}^{t} \rho_{n-k}(s)g'_{n-k}(s)\, ds$$

$$= \rho_{n-k}(t)g_{n-k}(t) - \rho_{n-k}(t_2)g_{n-k}(t_2) + \int_{t_2}^{t} \frac{\rho_{n-k-1}(s)}{\tau_{n-k}(s)}g_{n-k}(s)\, ds$$

$$= \frac{\rho_{n-k}(t)\tau_{n-k}(t)}{\rho_{n-k-1}(t)}\rho_{n-k-1}(t)g'_{n-k-1}(t) - \rho_{n-k}(t_2)g_{n-k}(t_2)$$

$$+ \int_{t_2}^{t} \rho_{n-k-1}(s)g'_{n-k-1}(s)\, ds$$

$$= -\frac{\rho_{n-k}(t)}{\rho'_{n-k}(t)}u'_k(t) + u_k(t) - \rho_{n-k}(t_2)g_{n-k}(t_2)$$

Therefore $u_k(t)$ satisfies the equation

$$u'(t)\frac{\rho_{n-k}(t)}{\rho'_{n-k}(t)} - u(t) + \phi_k(t) = 0 \qquad (5.7.8)$$

In the other form,

$$u'(t) - \frac{\rho'_{n-k}(t)}{\rho_{n-k}(t)}u(t) + \frac{\rho'_{n-k}(t)}{\rho_{n-k}(t)}\phi_k(t) = 0 \qquad (5.7.9)$$

where $\phi_k(t) = u_{k-1}(t) + \rho_{n-k}(t_2)g_{n-k}(t_2)$. It follows from (5.7.7) that $u_k(t_2) = 0$, while conditions (i) and (ii) imply $\rho_{n-k}(t) > 0$, $\rho'_{n-k}(t) < 0$, and $\lim_{t\to\infty}\rho_{n-k}(t) = 0$. Hence conditions (5.7.3) are satisfied for the functions $\phi_k(t)$ and $\rho_{n-k}(t)$. Applying Lemma 5.7.1 to equation (5.7.9), we have that if $\lim_{t\to\infty}u_{k-1}(t) = \pm\infty$, then also $\lim_{t\to\infty}u_k(t) = \pm\infty$.

From Lemma 5.7.2, the boundary $\lim_{t\to\infty}u_{k-1}(t)$ exists on R^* which implies that the boundary $\lim_{t\to\infty}u_k(t)$ also exists on R^*.

We multiply both sides of equation (5.7.1) by $\rho_{n-1}(t)$ and integrate over interval $[t_2, t]$. We get

$$\int_{t_2}^{t} \rho_{n-1}(s)g_{n-1}'(s)\, ds + \int_{t_2}^{t} \rho_{n-1}(s)f(s, y(\Delta(s, y(s))))\, ds = \int_{t_2}^{t} \rho_{n-1}(s)Q(s)\, ds$$

$$(5.7.10)$$

The following cases are possible:

$$\int_{t_2}^{\infty} \rho_{n-1}(t)f(t, y(\Delta(t, y(t))))\, dt = \infty \qquad (5.7.11)$$

or

$$\int_{t_2}^{\infty} \rho_{n-1}(t)f(t, y(\Delta(t, y(t))))\, dt < \infty \qquad (5.7.12)$$

Assume relation (5.7.11) holds. From (5.7.5) it can be seen that the right-hand side of (5.7.10) tends to a finite boundary as $t \to \infty$. Hence (5.7.10) and (5.7.11) imply

$$\lim_{t \to \infty} \int_{t_2}^{t} \rho_{n-1}(s)g_{n-1}'(s)\, ds = -\infty$$

that is,

$$\lim_{t \to \infty} u_0(t) = -\infty$$

Applying Lemma 5.7.1 to equation (5.7.9) for $k = 1$, we obtain $\lim_{t \to \infty} u_1(t) = -\infty$. Analogously we find that $\lim_{t \to \infty} u_{n-1}(t) = -\infty$. From (5.7.6) and (5.7.7) we have

$$u_{n-1}(t) = \int_{t_2}^{t} \rho_0(s)g_0'(s)\, ds = y(t)$$

Therefore $\lim_{t \to \infty} y(t) = -\infty$ which contradicts $y(t) > 0$. Hence relation (5.7.12) holds. Carrying out boundary transition in (5.7.10), from (5.7.5) and (5.7.12) it follows that the boundary $\lim_{t \to \infty} u_0(t)$ exists and is finite.

It follows from Lemma 5.7.2 applied to (5.7.8) for $k = 1$ that the boundary $\lim_{t \to \infty} u_1(t)$ exists on R^*. But this boundary must be finite. Indeed, if $\lim_{t \to \infty} u_1(t) = +\infty$ or $\lim_{t \to \infty} u_1(t) = -\infty$, then by reasoning analogous to the above it can be proved that $\lim_{t \to \infty} y(t) = +\infty$ or $\lim_{t \to \infty} y(t) = -\infty$. The former contradicts the boundedness of $y(t)$ and the latter contradicts $y(t) > 0$. Hence the boundary $\lim_{t\to\infty} u_1(t)$ exists and is finite. By similar reasoning we finally obtain

$$\lim_{t\to\infty} y(t) = \alpha \qquad \text{where} \quad \alpha \in R \tag{5.7.13}$$

We shall prove that

$$\liminf_{t\to\infty} y(\Delta(t, y(t))) = \liminf_{t\to\infty} y(t) = 0 \tag{5.7.14}$$

Because $y(\Delta(t, y(t)))) > 0$ for $t \geq t_2$, then $\liminf_{t\to\infty} y(\Delta(t, y(t))) = d \geq 0$. Assume that $d > 0$. Then there exists a number $t_3 \geq t_2$ such that $y(\Delta(t, y(t))) \geq d$ for $t \geq t_3$, and condition (iii) implies that

$$f(t, d) \leq f(t, y(\Delta(t, y(t)))) \qquad \text{for} \quad t \geq t_3$$

that is,

$$\rho_{n-1}(t)f(t, d) \leq \rho_{n-1}(t)f(t, y(\Delta(t, y(t)))) \qquad \text{for} \quad t \geq t_3$$

From the final inequality and (5.7.12) there follows $\int_{t_3}^{\infty} \rho_{n-1}(t)f(t, d)\, dt < \infty$, which contradicts (5.7.4). Therefore $d = 0$, i.e.,

$$\liminf_{t\to\infty} y(\Delta(t, y(t))) = 0 \tag{5.7.15}$$

Because $y(t) > 0$ we have $\liminf y(t) = d_1 \geq 0$. Assume that $d_1 > 0$. Then there exists a number $t_4 \geq t_1$ such that

$$y(t) \geq d_1 \qquad \text{for} \quad t \geq t_4 \tag{5.7.16}$$

and from (v) we get $\Delta(t, y(t)) \geq \Delta(t, d_1)$ for $t \geq t_4$. The last inequality and (v) imply that there exists a number $t_5 \geq t_4$ such that $\Delta(t, y(t)) \geq t_4$ for $t > t_5$. Hence (5.7.16) implies that $y(\Delta(t, y(t))) \geq d_1 > 0$ for $t \geq t_5$, which contradicts (5.7.15). Thus relation (5.7.14) is proved to hold. (5.7.13) and (5.7.14) imply $\lim y(t) = 0$. Thus Theorem 5.7.1 is proved.

EXAMPLE 5.7.1 Consider the equation

$$(t(t(t(t^2 y'(t))')')')' + y(\beta t + y^2(t)) = \frac{t^2}{\beta t^3 + 1} \tag{5.7.17}$$

where $\beta = \text{const} > 0$ and $t \geq t_0 > 0$. The functions $\tau_1(t) = t^2$, $\tau_2(t) = \tau_3(t) = t$, $\rho_1(t) = \rho_2(t) = \rho_3(t) = t^{-1}$, $f(t, u) = u$, $\Delta(t, u) = \beta t + u^2$, and $Q(t) = t^2/(\beta t^3 + 1)$ satisfy all conditions of Theorem 5.7.1. Therefore all bounded nonoscillating solutions of (5.7.17) tend to zero as $t \to \infty$. For instance $y(t) = t^{-1}$ is one of these solutions.

5.8 NOTES

Theorems 5.1.1 and 5.2.1 are from Zhang [299]. For related work see
also Philos and Sficas [223], Morgenthal [188], Terry [265], and Erbe [61].
Theorems 5.1.3 to 5.1.8 are from Kusano and Naito [138]; see also [133],
[241], and [263]. Lemmas 5.2.1 and 5.2.2 belong to Kuguradze [113];
Lemma 5.2.3 is adapted from Grammatikopoulis et al. [92]. Theorems
5.2.1 and 5.2.2 are from Zhang and Ladde [305]. For related work see
Grace and Lalli [71-74, 76], Dahiya [44, 47], Chen [37], and Liu [172].
Theorems 5.3.1 to 5.3.4 are due to Ladas and Stavroulakis [151, 153];
for related work see Kusano [139, 141]. Theorems 5.3.5 to 5.3.7 are taken
from Kusano [141]; see also Bogar [16] and Dahiya [47]. Theorems 5.3.8
to 5.3.12 are new; for related work see Ladas et al. [157], Lovelady [176],
and Sficas and Staikos [226]. Theorems 5.3.13 and 5.3.14 are taken from
Kusano [142]; see also Dahiya and Kusano [51], Dahiya [47], Graef and Seal
[89], Mahfoud [180], Grimmer [93], Gustafson [94], Nasyhova [195], and
Terry [269]. Theorems 5.4.1 to 5.4.4 are taken from Ohriska [201]. For
related work see [77, 84, 85, 90, 129, 130, 178, 194, 203, 227, 231, 230, 303].
Theorems 5.5.1 to 5.5.4 are direct extensions of the results given by
Ivanov et al. [102]. For related work, see Werbowski [286], Graef et al. [83],
Staikos and Sficas [253], and Staikos [258]. The asymptotic behavior of
solutions is discussed by Graef and Spikes [79], Graef [80], Grammatikop-
oulos [90], Chen [33], Chen and Yeh [36], and Chen, Yeh, and Yu [39].
Theorems 5.6.1 and 5.6.2 are from Ohriska [199]. For related work see
Marusiak [182], Naito [193], and Chen and Yeh [35]. Theorem 5.7.1 is from
Mishev and Bainov [187]. A survey of oscillation results for forced equations
prior to 1976 can be found in the paper by Kartsatos [108]. For recent work
see Dahiya [45, 50], Graef [80, 87], Graef and Spikes [79], Graef et al. [88],
Chen [32], Singh [242, 250], Onose [210], Foster [64], Johnson and Yan [104],
Kartsatos and Manougian [106], Kartsatos [107], Kartsatos and Toro [110],
Staikos and Sficas [254]. For oscillations on both sides see Fite [62] and
Norkin [198]. For results on comparison theorems see Kusano and Naito
[140].

6

Systems of Differential Equations

6.0 INTRODUCTION

We begin with two dimensional linear systems and develop several new results in Section 6.1. Nonlinear two dimensional systems are discussed in Sections 6.2 and 6.3. Higher order linear systems form the content of Section 6.4.

6.1 TWO-DIMENSIONAL LINEAR SYSTEMS

We consider the linear system of functional differential equations with retarded arguments,

$$y_1'(t) = a_{11}y_1(t) + a_{12}y_2(t - \tau_1)$$
$$y_2'(t) = a_{21}y_1(t - \tau_2) + a_{22}y_2(t) \tag{6.1.1}$$

where $a_{ij} \in R$ and $\tau_i > 0$ for $i, j = 1, 2$. The ordinary system of differential equations corresponding to (6.1.1) is

$$y_1'(t) = a_{11}y_1(t) + a_{12}y_2(t)$$
$$y_2'(t) = a_{21}y_1(t) + a_{22}y_2(t) \tag{6.1.2}$$

As in Section 1.2, a solution $(y_1(t), y_2(t))$ of (6.1.2) or (6.1.1) is said to be <u>oscillatory</u> if y_1 and y_2 both have arbitrarily large zeros in R_+. Otherwise (y_1, y_2) is said to be <u>nonoscillatory</u>.

It is easy to see that the oscillatory and nonoscillatory behavior of (6.1.2) depends on the characteristic roots corresponding to the characteristic equation associated with (6.1.2). In fact, every solution of (6.1.2) oscillates if

$$(a_{11} - a_{22})^2 + 4a_{12}a_{21} < 0 \tag{6.1.3}$$

and every solution of (6.1.2) is nonoscillatory if

$$(a_{11} - a_{22})^2 + 4a_{12}a_{21} \geq 0 \qquad (6.1.4)$$

In order to investigate the oscillatory and nonoscillatory properties of (6.1.1), we present the following lemmas.

LEMMA 6.1.1 Let $a_{ij} \in R$, and $\tau_i > 0$ for $i, j = 1, 2$. Further assume $(y_1(t), y_2(t))$ is a solution of (6.1.1). Then

$$y_1(t) = \exp\left[\frac{(a_{11} + a_{22})t}{2}\right] w_1(t)$$

$$y_2(t) = \exp\left[\frac{(a_{11} + a_{22})t}{2}\right] w_2(t) \qquad (6.1.5)$$

where $w_i(t)$ for $i = 1, 2$ is a solution of

$$w''(t) - \frac{(a_{22} - a_{11})^2}{4} w(t) - a_{12}a_{21} \exp\left[-\frac{\tau(a_{11} + a_{22})}{2}\right] w(t - \tau) = 0 \qquad (6.1.6)$$

where $\tau = \tau_1 + \tau_2$.

Proof: From (6.1.1) and (6.1.5), we have

$$w_1'(t) = \frac{(a_{11} - a_{22})}{2} w_1(t) + a_{12} \exp\left[-\frac{\tau_1(a_{11} + a_{22})}{2}\right] w_2(t - \tau_1)$$

$$w_2'(t) = \frac{(a_{22} - a_{11})}{2} w_2(t) + a_{21} \exp\left[-\frac{\tau_2(a_{11} + a_{22})}{2}\right] w_1(t - \tau_2) \qquad (6.1.7)$$

By differentiating $w_i(t)$ for $i = 1, 2$ twice and using (6.1.7), one can easily conclude that for $i = 1, 2$, $w_i(t)$ satisfies (6.1.6). This completes the proof of the lemma.

LEMMA 6.1.2 Let $w(t)$ be a solution of (6.1.6). Further assume that β is a solution of

$$a_{12}r^2 + (a_{11} - a_{22})r - a_{21} = 0 \qquad (6.1.8)$$

Then

$$y_1(t) = \exp\left[\frac{(a_{11} + a_{12})t}{2}\right] w(t)$$

$$y_2(t) = \beta \exp\left[\frac{(a_{11} + a_{22})t}{2}\right] w(t) \qquad (6.1.9)$$

is a solution of (6.1.1).

Proof: The proof of Lemma 6.1.2 follows directly from (6.1.8) and the definition of a solution.

REMARK 6.1.1 From Lemmas 6.1.1 and 6.1.2, it is obvious that the oscillatory and nonoscillatory properties of (6.1.1) are equivalent to the oscillatory and nonoscillatory properties of (6.1.6).

Now, we are ready to formulate sufficient conditions for oscillatory solutions of (6.1.1).

THEOREM 6.1.1 Assume that $a_{ij} \in R$, $\tau_i > 0$, $i, j = 1, 2$ and

$$a_{11} = a_{22} \quad \text{and} \quad a_{12} a_{21} < 0 \qquad (6.1.10)$$

Then every solution of (6.1.1) is oscillatory.

Proof: From Lemma 6.1.1 and (6.1.10) we have

$$y_1(t) = \exp(a_{11} t) w_1(t)$$
$$y_2(t) = \exp(a_{22} t) w_2(t) \qquad (6.1.11)$$

where $w_i(t)$ for $i = 1, 2$ are solutions of

$$w''(t) - a_{12} a_{21} \exp(-\tau a_{11}) w(t - \tau) = 0 \qquad (6.1.12)$$

Obviously, the characteristic equation of (6.1.12) has no real roots. Therefore every solution of (6.1.12) is oscillatory, and with Remark 6.1.1, the conclusion of the theorem follows immediately.

REMARK 6.1.2 We note that under condition (6.1.10), the oscillatory property of (6.1.2) is unaffected by the presence of past memory.

THEOREM 6.1.2 Assume that

$$a_{11} = a_{22} \quad \text{and} \quad a_{12} a_{21} > 0 \qquad (6.1.13)$$

and

$$\tau^2 a_{12} a_{21} \exp(-a_{11} \tau) \geq 2 \qquad (6.1.14)$$

where $\tau = \tau_1 + \tau_2$. Then every solution of the form

$$y_1(t) = 0(\exp(a_{11} t))$$
$$y_2(t) = 0(\exp(a_{11} t)) \qquad (6.1.15)$$

of (6.1.1) is oscillatory.

Proof: From Lemma 6.1.1 and (6.1.13), we arrive at equation (6.1.12) with the coefficient of $w(t - \tau)$ being a negative constant. Now, by application of Corollary 4.1.1, one can easily conclude that every bounded solution of (6.1.12) is oscillatory. This together with (6.1.11) and Remark 6.1.1 implies that every solution of (6.1.1) of the form (6.1.15) is oscillatory. This completes the proof of the theorem.

REMARK 6.1.3 We observe that condition (6.1.13) implies (6.1.4). Hence, every solution of (6.1.2) is nonoscillatory. Therefore, under the conditions of Theorem 6.1.2, the oscillation of solutions of (6.1.1) are generated by the delays.

In the following, we present a procedure that enables us to construct a linear system of functional differential equations having oscillatory behavior. However, its corresponding ordinary system is nonoscillatory. For this purpose, we consider the characteristic equation

$$\lambda^2 - a_{12}a_{21} \exp(-a_{11}\tau) \exp(-\tau\lambda) = 0 \tag{6.1.16}$$

corresponding to (6.1.12) satisfying condition (6.1.10). Assume that (6.1.16) has the purely imaginary root iy. Then (6.1.16) reduces to

$$-y^2 - a_{12}a_{21} \exp(-a_{11}\tau)(\cos y\tau + i \sin y\tau) = 0 \tag{6.1.17}$$

This implies that

$$\sin y\tau = 0, \qquad y = \frac{k\pi}{\tau}, \qquad k = 0, \pm 1, \pm 2, \ldots$$

This together with (6.1.17) yields

$$-\left(\frac{k\pi}{\tau}\right)^2 - a_{12}a_{21} \exp(-a_{11}\tau)(-1)^k = 0 \tag{6.1.18}$$

Assume that (6.1.18) has an odd integral root k. From this, we can conclude that (6.1.16) has a purely imaginary root $y = k\pi/\tau$. Moreover, $w(t) = c_1 \cos(k\pi/\tau)t + c_2 \sin(k\pi/\tau)t$ is a general solution of (6.1.12) satisfying the conditions (6.1.13) and (6.3.14). Now we are ready to present an example.

EXAMPLE 6.1.1 Consider the system

$$y_1'(t) = -y_1(t) + y_2(t)$$

$$y_2'(t) = -3e^{-\pi}y_1(t) - y_2(t) \tag{6.1.19}$$

We note that (6.1.19) is nonoscillatory. From (6.1.18), we can construct a corresponding delay system

$$y_1'(t) = -y_1(t) + y_2\left(t - \frac{\pi}{2}\right)$$

$$y_2'(t) = 3^2 e^{-\pi} y_1\left(t - \frac{\pi}{2}\right) - y_2(t) \tag{6.1.20}$$

which has an oscillatory solution

$$y_1(t) = e^{-t}(\cos 3t + \sin 3t)$$

$$y_2(t) = 3 \exp\ -\left(t + \frac{\pi}{2}\right)\ (\cos 3t + \sin 3t) \tag{6.1.21}$$

THEOREM 6.1.3 If $a_{12}a_{21} < 0$ and

$$\lambda_0 > -\frac{1}{\tau} + \sqrt{\left(\frac{1}{\tau}\right)^2 + \left(\frac{a_{11} - a_{22}}{2}\right)^2} \tag{6.1.22}$$

where λ_0 satisfies the equation

$$2\lambda_0 + \tau a_{12}a_{21} \exp\left[-\frac{\tau(a_{11} + a_{22})}{2}\right] \exp(-\lambda_0 \tau) = 0 \tag{6.1.23}$$

and $\tau = \tau_1 + \tau_2$, then every solution of (6.1.1) is oscillatory.

Proof: We consider the characteristic equation, corresponding to (6.1.6),

$$F(\lambda) = \lambda^2 - \frac{(a_{22} - a_{11})^2}{4} - a_{12}a_{21} \exp\left[-\frac{\tau(a_{11} + a_{22})}{2}\right] \exp(-\lambda\tau) = 0$$

Obviously,

$$F'(\lambda) = 2\lambda + \tau a_{12}a_{21} \exp\left[-\frac{\tau(a_{11} + a_{22})}{2}\right] \exp(-\lambda\tau)$$

$$F''(\lambda) = 2 - \tau^2 a_{12}a_{21} \exp\left[-\frac{\tau(a_{11} + a_{22})}{2}\right] \exp(-\lambda\tau) > 0$$

This implies that $F(\lambda)$ has a minimum value $F(\lambda_0)$, where λ_0 is a root of the equation $F'(\lambda) = 0$. If $F(\lambda_0) > 0$, then $F(\lambda) = 0$ has no real roots. According to $F'(\lambda_0) = 0$, we have

$$F(\lambda_0) = \lambda_0^2 - \left(\frac{a_{11} - a_{22}}{2}\right)^2 + \frac{2\lambda_0}{\tau}$$

The condition (6.1.22) guarantees $F(\lambda_0) > 0$. That is, equation (6.1.6) has no nonoscillatory solution. The proof is completed.

REMARK 6.1.4 We can compute the real root λ_0 of equation (6.1.23) using a graphic method or computer techniques. Therefore Theorem 6.1.3 is useful in application.

EXAMPLE 6.1.2 We consider

$$y_1'(t) = -y_1(t) + y_2\left(t - \frac{3}{2}\right)$$

$$y_2'(t) = -y_1\left(t - \frac{1}{2}\right) - y_2(t) \tag{6.1.24}$$

$\lambda_0 = 1$ is a root of equation (6.1.23). Obviously, it satisfies the condition (6.1.22), so every solution of equation (6.1.24) is oscillatory.

In the following, we present a result that establishes the existence of a nonoscillatory solution of (6.1.1).

THEOREM 6.1.4 Assume that $a_{11} a_{22} > 0$ and

$$\tau_1 + \tau_2 = \left[- \frac{2}{(a_{11} + a_{22})} \right] \ln \left(- \frac{(a_{11} - a_{22})^2}{4(a_{12} a_{21})} \right) \tag{6.1.25}$$

Then

$$y_1(t) = c_1 \exp \left[\frac{a_{11} + a_{22}}{2} \right] t$$

$$\tag{6.1.26}$$

$$y_2(t) = c_2 \exp \left[\frac{a_{11} + a_{22}}{2} \right] t$$

is a nonoscillatory solution of (6.1.1), whenever (c_1, c_2) is a nonzero solution of the following algebraic system:

$$\alpha \frac{a_{22} - a_{11}}{2} - \beta a_{12} \exp \left(- \frac{(a_{11} + a_{22}) \tau_1}{2} \right) = 0$$

$$\tag{6.1.27}$$

$$\alpha a_{21} \exp \left(- \frac{(a_{11} + a_{22}) \tau_2}{2} \right) - \beta \frac{a_{11} - a_{22}}{2} = 0$$

Proof: From (6.1.25), we conclude that

$$\frac{(a_{11} - a_{22})^2}{4} + a_{21} a_{12} \exp \left[-(a_{11} + a_{22}) \frac{\tau_1 + \tau_2}{2} \right] = 0$$

This implies that (6.1.27) has a nonzero solution, say, (c_1, c_2).
    To prove the function (6.1.26) is a solution of (6.1.1), we have

$$\frac{dy_1(t)}{dt} = c_1 \left(\frac{a_{11} + a_{22}}{2} \right) \exp \left(\frac{a_{11} + a_{22}}{2} \right) t$$

From (6.1.27), it is easy to see that

$$\frac{dy_1(t)}{dt} = a_{11} c_1 \exp \left(\frac{a_{11} + a_{22}}{2} \right) t + a_{12} c_2 \exp \left(\frac{a_{11} + a_{22}}{2} \right) (t - \tau_1)$$

$$= a_{11} y_1(t) + a_{12} y(t - \tau_1)$$

Similarly,

$$\frac{dy_2(t)}{dt} = c_2\left(\frac{a_{11} + a_{22}}{2}\right) \exp\left(\frac{a_{11} + a_{22}}{2}\right) t$$

$$= a_{21} c_1 \exp\left(\frac{a_{11} + a_{22}}{2}\right)(t - \tau_2) + a_{22} c_2 \exp\left(\frac{a_{11} + a_{22}}{2}\right) t$$

$$= a_{21} y_1(t - \tau_1) + a_{22} y_2(t)$$

The theorem is proved.

EXAMPLE 6.1.3 We consider

$$y_1'(t) = 2y_1(t) - y_2\left(t - \frac{2}{3} \ln 2\right)$$

$$y_2'(t) = y_1\left(t - \frac{2}{3} \ln 2\right) + y_2(t)$$

(6.1.28)

This has nonoscillatory solution $y_1(t) = \exp\left(\frac{3}{2} t\right)$, $y_2(t) = \exp\left(\frac{3}{2} t\right)$, but the corresponding system without delay,

$$y_1'(t) = 2y_1(t) - y_2(t)$$

$$y_2'(t) = y_1(t) + y_2(t)$$

(6.1.29)

is oscillatory.

REMARK 6.1.5 From (6.1.25), we note that the system (6.1.2) is oscillatory. This, together with Theorem 6.1.4, implies that the nonoscillations are generated by delay.

REMARK 6.1.6 We found that the oscillation of (6.1.1) depends on the total delay $\tau = \tau_1 + \tau_2$ only; it is independent of interchange of τ_1 and τ_2. This follows from Lemma 6.1.1. Hence, the mixed type system

$$y_1'(t) = a_{11} y_1(t) + a_{12} y_2(t - \tau_1)$$

$$y_2'(t) = a_{21} y_1(t + \tau_1) + a_{22} y_2(t)$$
$\quad \tau_i > 0, \quad i = 1, 2 \qquad$ (6.1.30)

if $\tau_1 - \tau_2 > 0$ is a delay system essentially according to Lemma 6.1.1. The results of this section remain valid for (6.1.30).

REMARK 6.1.7 We can use similar arguments to study the system with advanced arguments

$$y_1'(t) = a_{11} y_1(t) + a_{12} y_2(t + \tau_1)$$

$$y_2'(t) = a_{21} y_1(t + \tau_2) + a_{22} y_2(t)$$

(6.1.31)

where $\tau_i > 0$, $i = 1, 2$.

We consider the case of variable coefficients in (6.1.1),

$$y_1'(t) = a_{11}(t)y_1(t) + a_{12}(t)y_2(t - \tau_1)$$
$$y_2'(t) = a_{21}(t)y_1(t - \tau_2) + a_{22}(t)y_2(t) \tag{6.1.32}$$

where $a_{ij}(t)$ is continuous and $a_{ij}(t)$, $i \neq j$, is absolutely continuous, $i, j = 1, 2$.

Let

$$y_1(t) = u(t) \exp\left(\int_{t_0}^{t} a_{11}(s)\ ds\right)$$

$$y_2(t) = v(t) \exp\left(\int_{t_0}^{t} a_{22}(s)\ ds\right) \tag{6.1.33}$$

Under this transformation system (6.1.33) is transformed into

$$u'(t) = \exp\left[-\int_{t_0}^{t} a_{11}(s)\ ds\right] a_{12} \exp\left[\int_{t_0}^{t-\tau_1} a_{22}(s)\ ds\right] v(t - \tau_1)$$

$$v'(t) = \exp\left[-\int_{t_0}^{t} a_{22}(s)\ ds\right] a_{21} \exp\left[\int_{t_0}^{t-\tau_1} a_{11}(s)\ ds\right] u(t - \tau_2) \tag{6.1.34}$$

By defining

$$a_{ij}(t) = a_{ij}(t_0), \quad i, j = 1, 2, \quad t < t_0$$

system (6.1.34) can reduce a single second order differential equation with retarded arguments:

$$u''(t) = p_1(t)u'(t) + p_2(t)u(t - \tau_1 - \tau_2) \tag{6.1.35}$$

where

$$p_1(t) = -a_{11}(t) + \frac{a_{12}'(t)}{a_{12}(t)} + a_{22}(t - \tau_1)$$

$$p_2(t) = a_{12}(t)a_{21}(t - \tau_1) \exp\left[-\int_{t-\tau_1-\tau_2}^{t} a_{11}(s)\ ds\right] \tag{6.1.36}$$

or

$$v''(t) = R_1(t)v'(t) + R_2(t)v(t - \tau_1 - \tau_2) \tag{6.1.37}$$

where

$$R_1(t) = -a_{22}(t) + \frac{a_{21}'(t)}{a_{21}(t)} + a_{11}(t - \tau_2)$$

$$R_2(t) = a_{21}(t)a_{12}(t - \tau_2) \exp\left[-\int_{t-\tau_1-\tau_2}^{t} a_{22}(s)\,ds\right] \tag{6.1.38}$$

Of course, we assume that $a_{ij}(t) \neq 0$ if $i \neq j$. By the following transformation,

$$u(t) = w(t) \exp\left[\frac{1}{2}\int_{t_0}^{t} p_1(s)\,ds\right]$$

$$v(t) = z(t) \exp\left[\frac{1}{2}\int_{t_0}^{t} R_1(s)\,ds\right] \tag{6.1.39}$$

Eqs. (6.1.35) and (6.1.36) can be transformed into

$$w''(t) + \left[\frac{p_1'(t)}{2} - \frac{p_2'(t)}{2}\right]w(t) - p_2(t) \exp\left[\frac{1}{2}\int_{t_0}^{t-\tau_1-\tau_2} p_1(s)\,ds\right]w(t - \tau_1 - \tau_2) = 0 \tag{6.1.40}$$

Similarly, equations (6.1.37) and (6.1.38) are transformed into

$$z''(t) + \left[\frac{R_1'(t)}{2} - \frac{R_2'(t)}{4}\right]z(t) - R_2(t) \exp\left[\frac{1}{2}\int_{t_0}^{t-\tau_1-\tau_2} R_1(s)\,ds\right]z(t - \tau_1 - \tau_2) = 0 \tag{6.1.41}$$

Equations of type (6.1.40) and (6.1.41) were studied by Norkin [197]. It is easy to see that transformations (6.1.33) and (6.1.39) do not change the oscillation of the system (6.1.32).

The study of oscillation of equations of the form

$$w''(t) + N(t)w(t) + M(t)y(t - \tau) = 0 \tag{6.1.42}$$

is far from being complete. For example, when $N(t)$ and $M(t)$ admit sign change, we have not seen any results about the oscillation of equation (6.1.42).

Now we prefer using another transformation to reduce equation (6.1.35) and (6.1.37).

Let

$$r(t) = \exp\left(-\int^{t} p_1(s)\,ds\right) \tag{6.1.43}$$

Then equation (6.1.35) is transformed into

$$(r(t)u'(t))' = p_2(t)r(t)u(t - \tau_1 - \tau_2) \tag{6.1.44}$$

Similarly, let

$$q(t) = \exp\left(-\int^{t} R_1(s)\,ds\right) \tag{6.1.45}$$

Then equation (6.1.37) is transformed into

$$[q(t)v'(t)]' = R_2(t)q(t)v(t - \tau_1 - \tau_2) \tag{6.1.46}$$

First we present a theorem regarding the system (6.1.32).

THEOREM 6.1.5 Assume that $a_{ij}(t)$, $i \neq j$, have a constant sign for sufficiently large $t \geq 0$; then the components $y_1(t)$ and $y_2(t)$ in system (6.1.32) have the same oscillatory behavior.

Proof: In fact, let $y_2(t)$ be nonoscillatory. From the first equation of system (6.1.32), we have

$$y_1(t) = \exp\left[\int_{t_0}^{t} a_{11}(s)\,ds\right]\left[c + \int_{t_0}^{t} \exp\left[-\int_{t_0}^{s} a_{11}(s_1)\,ds_1\right] a_{12}(s)y_2(s - \tau_1)\,ds\right]$$

and it follows that $y_1(t)$ has same sign if t is sufficiently large. Thus $y_1(t)$ also is nonoscillatory. Similarly, if $y_1(t)$ is nonoscillatory then $y_2(t)$ is too. From Theorem 6.1.5, we can use any one of (6.1.35) ((6.1.44)) and (6.1.37) ((6.1.46)) to discuss the oscillation of the system (6.1.32).

For example, for equation (6.1.44) with $p_i(t) > 0$, we can use the results in Section 4.1 especially, to obtain the following results.

THEOREM 6.1.6 Assume that

(i) $r(t) > 0$ is nondecreasing and $R(t) = \int_{t_0}^{t} \dfrac{ds}{r(s)} \to \infty$ as $t \to \infty$.

(ii) $p_2(t) > 0$ and $\int_{t_0}^{\infty} R(t)p_2(t)r(t)\,dt = \infty$.

(iii) $\limsup_{t\to\infty}\left[\dfrac{1}{r(t - \tau)}\int_{t-\tau}^{t} p_2(s)r(s)(t - s)\,ds\right] > 1$.

Then system (6.1.32) has oscillatory solutions.

For equation (6.1.44) with $p_2(t) < 0$ we can use a theorem in [309] and obtain the following oscillatory result with respect to (6.1.32).

THEOREM 6.1.7 Assume that $r(t) \geq 0$, $R(t) = \int_{t_0}^{t} ds/r(s) \to \infty$ as $t \to \infty$, $p_2(t) < 0$, and

$$\int_{t_2}^{t} p_2(s)r(s)\left[c_1 - c_2\int_{t_1}^{s-\tau} R(s_1)r(s_1)p_2(s_1)\,ds_1\right]ds \to -\infty$$

as $t \to \infty$, $t - \tau \geq t_1$ for $t \geq t_2$, where c_1 and $c_2 > 0$ are arbitrary constants. Then every solution of (6.1.32) oscillates.

Proof: From the theorem in [309], under the conditions of Theorem 6.1.7 the equation (6.1.44) is oscillatory because of the fact that the transformation (6.1.33) preserves oscillation. The proof is completed.

EXAMPLE 6.1.4 We consider

$$y_1'(t) = \sin t \, y_1(t) + y_2\left(t - \frac{\pi}{2}\right)$$

$$y_2'(t) = -y_1\left(t - \frac{\pi}{2}\right) + \sin t \, y_2(t)$$

$$(6.1.47)$$

(6.1.47) satisfies the conditions of Theorem 6.1.7; therefore every solution is oscillatory.

6.2 NONLINEAR SYSTEMS OF SPECIAL FORM

We consider

$$y_1'(t) = a_{11}(t)y_1(t) + a_{12}(t)y_2(t)$$

$$y_2'(t) = f(t, y_1(g(t))) + a_{22}(t)y_2(t)$$

$$(6.2.1)$$

where $a_{ij}(t)$ and $g(t)$ are continuous on R_+, $f(t,y) \in C[R_+ \times R, R]$, $\lim_{t \to \infty} g(t) = \infty$, and $\sup_{t \geq T} |f(t, y)| > 0$ for any $T \geq 0$ and $y \neq 0$. As in Chapter 1, hereafter the term "solution" will be interpreted to mean a solution $(y_1(t), y_2(t))$ of (6.2.1) which exists at some ray $[T_0, \infty)$ and satisfies

$$\sup \{|y_1(t)| + |y_2(t)| : t \geq T\} > 0 \quad \text{for any} \quad T \geq T_0$$

DEFINITION 6.2.1 A solution (6.2.1) is said to be <u>oscillatory</u> (resp. <u>weakly oscillatory</u>) if each component (resp. at least one component) has arbitrarily large zeros. A solution of (6.2.1) is said to be <u>nonoscillatory</u> (resp. <u>weakly nonoscillatory</u>) if each component (resp. at least one component) is eventually of constant sign.

LEMMA 6.2.1 Let $(y_1(t), y_2(t))$ be a solution of (6.2.1); then

$$y_1(t) = u(t) \exp \int_{t_0}^{t} a_{11}(s) \, ds$$

$$(6.2.2)$$

$$y_2(t) = v(t) \exp \int_{t_0}^{t} a_{22}(s) \, ds$$

where $(u(t), v(t))$ satisfies the system

$$u'(t) = a_{12}(t) \exp \int_{t_0}^{t} (a_{22}(s) - a_{11}(s))\, ds\; v(t)$$

$$v'(t) = \exp\left(-\int_{t_0}^{t} a_{22}(s)\, ds\right) f\left(t,\; u(g(t)) \exp\left(\int_{t_0}^{g(t)} a_{11}(s)\, ds\right)\right) \tag{6.2.3}$$

Proof: The conclusion of Lemma 6.2.1 can be obtained directly.

From Lemma 6.2.1, without loss of generality, we consider the system of the form

$$y_1'(t) = p(t)y_2(t)$$

$$y_2'(t) = -f(t,\; y_1(g(t))) \tag{6.2.4}$$

It is convenient to distinguish the two cases

$$\int_a^{\infty} p(t)\, dt = \infty \quad \text{and} \quad \int_a^{\infty} p(t)\, dt < \infty$$

which will be examined separately.

We consider only the case $\int_a^{\infty} p(t)\, dt = \infty$ here. Set

$$P(t) = \int_a^{t} p(s)\, ds, \quad g_*(t) = \min(g(t), t)$$

LEMMA 6.2.2 Assume that

(i) $p \in C[R_+, R]$ and $p(t)$ are not identically zero on any infinite subinterval of $[a, \infty)$.

(ii) $g \in C[R_+, R]$ and $\lim_{t \to \infty} g(t) = \infty$.

(iii) $f \in C(R_+ \times R, R]$, $yf(t, y) \geq 0$ and $\sup_{t \geq T} |f(t, y)| > 0$ for any $T > a$ and $y \neq 0$.

Let $(y_1(t), y_2(t))$ be a weakly nonoscillatory solution of (6.2.4). Then it is nonoscillatory and there exist constants $T > a$, $k_1 > 0$, and $k_2 > 0$ such that

$$y_1(t)y_2(t) > 0 \quad \text{for } t \geq T \tag{6.2.5}$$

$$k_1 P(t)|y_2(t)| \leq |y_1(t)| \leq k_2 P(t) \quad \text{for } t \geq T \tag{6.2.6}$$

Proof: If $y_1(t)$ has arbitrarily large zeros, then so does $y_2(t)$. Assuming the contrary, the first equation of (6.2.4) would imply that $y_1(t)$ is a monotone function and this contradicts the fact that $y_1(t)$ is oscillatory. It follows that for a weakly nonoscillatory solution $(y_1(t), y_2(t))$ the first component

is eventually of constant sign. We may suppose that $y_1(t)$ is eventually positive, since a similar argument holds if $y_1(t)$ is eventually negative. There
exists $T > a$ such that $y_1(g_*(t)) > 0$ for $t \geq T$. From the second equation of
(6.2.4) we find that $y_2(t)$ is decreasing on $[T, \infty)$. If $y_2(t_0) < 0$ for some
$t_0 > T$, then $y_2(t) \leq y_2(t_0) < 0$ for $t \geq t_0$. Taking this into account and integrating the first equation of (6.2.4), we have

$$y_1(t) \leq y_1(t_0) + y_2(t_0) \int_{t_0}^{t} p(s)\, ds \qquad t \geq t_0$$

which implies that $y_1(t) \to -\infty$ as $t \to \infty$. This is a contradiction. Therefore,
we must have $y_1(t) > 0$ for $t \geq T$. Integrating again the first equation of
(6.2.4) and using the decreasing nature of $y_2(t)$, we obtain

$$y_2(t) \int_{T}^{t} p(s)\, ds \leq y_1(t) \leq y_1(T) + y_2(T) \int_{T}^{t} p(s)\, ds \qquad t \geq T$$

from which (6.2.6) follows immediately. This completes the proof.

Necessary conditions for the oscillation of all solutions of (6.2.4) will
be derived from the following theorems, which give conditions guaranteeing
the existence of two extreme kinds of nonoscillatory solution of (6.2.4).

THEOREM 6.2.1   Assume that conditions (i)–(iii) of Lemma 6.2.2 are
satisfied. Let f be either superlinear or sublinear, and

$$\int^{\infty} |f(t, kP(g(t)))|\, dt < \infty \qquad \text{for some } k \neq 0 \tag{6.2.7}$$

Then (6.2.4) has a nonoscillatory solution $(y_1(t), y_2(t))$ with the properties

$$\lim_{t \to \infty} \frac{y_1(t)}{P(t)} = k \qquad \lim_{t \to \infty} y_2(t) = k \tag{6.2.8}$$

Proof: We give a proof for the case where f is sublinear and $k > 0$. The
remaining cases can be treated similarly.
   Take $T > a$ so large that

$$\int_{T}^{\infty} f(t, kP(g(t)))\, dt \leq \frac{k}{2} \qquad \text{and} \qquad T_0 = \inf_{t \geq T} g_*(t) > a$$

Let C_D denote the linear space of all continuous vector functions $\xi(t) =
(y_1(t), y_2(t))$ on $[T_0, \infty)$ such that

$$\|\xi\| = \max\left\{ \sup_{t \geq T_0} P^{-2}(t)|y_1(t)|, \ \sup_{t \geq T_0} |y_2(t)| \right\} < \infty \tag{6.2.9}$$

It is clear that C_p becomes a Banach space under the norm defined by (6.2.9). Define a set F by

$$F = \{(y_1, y_2) \in C_p: kP(t) \le y_1(t) \le 2kP(t),\ k \le y_2(t) \le 2k,\ t \ge T_0\}$$

Obviously, F is a bounded, convex, and closed subset of C_p.

Let Φ designate the operator which assigns to every element $\xi = (y_1, y_2)$ of F a vector function $\Phi\xi = (\Phi y_1,\ \Phi y_2)$ defined by

$$(\Phi y_1)(t) = y(T_0)\int_a^{T_0} p(s)\ ds + \int_{T_0}^t p(s)y_2(s)\ ds \qquad t \ge T_0$$

$$(\Phi y_2)(t) = \begin{cases} k + \displaystyle\int_t^\infty f(s,\ y_2(g(s)))\ ds & t \ge T \\[2em] k + \displaystyle\int_T^\infty f(s,\ y_2(g(s)))\ ds & T_0 \le t \le T \end{cases}$$

(i) Φ maps F into F.

The following inequalities are obvious:

$$kP(t) \le (\Phi y_1)(t) \le 2kP(t) \qquad \text{for } t \ge T_0$$

$$(\Phi y_2)(t) \ge k \qquad\qquad\qquad \text{for } t \ge T_0$$

Using the sublinearity of f, we see that

$$(\Phi y_2)(t) \le k + \int_T^\infty y_1(g(s))f(s,\ y_2(g(s)))/y_1(g(s))\ ds$$

$$\le k + \int_T^\infty 2kP(g(s)) \cdot f(s,\ kP(g(s)))/kP(g(s))\ ds$$

$$= k + 2 \int_T^\infty f(s,\ kP(g(s)))\ ds \le 2k \qquad t \ge T_0$$

(ii) Φ is continuous.

Let $\xi_n = (y_{1n}, y_{2n})$ be a sequence of elements of F converging to an element $\xi = (y_1, y_2)$ of F: $\lim_{n\to\infty} \|\xi_n - \xi\| = 0$. It is easy to verify that for $t \ge T_0$,

$$P^{-2}(t)\,|(\Phi y_{1n})(t) - (\Phi y_1)(t)| \le P^{-1}(T_0)\sup_{s \ge T_0}|y_{2n}(s) - y_2(s)| \qquad (6.2.10)$$

and

$$| (\Phi y_{2n})(t) - (\Phi y_2)(t) | \leq \int_T^\infty F_n(s)\, ds \qquad (6.2.11)$$

where $F_n(s) = | f(s, y_{1n}(g(s))) - f(s, y_1(g(s))) |$. Evidently, the right-hand side of (6.2.10) tends to zero as $n \to \infty$. Since $F_n(s) \leq 4f(s, kP(g(s)))$ and $F_n(s) \to 0$ as $n \to \infty$ for $s \geq T_0$, the Lebesgue dominated convergence theorem implies that the right-hand side of (6.2.11) tends to zero as $n \to \infty$. It follows that $\lim_{n \to \infty} \| \Phi \xi_n - \Phi \xi \| = 0$.

(iii) ΦF is precompact.

By a theorem of Levitan it is sufficient to show that when (y_1, y_2) ranges over F, the family of functions $\{ P^{-2} \Phi y_1 \}$ and $\{ \Phi y_2 \}$ are uniformly bounded and equicontinuous on $[T_0, \infty)$. Since the uniform boundedness is clear, we need only to demonstrate the equicontinuity. This will be done if it is shown that, for any given $\epsilon > 0$, the interval $[T_0, \infty)$ can be decomposed into a finite number of subintervals in such a way that on each subinterval all functions of these families have oscillations less than ϵ.

Let $(y_1, y_2) \in F$. Then, we have for $t_2 > t_1 \geq T$,

$$| (P^{-2}\Phi y_1)(t_2) - (P^{-2}\Phi y_1)(t_1) | \leq P^{-2}(t_2) \int_T^{t_2} p(s) y_2(s)\, ds + P^{-2}(t_1) \int_T^{t_1} p(s) y_2(s)\, ds$$

$$\leq 4kP^{-1}(t_1)$$

$$| (\Phi y_2)(t_2) - (\Phi y_2)(t_1) | \leq \int_{t_2}^{t_1} f(s, y_1(g(s)))\, ds \leq 2 \int_{t_1}^\infty f(s, kP(g(s)))\, ds$$

Therefore, for any given $\epsilon > 0$, there exists $T^* > T$ such that

$$| (P^{-2}\Phi y_1)(t_2) - (P^{-2}\Phi y_1)(t_1) | < \epsilon$$

$$| (\Phi y_2)(t_2) - (\Phi y_2)(t_1) | < \epsilon \qquad (6.2.12)$$

provided that $t_3 > t_1 \geq T^*$. Now, if $T_0 \leq t_1 < t_2 \leq T^*$, we have

$$| (P^{-2}\Phi y_1)(t_2) - (P^{-2}\Phi y_1)(t_1) | \leq 2kP^{-2}(T_0) \int_{t_1}^{t_2} p(s)\, ds + 2k|P^{-2}(t_2)$$

$$- P^{-2}(t_1)| \int_T^{T^*} p(s)\, ds$$

$$| (\Phi y_2)(t_2) - (\Phi y_1)(t_1) | \leq 2 \int_{t_1}^{t_2} f(s, kP(g(s)))\, ds$$

The above inequalities ensure that there exists a $\delta = \delta(\epsilon) > 0$ such that (6.2.12) holds for any t_1, $t_2 \in [T_0, T^*]$ with $0 < t_2 - t_1 < \delta$. Consequently the interval $[T_0, \infty)$ admits the required decomposition.

We now apply the Schauder fixed point theorem to the operator Φ, concluding that Φ has a fixed point $\xi = (y_1, y_2) \in F$. It is easily checked that this fixed point provides a solution of the system (6.2.4) with the asymptotic property (6.2.8). The proof is complete.

THEOREM 6.2.2 Assume that conditions (i)-(iii) of Lemma 6.2.2 are satisfied. Let f be either superlinear or sublinear and

$$\int_{t_0}^{\infty} p(t) \,|\, f(t,k) \,|\, dt < \infty \qquad \text{for some } k \neq 0 \tag{6.2.13}$$

Then (6.2.4) has a nonoscillatory solution $(y_1(t), y_2(t))$ with the properties

$$\lim_{t \to \infty} y_1(t) = k, \qquad \lim_{t \to \infty} y_2(t) = 0 \tag{6.2.14}$$

Proof: The proof is similar to the proof of Theorem 6.2.1, as long as an operator Φ is defined which assigns to every $\xi = (y_1, y_2) \in F$ a vector function $\Phi\xi = (\Phi y_1, \Phi y_2)$ given by

$$(\Phi y_1)(t) = \begin{cases} k - \displaystyle\int_t^{\infty} p(s)y_2(s)\, ds, & t \geq T \\[2em] k - \displaystyle\int_T^{\infty} p(s)y_2(s)\, ds, & T_0 \leq t \leq T \end{cases}$$

$$(\Phi y_2)(t) = \begin{cases} \displaystyle\int_t^{\infty} f(s, y_1(g(s)))\, ds, & t \geq T \\[2em] \displaystyle\int_T^{\infty} f(s, y_1(g(s)))\, ds, & T_0 \leq t \leq T \end{cases}$$

We now present sufficient conditions for the oscillation of all solutions of (6.2.4) by limiting ourselves to the strongly superlinear and strongly sublinear cases.

THEOREM 6.2.3 Assume that conditions (i)-(iii) of Lemma 6.2.1 hold, and f is strongly superlinear. Suppose there exists a differentiable function $h_*(t)$ on $[a, \infty)$ such that

$$0 \leq h_*(t) \leq g_*(t), \qquad h_*'(t) \geq 0, \qquad \lim_{t \to \infty} h_*(t) = \infty \tag{6.2.15}$$

and

$$\int_{t_0}^{\infty} P(h_*(t))\,|f(t,k)|\,dt = \infty \qquad \text{for all } k \neq 0 \tag{6.2.16}$$

Then all solutions of (6.2.4) are oscillatory.

Proof: Otherwise there exists a weakly nonoscillatory solution $(y_1(t), y_2(t))$. By Lemma 6.2.2, $(y_1(t), y_2(t))$ is nonoscillatory, and without loss of generality we may suppose that $y_1(t) > 0$ and $y_2(t) > 0$ eventually, say for $t \geq T > a$. Take $t_0 \geq T$ so that $h_*(t) \geq T$ for $t \geq t_0$. Let $\alpha > 1$ be the superlinearity constant of (6.2.4). We compute

$$[P(h_*(t))y_2(t)y_1^{-\alpha}(h_*(t))]'$$

$$\leq h_*'(t)p(h_*(t))y_2(t)y_1^{-\alpha}(h_*(t)) - P(h_*(t))f(t,\; y_1(g(t)))y_1^{-\alpha}(h_*(t)))$$

$$\leq h_*'(t)p(h_*(t))y_2(h_*(t))y_1^{-\alpha}(h_*(t)) - P(h_*(t))f(t,\; y_1(g(t)))y_1^{-\alpha}(g(t))$$

$$\leq h_*'(t)y_1'(h_*(t))y_1^{-\alpha}(h_*(t)) - P(h_*(t))f(t,k)k^{-\alpha} \tag{6.2.17}$$

for $t \geq t_0$, where $k = y_1(T) > 0$. In deriving (6.2.17) we have used the equations in (6.2.4), the strong superlinearity, and the fact that $y_1(t)$ is increasing and $y_2(t)$ is decreasing. Integrating (6.2.17), we obtain

$$k^{-\alpha}\int_{t_0}^{t} P(h_*(s))f(s,k)\,ds < \frac{y_1^{1-\alpha}(h_*(t_0))}{\alpha - 1} + P(h_*(t_0))y_2(t_0)y_1^{-\alpha}(h_*(t_0))$$

which gives a contradiction to (6.2.16) in the limit as $t \to \infty$.

THEOREM 6.2.4 Let (6.2.4) be strongly sublinear with sublinearity constant $0 < \beta < 1$, and assume conditions (i)-(iii) of Lemma 6.2.1 are satisfied. Further assume

$$\int^{\infty} \left[\frac{P(g_*(t))}{P(g(t))}\right]^{\beta} |f(t,\; kP(g(t)))|\,dt = \infty \tag{6.2.18}$$

for all $k \neq 0$. Then all solutions of (6.2.4) are oscillatory.

Proof: Let $(y_1(t), y_2(t))$ be a weakly nonoscillatory solution of (6.2.4). We may suppose that $y_1(t) > 0$ and $y_2(t) > 0$ for $t \geq T$. By Lemma 6.2.2 there exist positive constants k_1 and k_2 such that

$$k_1 P(t)y_2(t) \leq y_1(t) \leq k_2 P(t) \qquad \text{for } t \geq T \tag{6.2.19}$$

Using the first inequality of (6.2.19) and the monotone nature of $y_1(t)$ and $y_2(t)$, we have

$$y_2^{-\beta}(t) \geq y_2^{-\beta}(g_*(t)) \geq k_1^\beta P^\beta(g_*(t))y_1^{-\beta}(g_*(t))$$

$$\geq k_1^\beta P^\beta(g_*(t))y_1^{-\beta}(g(t)) , \qquad t \geq t_0 \tag{6.2.20}$$

where $t_0 \geq T$ is chosen so that $g_*(t) \geq T$ for $t \geq t_0$. From (6.2.20), the second inequality of (6.2.19), and the strong sublinearity of f we see that

$$-(y_2^{1-\beta}(t))' = (1 - \beta)y_2^{-\beta}(t)f(t, y_1(g(t)))$$

$$\geq (1 - \beta)k_1^\beta P^\beta(g_*(t))y_1^{-\beta}(g(t))f(t, y_1(g(t)))$$

$$\geq (1 - \beta)\left(\frac{k_1}{k_2}\right)^\beta \left[\frac{P(g_*(t))}{P(g(t))}\right]^\beta f(t, k_2 P(g(t))) \tag{6.2.21}$$

for $t \geq t_0$. An integration of (6.2.21) yields a contradiction to (6.2.18). This completes the proof.

Combining the foregoing results we are able to obtain necessary and sufficient conditions for the oscillation of all solutions of certain classes of nonlinear differential systems of the form (6.2.4).

THEOREM 6.2.5 Let (6.2.4) be strongly superlinear and hypotheses (i)–(iii) of Lemma 6.2.1 be satisfied. Suppose there is a differentiable function $h_*(t)$ on $[a,\infty)$ satisfying (6.2.15) and

$$\liminf_{t \to \infty} \frac{P(h_*(t))}{P(t)} > 0 \tag{6.2.22}$$

Then all solutions of (6.2.4) are oscillatory if and only if

$$\int^{\infty} P(t)|f(t,k)| \, dt = \infty \qquad \text{for all } k \neq 0 \tag{6.2.23}$$

Proof: The result follows from Theorems 6.2.2 and 6.2.3.

THEOREM 6.2.6 Let (6.2.4) be strongly sublinear. Suppose that

$$\liminf_{t \to \infty} \frac{P(g_*(t))}{P(g(t))} > 0 \tag{6.2.24}$$

Then all solutions of (6.2.4) are oscillatory if and only if

$$\int^{\infty} |f(t, kP(g(t)))| \, dt = \infty \tag{6.2.25}$$

for all $k \neq 0$.

Proof: The result follows from Theorems 6.2.1 and 6.2.4.

REMARK 6.2.1 It is easy to see that (6.2.22) holds if $g(t) \geq t$, and (6.2.24) holds if $g(t) \leq t$. Thus, we have characterized the oscillation situation for the strongly sublinear retarded system (6.2.4) as well as for the strongly superlinear advanced system (6.2.4).

EXAMPLE 6.2.1 Consider the system

$$y_1'(t) = (1 + \cos(\ln t))y(t)$$

$$y_2'(t) = -\frac{y_1^3(\ln t)}{t^\nu(\ln(\ln t) + \sin(\ln(\ln t)))^3}$$

(6.2.26)

where ν is a constant. Here we can take $P(t) = t$ and $h_*(t) = \ln t$. If $\nu \leq 1$, then (6.2.16) is satisfied, so that all solutions of (6.2.26) are oscillatory, by Theorem 6.2.3. If $\nu > 2$, then (6.2.9) holds for every $k \neq 0$, and so on account of Theorem 6.2.2, Eq. (6.2.26) has a nonoscillatory solution $(y_1(t), y_2(t))$ such that $\lim_{t \to \infty} y_1(t) = k$, $\lim_{t \to \infty} y_2(t) = 0$ for every $k \neq 0$. Our results do not apply to the case $1 < \nu < 2$. Notice that for $\nu = 2$, (6.2.23) is satisfied but (6.2.22) is violated, and that (6.2.26) has a nonoscillatory solution $(y_1(t), y_2(t))$ with

$$y_1(t) = \ln t + \sin(\ln t), \qquad y_2(t) = \frac{1}{t}$$

This observation tells us that there is, in general, a gap between the sufficient condition (6.2.16) and the necessary condition (6.2.23) for the oscillation of all solutions of the strongly superlinear system (6.2.4).

In [118], the author obtained several results with respect to the oscillation of (6.2.4) for the case when $\int^\infty p(t)\,dt < \infty$. The interested reader can refer to that paper.

We now consider a system of the form

$$y_i'(t) = f_i(t, y_1(\tau_1(t)), y_2(\tau_1(t)), y_1(\tau_2(t)), y_2(\tau_2(t)), \qquad t \geq t_0, \quad i = 1, 2 \quad (6.2.27)$$

THEOREM 6.2.7 Assume that

(i) $f_i(t, u_1, u_2, v_1, v_2)$ are continuous in all arguments, and

$$f_1(t, u_1, u_2, v_1, v_2)\, \text{sgn}\, v_2 \geq a_1(t)|v_2|^{\alpha_1}$$

$$f_2(t, u_1, u_2, v_1, v_2)\, \text{sgn}\, u_1 \leq -a_2(t)|u_1|^{\alpha_2}$$

(6.2.28)

where $a_i(t) \geq 0$ are continuous and α_i is positive, $i = 1, 2$.

(ii) $\tau_k(t) \in C[R_+, R]$, $\tau_k(t) \leq t$,   and $\lim_{t \to \infty} \tau_k(t) = \infty$, $k = 1, 2$

Further assume that $\tau_1(t)$ is an absolutely continuous nondecreasing function, and

$$\int_{t_0}^{\infty} a_1(t) \, dt = \infty \tag{6.2.29}$$

Set $A(x) = \int_{t_0}^{x} a_1(s) \, ds$, and

$$\int^{\infty} (A[\tau_1(t)])^{\nu} a_2(t) \, dt = \infty \tag{6.2.30}$$

where

$$\nu = \begin{cases} \alpha_2 - \epsilon \, (0 < \epsilon \leq \alpha_2), & \text{for } \alpha_1 \alpha_2 = 1 \\ \alpha_2, & \text{for } \alpha_1 \alpha_2 < 1 \\ \alpha_1, & \text{for } 0 < \alpha_1 \leq 1, \ \alpha_2 > 1 \end{cases} \tag{6.2.31}$$

Then each solution $(y_1(t), y_2(t))$ of system (6.2.27) oscillates in the sense of Definition 6.2.1.

Proof: Otherwise, assume that system (6.2.27) has a solution $(y_1(t), y_2(t))$ that is not oscillatory. By Lemma 6.2.2, from condition (6.2.28) it follows that both the components of this solution are nonoscillatory and $y_1(t)y_2(t) > 0$ for $t \geq T$ if T is a sufficiently large number. For definiteness, let us suppose that $y_1(t) > 0$ and $y_2(t) > 0$ $(t \geq t_1)$. Then it follows from (6.2.27) and (6.2.28) that $y_1(t)$ monotonically increases and $y_2(t)$ monotonically decreases, and consequently, $y_2(\tau_2(t)) > y_2(t)$ for $t \geq t_2 \geq T$. By virtue of the condition (6.2.28) we have

$$y_1'(t) \geq a_1(t)[y_2(t)]^{\alpha_1} \tag{6.2.32}$$

$$y_2'(t) \leq -a_2(t)[y_1(\sigma_1(t))]^{\alpha_2}, \quad t \geq t_2 \tag{6.2.33}$$

Integrating (6.2.32) over $[t_2, t]$, we get

$$y_1(t) \geq \int_{t_2}^{t} a_1(s) \, [y_2(s)]^{\alpha_1} \, ds \quad \text{for } t \geq t_2$$

and for $t \geq t_3 \geq t_2$, we have

$$y_1(\tau_1(t)) \geq \int_{t_2}^{\tau_1(t)} a_1(s)[y_2(s)]^{\alpha_1} \, ds \tag{6.2.34}$$

Suppose that $\alpha_1 \alpha_2 = 1$. Since $y_1(t)$ is nondecreasing, it follows that $y_1(\tau_1(t)) \geq k$ (k is a positive constant). Therefore

$$\frac{y_1(\tau_1(t))}{k} \geq \left[\frac{y_1(\tau_1(t))}{k}\right]^{\beta} \qquad \text{for } \beta \in [0, 1) \tag{6.2.35}$$

Because $\tau_1(t)$ is nondecreasing and $y_2(t)$ is nonincreasing, by virtue of (6.2.34), (6.2.35), and the condition $\alpha_1 \alpha_2 = 1$, we have

$$y_2'(t) \leq -Ma_2(t)\,[y_2(t)]^{\beta}\left[\int_{t_2}^{\tau_1(t)} a_1(s)\,ds\right]^{\alpha_2 \beta}$$

for $t \geq t_2$ (M > 0 is a constant). Dividing both the sides of this inequality by $[y_2(t)]^{\beta}$, integrating, and setting $\beta = \alpha_2 - \epsilon/\alpha_2$ ($0 < \epsilon \leq \alpha_2$), we are led to a contradiction.

For $\alpha_1 \alpha_2 < 1$, from (6.2.33) and (6.2.34) we have

$$y_2'(t) \leq -a_2(t)\left[\int_{t_3}^{\tau_1(t)} a_1(s)\,[y_2(s)]^{\alpha_1}\,ds\right]^{\alpha_2}, \qquad t \geq t_3$$

By virtue of the monotonicity property of $\tau_1(t)$ and $y_2(t)$, we have

$$y_2'(t) \leq -a_2(t)\,[y_2(t)]^{\alpha_1 \alpha_2}\left[\int_{t_3}^{\tau_1(t)} a_1(s)\,ds\right]^{\alpha_2}$$

Dividing both the sides of this inequality by $[y_2(t)]^{\alpha_1 \alpha_2}$ and integrating over $[t_3, t]$, we have

$$\frac{[y_2(t)]^{1-\alpha_1\alpha_2}}{1-\alpha_1\alpha_2} - \frac{[y_2(t_3)]^{1-\alpha_1\alpha_2}}{1-\alpha_1\alpha_2} \leq -\int_{t_3}^{t} a_2(x)\left[\int_{t_3}^{\tau_1(x)} a_1(s)\,ds\right]^{\alpha_2} dx$$

Letting t tend to infinity, we get

$$\int_{t_3}^{\infty} a_2(x)\left[\int_{t_3}^{\tau_1(x)} a_1(s)\,ds\right]^{\alpha_2} dx < \infty$$

This contradicts condition (6.2.30).

For $0 < \alpha_1 \leq 1$ and $\alpha_2 > 1$, as before there is a nonoscillatory solution $y_1(t) > 0$, $y_2(t) > 0$, $t \geq t_1$. We use identity

$$uv\Big|_{t_3}^{t} - \int_{t_3}^{t} v\,du - \int_{t_3}^{t} u\,dv \equiv 0$$

where

$$u(t) = \frac{[y_2(t)]^{\alpha_1}}{[y_1(\tau_1(t))]^{\alpha_2}}, \qquad v(t) = \left[\int_0^{\tau_1(t)} a_1(s)\, ds\right]^{\alpha_1}$$

Therefore, we have

$$c \geq -\alpha_1 \int_{t_3}^t \frac{[y_2(s)]^{\alpha_1-1}\, y_2'(s)}{y_1(\tau_1(s))^{\alpha_2}} \left[\int_0^{\tau_1(s)} a_1(s_1)\, ds_1\right]^{\alpha_1} ds$$

$$- \alpha_1 \int_{t_3}^t \frac{[y_2(s)]^{\alpha_1}}{[y_1(\tau_1(s))]^{\alpha_2}}\, a_1(\tau_1(s))\, \tau'(s) \left(\int_0^{\tau_1(s)} a_1(s_1)\, ds_1\right)^{\alpha_1-1} ds \tag{6.2.36}$$

where $c \geq 0$. From (6.2.36), (6.2.32), and (6.2.33), we have

$$c \geq \alpha_1 \int_{t_3}^t a_2(s)[y_2(s)]^{\alpha_1-1}\left[\int_0^{\tau_1(s)} a_1(s_1)\, ds_1\right]^{\alpha_1} ds$$

$$- \alpha_1 \int_{t_3}^t \frac{y_1'(\tau_1(s))\, \tau_1'(s)}{[y_1(\tau_1(s))]^{\alpha_2}}\left[\int_0^{\tau_1(s)} a_1(s_1)\, ds_1\right]^{\alpha_1-1} ds$$

Because $y_1(t)$ is nondecreasing, it follows that

$$c \geq c_1 \int_{t_3}^t a_2(s)\left[\int_0^{\tau_1(s)} a_1(s_1)\, ds_1\right]^{\alpha_1} ds - c_2[y_1(\tau_1(s))]^{1-\alpha_1}\Big|_{t_3}^{t} \tag{6.2.37}$$

where

$$c_1 = \alpha_1[y_2(t_3)]^{\alpha_1-1}, \qquad c_2 = \frac{\alpha_1}{1-\alpha_2}\left[\int_0^{\tau_1(t_3)} a_1(s)\, ds\right]^{\alpha_1-1} < 0$$

From (6.2.37), we obtain

$$c \geq c_1 \int_{t_3}^t a_2(s)\left[\int_0^{\tau_1(s)} a_1(s_1)\, ds_1\right]^{\alpha_1} ds + c_2[y_1(\tau_1(t_3))]^{1-\alpha_2}$$

Letting $t \to \infty$, we have

$$\int_{t_3}^{\infty} a_2(t)\left[\int_0^{\tau_1(t)} a_1(s)\, ds\right]^{\alpha_1} dt < \infty$$

This is a contradiction. The proof is complete.

REMARK 6.2.1 A particular case of system (6.2.27) is

$$y_1'(t) = a_1(t)y_2(\tau_2(t))$$
$$y_2'(t) = -a_2(t)f(y_1(\tau_1(t)))$$
(6.2.38)

where $a_i \in C[R_+, R_+]$, $f \in C[R, R_+]$, $uf(u) > 0$ for $u \neq 0$, and $f(u)$ is a nondecreasing function.

6.3 LINEAR DELAY SYSTEMS

Consider

$$y_i'(t) = \sum_{j=1}^{n} a_{ij} y_j(t - \tau), \qquad t > 0, \quad i = 1, 2, \ldots, n \tag{6.3.1}$$

where a_{ij} and τ are real constants with $\tau > 0$.

LEMMA 6.3.1 Assume that in (6.3.1) a_{ij} and τ are real constants with $\tau > 0$. Define a solution vector

$$\tilde{y}(t) = (y_1(t), \ldots, y_n(t))^T$$

Then

$$y(t) = \sum_j \tilde{p}_j(t) e^{\lambda_j t}, \qquad t > 0 \tag{6.3.2}$$

where λ_j is a root of $H(\lambda) = 0$ and

$$H(\lambda) = \det [\lambda I - Ae^{-\lambda \tau}] \tag{6.3.3}$$

where I denotes the $n \times n$ identity matrix and A is $n \times n$ coefficient matrix of (6.3.1). The polynomial vector $\tilde{p}_j(t)$ is determined by

$$\tilde{p}_j(t) = \text{residue of } \left\{ e^{\lambda t} \text{ adj} [\lambda I - Ae^{-\lambda \tau}] \left[\tilde{\phi}(0) + A \int_{-\tau}^{0} \tilde{\phi}(s) \, ds \right] \middle/ H(\lambda) \right\}$$

at a root λ_j of $H(\lambda) = 0$, where
(6.3.4)

$$\tilde{\phi}(s) = (\phi_1(s), \ldots, \phi_n(s))^T, \qquad s \in [-\tau, 0]$$

is an initial vector.

Proof: We can rewrite (6.3.1) in vector matrix notation as follows:

$$\tilde{y}'(t) = A\tilde{y}(t - \tau), \qquad t > 0 \tag{6.3.5}$$

If $\tilde{\phi}$ is continuous on $[-\tau, 0]$ then one can show that solution of the initial value problem of (6.3.5) exists on $[0, \infty)$. In fact, we have

$$\tilde{y}(t) = \tilde{\phi}(0) + A \int_0^t \tilde{y}(s - \tau) \, ds, \qquad t > 0$$

$$= \tilde{\phi}(0) + A \int_{-\tau}^{t-\tau} y(\eta) \, d\eta$$

and hence

$$\|\tilde{y}(t)\| \leq \|\tilde{\phi}(0)\| + \|A\| \int_{-\tau}^0 \|\tilde{\phi}(\eta)\| \, d\eta + \|A\| \int_0^t \|\tilde{y}(\eta)\| \, d\eta$$

From this, by Gronwall's inequality it follows that

$$\|\tilde{y}(t)\| \leq \left[\|\tilde{\phi}(0)\| + \|A\| \int_{-\tau}^0 \|\tilde{\phi}(\eta)\| \, d\eta \right] \exp\left[\|A\|t\right]$$

showing that solutions of (6.3.5) are of exponential order. Thus one can use the Laplace transform method for the study of equations of the form (6.3.5).

Let $\tilde{y}(\lambda)$ denote the Laplace transform of a solution vector $\tilde{y}(t)$ of (6.3.5) defined by

$$\tilde{y}(\lambda) = (y_1(\lambda), \ldots, y_n(\lambda))^T$$

$$y_j(\lambda) = \int_0^\infty y_j(t) e^{-\lambda \tau} \, dt \qquad\qquad\qquad (6.3.6)$$

It follows from elementary properties of Laplace transforms that

$$\tilde{y}(\lambda) = [\lambda I - Ae^{-\lambda t}]^{-1} \left[\tilde{\phi}(0) + A \int_{-\tau}^0 \tilde{\phi}(s) \, ds \right]$$

$$= \mathrm{adj}[\lambda I - Ae^{-\lambda t}] \left[\tilde{\phi}(0) + A \int_{-\tau}^0 \tilde{\phi}(s) \, ds \right] \Big/ H(\lambda) \qquad (6.3.7)$$

where $H(\lambda)$ is defined by (6.3.3). By the inversion theorem on Laplace transforms we have from (6.3.5), (6.3.6), and (6.3.3) that any solution of (6.3.5) is given by the integral representation,

$$\tilde{y}(t) = \frac{1}{2\pi i} \int_{\sigma - i\infty}^{\sigma + i\infty} \left[e^{\lambda t} \, \mathrm{adj}[\lambda I - Ae^{-\lambda \tau}] \left[\tilde{\phi}(0) + A \int_{-\tau}^0 \tilde{\phi}(s) \, ds \right] \Big/ H(\lambda) \right] d\lambda$$

$$(6.3.8)$$

where σ is any real number greater than the real parts of roots of $H(\lambda) = 0$. The existence of such a real number σ is well known (Hale [98]). The integral in (6.3.8) can be evaluated using residue calculus, so that

$$\tilde{y}(t) = \sum_j \tilde{p}_j(t) e^{\lambda_j t}, \quad t > 0$$

where the polynomial vector $\tilde{p}_j(t)$ is defined by (6.3.4). The convergence of the series representation of the type in (6.3.2) has been established by Banks and Manitius [9]. The proof is complete.

THEOREM 6.3.1 Suppose that the matrix A of real coefficients a_{ij} (i, j = 1, 2, $\ldots$, n) in (6.3.1) is such that

(i) $\det A \neq 0$

(ii) The eigenvalues α_1, α_2, $\ldots$, α_n (real or complex) satisfy

$$|\alpha_j| \tau e > 1, \quad i = 1, 2, \ldots, n \tag{6.3.9}$$

Then all bounded solutions of (6.3.1) are oscillatory.

Proof: From Lemma 6.3.1, the solutions of (6.3.1) are representable as in (6.3.2). It follows that a necessary and sufficient condition for all bounded solutions of (6.3.1) to be oscillatory is that the characteristic equation $H(\lambda) = 0$ have no real nonpositive root. Since $\alpha_1, \ldots, \alpha_n$ are the eigenvalues of A, we have immediately that

$$\det[\lambda I - A e^{-\lambda \tau}] = 0 \Longleftrightarrow \sum_{j=1}^{n} [\lambda - \alpha_j e^{-\lambda \tau}] = 0 \tag{6.3.10}$$

Thus we are led to an investigation of the nature of the roots of

$$\lambda = \alpha_j e^{-\lambda \tau}, \quad j = 1, 2, \ldots, n \tag{6.3.11}$$

Suppose now that there exists a bounded nonoscillatory solution of (6.3.1). This implies that there exists a real nonpositive root, say λ^*, such that

$$\lambda^* = \alpha_j e^{-\lambda^* \tau} \quad \text{for some } j \in \{1, 2, \ldots, n\} \tag{6.3.12}$$

Since $\det A \neq 0$, then $\alpha_j \neq 0$, j = 1, 2, $\ldots$, n, and hence $\lambda^* \neq 0$; thus $\lambda^* < 0$. It will then follow from (6.3.12) that

$$1 = \tau |\alpha_j| \frac{e^{|\lambda^*| \tau}}{|\lambda^*| \tau} \quad \text{for } j \in \{1, 2, \ldots, n\}$$

$$\geq \tau |\alpha_j| e \quad \text{for some } j \in \{1, 2, \ldots, n\} \tag{6.3.13}$$

But (6.3.13) contradicts (6.3.9) and hence (6.3.1) cannot have a bounded nonoscillatory solution when conditions (i) and (ii) hold, and the proof is complete.

REMARK 6.3.1 Assume that eigenvalues $\alpha_1, \ldots, \alpha_n$ are complex numbers. Then (6.3.10) has no real roots. Therefore every solution of (6.3.1) is oscillatory. In other words, the oscillatory behavior of the linear system

$$\tilde{x}'(t) = A\tilde{x}(t) \tag{6.3.14}$$

is unaffected by the delay in that case.

REMARK 6.3.2 If there exists a positive eigenvalue α_j of A, then (6.3.1) has a bounded nonoscillatory solution. In fact from Theorem 2.3.1 and (6.3.11), we find that (6.3.1) has an unbounded nonoscillatory solution.

REMARK 6.3.3 Suppose that the matrix A in (6.3.1) has at least one real negative eigenvalue, say β, which is such that

$$0 < |\beta| \tau e \leq 1$$

Then (6.3.1) has at least one bounded nonoscillatory solution.

In fact, the portion of the characteristic equations of (6.3.1) corresponding to β is given by

$$\lambda = \beta e^{-\lambda \tau}$$

which is equivalent to $\mu = |\beta| e^{\mu \tau}$, where $\mu = -\lambda$. As in Section 5.4, it is easy to see that there exist positive real numbers μ such that $\mu = |\beta| e^{\mu \tau}$ when $|\beta| \tau e \leq 1$, and corresponding to such μ, we will have a solution of (6.3.1) in the form $p_\mu(t) e^{-\lambda t}$ where $p_\mu(t)$ is a polynomial in t. This solution is nonoscillatory and bounded.

We now consider a linear delay differential system of the form

$$\frac{d\tilde{y}(t)}{dt} = B\tilde{y}(t) + A\tilde{y}(t - \tau), \quad t > 0 \tag{6.3.15}$$

where A and B denote real constant $n \times n$ matrices with elements a_{ij}, b_{ij} (i, j = 1, 2, $\ldots$, n), respectively, and $\tau > 0$ is a constant.

DEFINITION 6.3.1 The <u>measure</u> $\mu(B)$ of the matrix B is defined by

$$\mu(B) = \lim_{\theta \to 0^+} \frac{\|I + \theta B\| - 1}{\theta} \tag{6.3.16}$$

where $\|\cdot\|$ is the norm of the matrix

It is well known, if we adopt the following norms of vectors and matrices:

$$\|\tilde{y}(t)\| = \sum_{i=1}^{n} |y_i(t)|$$

$$\|A\| = \max_{j} \sum_{i=1}^{n} |a_{ij}| , \qquad \|B\| = \max_{j} \sum_{j=1}^{n} |b_{ij}|$$

that the corresponding measure (6.3.16) is

$$\mu(B) = \max_{j} \left[b_{jj} + \sum_{\substack{i=1 \\ i \neq j}}^{n} |b_{ij}| \right]$$

THEOREM 6.3.2 Assume the following for the system (6.3.15):

(i) $\det A \neq 0$

(ii) $\|A\| > |\mu(B)|$

Then all bounded solutions of (6.3.15) corresponding to continuous initial conditions on $[-\tau, 0]$ are oscillatory on $[0, \infty)$.

Proof: Assume that there exists a solution, say $y(t) = (y_1(t), \ldots, y_n(t))^T$, of (6.3.15) which is bounded and nonoscillatory on $[0, \infty)$. It then follows that there exists a $t^* > 0$ such that no component of $y(t)$ has a zero for $t > t^* + \tau$, and as a consequence we have

$$\frac{d|y_i(t)|}{dt} \leq b_{ii}|y_i(t)| + \sum_{\substack{j=1 \\ j \neq i}}^{n} |b_{ij}| |y_j(t)| + \sum_{j=1}^{n} |a_{ij}| |y_j(t - \tau)| \qquad (6.3.17)$$

for $t \geq t^* + 2\tau$, $i = 1, 2, \ldots, n$. Set

$$u(t) \equiv \sum_{i=1}^{n} |y_i(t)|$$

According to the above assumption, $u(t) > 0$ for $t \geq t^* + \tau$, and

$$\frac{du(t)}{dt} \leq \mu(B)u(t) + \|A\|u(t - \tau) , \qquad t \geq t^* + 2\tau \qquad (6.3.18)$$

Consider now the scalar delay differential equation

$$\frac{dv(t)}{dt} = \mu(B)v(t) + \|A\|v(t - \tau) , \qquad t \geq t^* + 2\tau \qquad (6.3.19)$$

with $v(s) = u(s)$, $s \in [t^*, t^* + \tau]$. It is an elementary exercise to show that

$$u(t) \leq v(t) \qquad \text{for } t \geq t^* + 2\tau \tag{6.3.20}$$

We now claim that all bounded solutions of (6.3.19) are oscillatory on $[t^* + 2\tau, \infty)$. We consider the characteristic equation associated with (6.3.19), given by

$$\lambda = \mu(B) + \|A\| e^{-\lambda \tau} \tag{6.3.21}$$

Set

$$F(\lambda) = \lambda - \mu(B) - \|A\| e^{-\lambda \tau}$$

It is obvious that $F(0) = -\mu(B) - \|A\| < 0$, and

$$F'(\lambda) = 1 + \tau \|A\| e^{-\lambda \tau} > 0$$

Therefore $F(\lambda) = 0$ has no negative root, i.e., (6.3.21) has no negative root. This implies that every bounded solution of (6.3.19) is oscillatory. This contradicts the fact $u(t) > 0$. The proof is complete.

REMARK 6.3.4 If $\mu(B) < 0$ in (6.3.15), then (6.3.15) has a bounded oscillatory solution. In fact, let $\lambda = \alpha + i\beta$ in (6.3.21); then

$$\alpha + i\beta = \mu(B) + \|A\| e^{-\alpha \tau}(\cos \beta\tau - i \sin \beta\tau)$$

$$\alpha = \mu(B) + \|A\| e^{-\alpha \tau} \cos \beta\tau \tag{6.3.22}$$

$$\beta = -\|A\| e^{-\alpha \tau} \sin \beta\tau \tag{6.3.23}$$

Combining (6.3.22) and (6.3.23), we have

$$\alpha = \mu(B) - \beta \cot \beta\tau \tag{6.3.24}$$

Taking β sufficiently small and noting $\mu(B) < 0$, then $\alpha < 0$ from (6.3.24). We have proved that (6.3.19) has a bounded oscillatory solution when $\mu(B) < 0$ and $\tau > 0$. By virtue of (6.3.20) we observe that if (6.3.18) has a bounded oscillatory solution, then (6.3.15) has one too.

THEOREM 6.3.3 Let a_{ij}, τ_{ij} ($i, j = 1, 2, \ldots, n$), denote real constants such that $a_{ii} \neq 0$, $\tau_{ii} > 0$, and $\tau_{ij} \geq 0$ if $i \neq j$, and consider the system

$$y_i'(t) = \sum_{j=1}^{n} a_{ij} y_j(t - \tau_{ij}), \qquad t > 0 \tag{6.3.25}$$

If

(i) $\det A \neq 0$

(ii) $|a_{ii}| \tau_{ii} e > 1 + \sum_{\substack{j=1 \\ j \neq i}}^{n} |a_{ij}| \tau_{ij}$

then all bounded solutions of (6.3.25) corresponding to continuous initial conditions defined on $[-\tau, 0]$, $\tau = \max_{1 \le i, j \le n} \tau_{ij}$, are oscillatory on $[0, \infty)$.

Proof: The characteristic equation corresponding to (6.3.25) is given by

$$\det |\lambda I - a_{ij} e^{-\lambda \tau_{ij}}| = 0 \tag{6.3.26}$$

Suppose (6.3.25) has a bounded nonoscillatory solution. Then (6.3.26) has a real nonpositive root, say δ, such that

$$\det [\delta I - a_{ij} e^{-\delta \tau_{ij}}] = 0$$

Since $\det A \neq 0$, then $\delta \neq 0$ and $\delta < 0$ is an eigenvalue of the matrix with entries $a_{ij} \exp(-\delta \tau_{ij})$, $i, j = 1, 2, \ldots, n$. By Gershgorin's theorem, δ satisfies

$$|\delta - a_{ii} e^{-\delta \tau_{ii}}| \le \sum_{\substack{j=1 \\ j \neq i}}^{n} |a_{ij}| e^{-\delta \tau_{ij}} \quad \text{for some } i \in \{1, \ldots, n\}$$

It then follows from

$$|\delta| = |a_{ii} e^{-\delta \tau_{ii}} + \delta - a_{ii} e^{-\delta \tau_{ii}}|$$

$$\ge |a_{ii}| e^{-\delta \tau_{ii}} - |\delta - a_{ii} e^{-\delta \tau_{ii}}|$$

$$\ge |a_{ii}| e^{\tau_{ii}|\delta|} - \sum_{\substack{j=1 \\ j \neq i}}^{n} |a_{ij}| e^{\tau_{ij}|\delta|}$$

that

$$|\delta| + \sum_{\substack{j=1 \\ j \neq i}}^{n} |a_{ij}| e^{\tau_{ij}|\delta|} \ge |a_{ii}| e^{\tau_{ii}|\delta|}$$

Thus

$$\inf_{|\delta| > 0} \left\{ 1 + \sum_{\substack{j=1 \\ j \neq i}}^{n} \tau_{ij} |a_{ij}| \left(\frac{e^{|\delta| \tau_{ij}}}{|\delta| \tau_{ij}} \right) \right\} \ge \inf_{|\delta| > 0} \left\{ |a_{ii}| \tau_{ii} \left(\frac{e^{|\delta| \tau_{ij}}}{|\delta| \tau_{ii}} \right) \right\}$$

which leads to

$$(1 + e \sum_{\substack{j=1 \\ j \neq i}}^{n} \tau_{ij} |a_{ij}|) \geq |a_{ii}| \tau_{ii} e \quad \text{for some } i \in \{1, \ldots, n\} \quad (6.3.27)$$

But (6.3.27) contradicts condition (ii). Thus (6.3.25) cannot have a bounded nonoscillatory solution when the conditions of the theorem hold.

We now consider the system with variable coefficients

$$y_i'(t) = \sum_{j=1}^{n} p_{ij}(t) y_j(t - \tau(t)), \quad t > 0, \text{ for } i \in \{1, 2, \ldots, n\} \quad (6.3.28)$$

THEOREM 6.3.4   Let p_{ij} be continuous on $[0, \infty)$, $i, j = 1, 2, \ldots, n$. Let $p(t)$ denote the $n \times n$ matrix with elements $p_{ij}(t)$, assume $\tau(t) > 0$ is continuous on $[0, \infty)$ and $\lim_{t \to \infty} (t - \tau(t)) = \infty$. Let $\mu(\cdot)$ denote the matrix measure. Assume that $\mu(p(t))$ is negative for sufficiently large t, and

$$\lim_{t \to \infty} \int_{t-\tau(t)}^{t} (-\mu(p(s))) \, ds > \frac{1}{e} \quad (6.3.29)$$

Then every solution of (6.3.28) oscillates.

Proof: Suppose that there exists a solution, say $\tilde{y}(t) = (y_1(t), \ldots, y_n(t))$, of (6.3.28) which is nonoscillatory. Then it follows that there exists a $t^* > 0$ such that for $t > t^*$, no component of $\tilde{y}$ has a zero. In such a case we have

$$\frac{d}{dt} \sum_{i=1}^{n} |y_i(t)| \leq \mu(p(t)) \sum_{i=1}^{n} |y_i(t - \tau(t))|$$

for sufficiently large $t \geq T > t^*$, and

$$\frac{du(t)}{dt} \leq \mu(p(t)) u(t - \tau(t)) \quad (6.3.30)$$

where $\mu(p(t)) < 0$ and

$$u(t) = \sum_{i=1}^{n} |y_i(t)| > 0$$

$$u(t - \tau(t)) > 0$$

for $t \geq T$. This implies that (6.3.30) has a positive solution, which contradicts the conclusion of Theorem 2.1.1.

Finally, we consider a kind of nonlinear system

$$\frac{d\tilde{y}(t)}{dt} = p(t)\tilde{y}(t - \tau(t)) + \tilde{f}(t, \ \tilde{y}(t - \tau(t))) \tag{6.3.31}$$

THEOREM 6.3.5 Assume that p, $\tau > 0$, and f are continuous on their domain, $\lim_{t \to \infty} t - \tau(t) = \infty$, and

$$\| \tilde{f}(t, \ \tilde{y}(t - \tau(t))) \| \ \leq \ k(t) \, \| \tilde{y}(t - \tau(t)) \| \tag{6.3.32}$$

If $\mu(p(t)) + k(t) < 0$ for sufficiently large t, and

$$\lim_{t \to \infty} - \int_{t-\tau(t)}^{t} (\mu(p(s)) + k(s)) \ ds \ > \ \frac{1}{e} \tag{6.3.33}$$

then all solutions of (6.3.31) are oscillatory.

Proof: As before, we can derive

$$\frac{du(t)}{dt} \leq (\mu(p(t)) + k(t))u(t - \tau(t)) \tag{6.3.34}$$

where $u(t) = \| \tilde{y}(t) \|$. Assume that (6.3.31) has a nonoscillatory solution $\tilde{y}(t)$. This implies that $u(t) > 0$ for sufficiently large t. This fact together with (6.3.34) and (6.3.33) implies a contradiction. The proof is complete.

REMARK 6.3.5 The argument here can be applied to more general systems so that the results in Chapter 2 can be used here.

6.4 OSCILLATION OF HIGH ORDER
NONLINEAR SYSTEMS

We consider

$$y_k'(t) = f_k(t, \ y_1(g_1(t)), \ \cdots, \ y_n(g_n(t))) \, , \quad k = 1, \ 2, \ \ldots, \ n \tag{6.4.1}$$

LEMMA 6.4.1 Assume that $f_k \in C[R_+ \times R^n, \ R]$, $g_k \in C[R_+, \ R]$, $\lim_{t \to \infty} g_k(t) = \infty$ $(k = 1, \ 2, \ \ldots, \ n)$, and

$$y_{k+1} f_k(t, \ y_1, \ \cdots, \ y_n) \begin{cases} > 0 & \text{if } k = 1, \ \ldots, \ n - 1 \\ \\ < 0 & \text{if } k = n, \ y_{n+1} = y_1 \end{cases} \tag{6.4.2}$$

for $y_{k+1} \neq 0$ $(k = 1, \ 2, \ \ldots, \ n)$.

Then the oscillation (nonoscillation) of one of the components of any solution of system (6.4.1) implies the oscillation (nonoscillation) of every component of this solution $(y_1, \ \cdots, \ y_n)$.

Proof: Assume that the component $y_k(t)$ is nonoscillatory. Hence $y_k(t) \neq 0$

for all $t \geq t_1 \geq 0$. From (6.4.1) and (6.4.2) it follows that $y'_{k-1}(t) \neq 0$ for all $t \geq t_1$. Hence, there is a point $t_2 \geq t_1$ such that $y_{k-1}(t) \neq 0$ for $t \geq t_2$. Step by step we can find a sequence

$$t_1 \leq t_2 \leq \cdots \leq t_n$$

such that for $t \geq t_k$ ($k = 1, 2, \ldots, n$) the functions $y_k, y_{k-1}, \ldots, y_2, y_1, y_n, \ldots, y_{k+1}$ are monotonic and of constant sign. Hence, for $t > T$ all the functions $y_k(t)$ ($k = 1, 2, \ldots, n$) are monotonic and of constant sign. The proof is complete.

LEMMA 6.4.2 Assume that the conditions of Lemma 6.4.1 hold. In addition, assume that

$$|f_k(t, y_1, \ldots, y_n)| \geq a_{k+1}(t) |F_{k+1}(y_{k+1})| \quad k = 1, 2, \ldots, n, \quad y_{n+1} = y_1$$

$$(6.4.3)$$

where $a_k(t) \geq 0$ for $t \in R^+$ and $F_k \in C[R, R]$. Moreover,

$$u F_k(u) \begin{cases} > 0 & \text{if } k = 2, \ldots, n \\[2mm] < 0 & \text{if } k = n + 1 \end{cases}$$

$$(6.4.4)$$

and

$$|F_k(u_1)| \leq |F_k(u_2)|, \quad \text{for } |u_1| \leq |u_2|$$

$$(6.4.5)$$

and

$$\int_T^\infty a_k(t)\, dt = \infty \quad k = 2, \ldots, n$$

$$(6.4.6)$$

If some component y_i ($1 \leq i \leq n$) of a solution $(y_1, \ldots, y_n)$ of system (6.4.1) has the property

$$\liminf_{t \to \infty} |y_i(t)| = L_i$$

$$(6.4.7)$$

then

$$\lim_{t \to \infty} y_k(t) = +\infty \ (-\infty), \quad k = 1, 2, \ldots, i - 1$$

$$(6.4.8)$$

when $L_i > 0$ and $i > 1$, and

$$\liminf_{t \to \infty} |y_k(t)| = 0, \quad k = i + 1, \ldots, n$$

$$(6.4.9)$$

when $L_i < \infty$ and $i < n$.

Proof: Let $L_i > 0$. Then for sufficiently large $t \geq T \geq 0$, we have

$$|y_i(g_i(t))| \geq L = \begin{cases} \frac{1}{2}L_i & \text{when } L_i < \infty \\ 1 & \text{when } L_i = \infty \end{cases} \qquad (6.4.10)$$

From Lemma 6.4.1, it follows that the solution $(y_1, \ldots, y_n)$ is nonoscillatory. Let us suppose that $y_i(t) > 0$ for $t \geq T$. Integrating the $(i-1)$th equation of system (6.4.1) and then using (6.4.2), (6.4.3), and (6.4.10), we have

$$y_{i-1}(t) - y_{i-1}(T) = \int_T^t f_{i-1}(s, y_1, \ldots, y_n)\, ds$$

$$\geq \int_T^t a_i(s) F_i(y_i(g_i(s)))\, ds$$

$$\geq F_i(L) \int_T^t a_i(s)\, ds$$

From condition (6.4.6) and the above inequality it follows that $\lim_{t \to \infty} y_{i-1} = \infty$. Hence $y_{i-1}(g_{i-1}(t)) > 1$ for sufficiently large $t \geq T_1 \geq T$. Integrating now the $(i-2)$th equation we obtain

$$y_{i-2}(t) - y_{i-2}(T_1) = \int_{T_1}^t f_{i-2}(s, y_1, \ldots, y_n)\, ds$$

$$\geq F_{i-1}(1) \int_{T_1}^t a_{i-1}(s)\, ds$$

Hence we get $\lim_{t \to \infty} y_{i-2}(t) = \infty$.

Proceeding as before, we prove that $\lim_{t \to \infty} y_k(t) = \infty$ $(k = 1, 2, \ldots, i - 1)$. In the case when $y_i(t) < 0$ the proof is analogous.

Let $L_i < \infty$. We will prove that $\liminf_{t \to \infty} |y_k(t)| = 0$ $(k = i+1, \ldots, n)$. Suppose that it is otherwise, i.e., there is such a component $y_m(t)$ $(i < m \leq n)$ of the solution that

$$\liminf_{t \to \infty} |y_m(t)| = L_m > 0$$

Then from the first part of the proof it follows that

$$\lim_{t \to \infty} |y_k(t)| = L_k = \infty \qquad (k = 1, \ldots, i, \ldots, m - 1)$$

which contradicts $L_i < \infty$. Hence (6.4.9) holds.

LEMMA 6.4.3 Assume that the conditions of Lemma 6.4.2 hold. Then for every nonoscillatory solution $(y_1, \ldots, y_n)$ of system (6.4.1) there is a number i $(1 \leq i \leq n)$ such that $n + i$ is even, and for sufficiently large t, we have

$$y_1(t)y_k(t) > 0 \quad (k = 1, \ldots, i)$$

$$(-1)^{n+k} y_1(t)y_k(t) > 0 \quad (k = i + 1, \ldots, n) \tag{6.4.11}$$

Proof: Let $(y_1, \ldots, y_n)$ be a nonoscillatory solution of system (6.4.1). Then from Lemma 6.4.1 follows the existence of a point $T \geq 0$ such that $y_k(t)$ and $y_k(g_k(t))$ $(k = 1, 2, \ldots, n)$ are monotonic and of a constant sign, for $t \geq T$. Let us assume that $y_1(t) > 0$ for $t \geq T$ (in the case $y_1(t) < 0$ the proof is analogous). We show that $y_n(t) > 0$ for $t \geq T$. From (6.4.1) and (6.4.2) it follows that $y_n'(t) < 0$ for $t \geq T$. If $y_n(t) < 0$, then $\lim_{t\to\infty} y_n(t) = L_n < 0$. From Lemma 6.4.2, we have $\lim_{t\to\infty} y_k(t) = -\infty$ $(k = 1, \ldots, n - 1)$, contradicting the assumption $y_1(t) > 0$ for $t \geq T$. Therefore $y_n(t) > 0$ for $t \geq T$. Now, consider the sign of $y_{n-1}(t)$. There are two possibilities: either $y_{n-1}(t) > 0$, or $y_{n-1}(t) < 0$ for $t \geq T$. In the case $y_{n-1} \geq 0$, $y_{n-1}'(t) > 0$ from (6.4.2). Hence $\lim_{t\to\infty} y_{n-1}(t) = L_{n-1} > 0$, and from Lemma 6.4.2 we obtain $\lim_{t\to\infty} y_k(t) = \infty$ $(k = 1, 2, \ldots, n - 2)$. Therefore (6.4.11) are true for $i = n$.

In the case when $y_{n-1}(t) < 0$ for $t \geq T$ we show that $y_{n-2}(t) > 0$ for $t \geq T$. If $y_{n-2}(t) < 0$ for $t \geq T$, then from (6.4.1) and (6.4.2) it follows that $y_{n-2}(t) < 0$, which implies $\lim_{t\to\infty} y_{n-2}(t) = L_{n-2} < 0$. From Lemma 6.4.2, we have $\lim_{t\to\infty} y_k(t) = -\infty$ $(k = 1, \ldots, n - 3)$, which contradicts the fact that $y_1(t)$ is a positive function. Therefore $y_{n-2}(t) > 0$ for $t \geq T$. If now $y_{n-3}(t) > 0$ for $t \geq T$, then the inequalities (6.4.11) are true for $i = n - 2$. In the case $y_{n-3}(t) < 0$ we show that $y_{n-4}(t) > 0$ for $t \geq T$. Then the inequalities (6.4.11) are true for $i = n - 4$. Proceeding as above we prove that the inequalities (6.4.11) are true for i $(1 \leq i \leq n)$ where $i + n$ is even.

THEOREM 6.4.1 Assume that the conditions of Lemma 6.4.3 hold. Moreover we suppose that

$$\int_{g_i(T)}^{\infty} a_{i+1}(t_1) \left| F_{i+1} \left\{ \int_{g_{i+1}(t_1)}^{\infty} a_{i+2}(t_2) F_{i+2} \left[\int_{g_{i+2}(t_2)}^{\infty} a_{i+3}(t_3) F_{i+3} \left(\cdots \right. \right. \right. \right.$$

$$\times F_n \int_{g_n(t_n+1)}^{\infty} a_{n+1}(t_{n-i+1}) F_{n+1} \left(\int_{T}^{g_1(t_{n-1+i})} a_2(t_{n-i+2}) F_2 \left(\int_{T}^{g_2(t_{n-i-2})} a_3(t_{n-i+3}) \right. \right.$$

$$\times F_3\left(\cdots\left(F_{i-3}\left(\int_T^{g_{i-3}(t_{n-3})} a_{i-2}(t_{n-2})\,F_{i-2}\right.\right.\right.$$

$$\times\left(\alpha\int_T^{g_{i-2}(t_{n-2})} a_{i-1}(t_{n-1})\,dt_{n-1}\right)dt_{n-2}\right)\cdots\Bigg]dt_2\Bigg\}\Bigg|\,dt_1 = \infty \qquad (6.4.12)$$

$i = 2, \ldots, n$, for an arbitrary constant $\alpha \neq 0$. Then every solution of system (6.4.1) is

- (a) Oscillatory for even n
- (b) Either oscillatory or tending monotonically to zero as $t \to \infty$ for odd n

Proof: Assume that (6.4.1) has a nonoscillatory solution $(y_1, \ldots, y_n)$. Let $T_0 \geq 0$ be such that the functions $y_k(t)$ and $y_k(g_k(t))$ $(k = 1, 2, \ldots, n)$ have a constant sign for $t \geq T_0$. Next we choose $T_1 \geq T_0$ such that $g_k(t) \geq T_0$ $(k = 1, 2, \ldots, n)$ for $t \geq T_1$.

(a) For an even number n, from Lemma 6.4.3 it follows that there is an even number i $(2 \leq i \leq n)$ such that for $t \geq T_0$, (6.4.11) holds. From (6.4.1), (6.4.2), and (6.4.11), for $t \geq T_0$ we obtain the inequalities

$$y_{k+1}(t)y_k'(t) > 0, \quad k = 1, \ldots, i-1$$

$$(-1)^{n+k-1}y_{k+1}y_k'(t) > 0, \quad k = i, i+1, \ldots, n \qquad (6.4.13)$$

From (6.4.13) we have $\lim_{t\to\infty} y_i(t) = L_i$, L_i is a constant, and

$$|y_{i-1}(g_{i-1}(t)))| \geq |y_{i-1}(T_0)| > 0, \quad \text{for } t \geq T$$

hence from Lemma 6.4.2, we get

$$\lim_{t\to\infty} |y_k(t)| = \begin{cases} \infty & k = 1, \ldots, i-2 \\ 0 & k = i+1, \ldots, n \end{cases}$$

Integrating the ith equation from $g_i(T_1)$ to ∞, we have

$$\infty > |y_i(g_i(T_1)))| - |L_i| = \int_{g_i(T_1)}^{\infty} |f_i|\,dt$$

$$\geq \int_{g_i(T_1)}^{\infty} a_{i+1}(t)\,|F_{i+1}(y_{i+1}(g_{i+1}(t)))|\,dt \qquad (6.4.14)$$

Integrating the remaining equations from $g_k(t)$ to ∞ for $k = i+1, \ldots, n$, and from T_1 to $g_k(t)$ for $k = 1, \ldots, i - 2$, respectively, we get

$$|y_k(g_k(t))| = \int_{g_k(t)}^{\infty} |f_k| \, dt$$

$$\geq \int_{g_k(t)}^{\infty} a_{k+1} |F_{k+1}(y_{k+1}(g_{k+1}(s)))| \, ds \qquad (6.4.15)$$

for $k = i + 1, \ldots, n$ and

$$|y_k(g_k(t))| \geq |y_k(g_k(t))| - |y_k(T_1)|$$

$$= \int_{T_1}^{g_k(t)} |f_k| \, ds \geq \int_{T_1}^{g_k(t)} a_{k+1}(s) |F_{k+1}(g_{k+1}(s))| \, ds \qquad (6.4.16)$$

for $k = 1, \ldots, i - 2$.

Applying $(6.4.14)-(6.4.16)$, we obtain

$$\infty > \int_{g_i(t)}^{\infty} a_{i+1}(t_1) |F_{i+1}(y_{i+1}(g_{i+1}(t_1)))| \, dt_1$$

$$\geq \int_{g_i(T_1)}^{\infty} a_{i+1}(t_1) \left| F_{i+1} \left\{ \int_{g_{i+1}(t_1)}^{\infty} a_{i+2}(t_2) F_{i+2} \right. \right.$$

$$\times \left[\int_{g_{i+2}(t_2)}^{\infty} a_{i+3}(t_3) F_{i+3} \Bigl(\cdots \Bigl(F_n \Bigl(\int_{g_n(t_{n-1})}^{\infty} a_{n+1}(t_{n-i+1}) \right.$$

$$\times \; F_{n+1} \int_{T}^{g_1(t_{n-i+1})} a_2(t_{n-i+2}) F_2$$

$$\times \Bigl(\int_{T}^{g_2(t_{n-i+2})} a_3(t_{n-i+3}) F_3 \Bigl(\cdots F_{i-3} \Bigl(\int_{T}^{g_{i-3}(t_{n-3})} a_{i-2}(t_{n-2}) F_{i-2}$$

$$\times \Bigl(\alpha \int_{T}^{g_{i-2}(t_{n-2})} a_{i-1}(t_{n-1}) \, dt_{n-1} \Bigr) dt_{n-2} \Bigr) \cdots \Bigr] dt_2 \left. \right\} \right| dt_1 = \infty$$

for $i = 2, \ldots, n$, where $\alpha = -F_{i-1}(y_{i-1}(T_0))$. The above inequality contradicts (6.4.12). Hence in the case when n is an even number, every solution of the system (6.4.1) is oscillatory.

(b) For odd number n we shall prove that every nonoscillatory solution tends to zero as $t \to \infty$. From Lemma 6.4.3, it follows that inequalities (6.4.11) are satisfied when i $(1 \leq i \leq n)$ is odd. As in the first part, we can prove that the case $1 < i \leq n$ is impossible. Therefore inequalities (6.4.11) hold only for $i = 1$. Hence $\lim_{t \to \infty} y_1(t) = L_1$, $|L_1| < \infty$. From Lemma 6.4.2 we have $\lim_{t \to \infty} y_k(t) = 0$, $k = 2, \ldots, n$.

We demonstrate that $L_1 = 0$. If $|L_1| > 0$, then for sufficiently large $t \geq T_2 \geq T$, the inequality $|y_1(g_1(t))| \geq |L_1|$ holds. Integrating the second equation from $g_2(T_2)$ to ∞ and the remaining ones from $g_k(t)$ to ∞, we get

$$\infty > |y_2(g_2(T_2))| = \int_{g_2(T_2)}^{\infty} |f_2(t, y_1, \ldots, y_n)| \, dt$$

$$\geq \int_{g_2(T_2)}^{\infty} a_3(t) |F_3(y_3(g_3(t)))| \, dt$$

and

$$|y_k(g_k(t))| = \int_{g_k(t)}^{\infty} |f_k| \, dt \geq \int_{g_k(t)}^{\infty} a_{k+1}(t) |F_{k+1}(y_{k+1}(g_{k+1}(t)))| \, dt$$

for $k = 3, \ldots, n$, and $y_{n+1} = y_1$.

From the above-mentioned inequalities and from the fact that the functions F_k are monotonic we have

$$\infty > \int_{g_2(T_2)}^{\infty} a_3(t) |F_3(y_3(g_3(t)))| \, dt$$

$$\geq \int_{g_2(T_2)}^{\infty} a_3(t_1) \left| F_3 \left\{ \int_{g_3(t_1)}^{\infty} a_4(t_2) F_4 \left[\cdots \right. \right. \right.$$

$$\left. \left. \left. \cdots F_n \left(\int_{g_n(t_{n-2})}^{\infty} a_{n+1}(t_{n-1}) F_{n+1}(y_1(g_1(t_{n-1}))) \, dt_{n-1} \right) \cdots \right] dt_2 \right\} \right| dt_1$$

$$\geq \int_{g_2(T_2)}^{\infty} a_3(t_1) \left| F_3 \left\{ \int_{g_3(t_1)}^{\infty} a_4(t_2) F_4 \left(\int_{g_4(t_2)}^{\infty} a_5(t_3) \right. \right. \right.$$

$$\left. \left. \left. \times F_5 \left(\ldots F_n \left(\alpha \int_{g_n(t_{n-2})}^{\infty} a_{n+1}(t_{n-1}) \, dt_{n-1} \right) \ldots \right] dt_2 \right\} \right) dt_1$$

where $\alpha = -A_{n+1}(L_1) > 0$. The above inequality contradicts (6.4.12). There-fore L_1 must be equal to zero. Then every nonoscillatory solution of system (6.4.1) tends to zero as $t \to \infty$, for any odd n.

Similarly, we can prove the next theorem.

THEOREM 6.4.2 Assume that the conditions of Lemma 6.4.3 hold. Moreover we assume that

$$\int_T^{\infty} a_2(t_2) \left| F_2 \left\{ \int_{g_2(t_2)}^{\infty} a_3(t_3) F_3 \left[\int_{g_3(t_3)}^{\infty} a_4(t_4) F_4 \left(\ldots F_{n-1} \left(\int_{g_{n-1}(t_{n-1})}^{\infty} a_n(t_n) F_n \right. \right. \right. \right. \right.$$

$$\left. \left. \left. \left. \times \left(\alpha \int_{g_n(t_n)}^{\infty} a_{n+1}(t_{n+1}) \, dt_{n+1} \, dt_n \right) \ldots \right] dt_3 \right\} \right| dt_2 = \infty$$

for an arbitrary constant $\alpha \neq 0$. Then every bounded solution of system (6.4.1) is

a) Oscillatory for even n
b) Either oscillatory or tending monotonically to zero as $t \to \infty$ for odd n

6.5 NOTES

The survey paper [237] contains the results on oscillation of solutions of systems of ODE with or without deviating argument up to the 70s. The results of Section 6.1 are taken from Ladde and Zhang [163]. Lemma 6.2.2 and Theorem 6.2.6 are based on the work of Kitamura and Kusano [115, 117, 118]. Theorem 6.2.7 belongs to Varekh and Shevelo [281]. Theorems 6.3.1 and 6.3.3 are taken from Gopalsamy [69]. Theorem 6.3.2 is an improved form of a result in [69]. Theorems 6.3.4 and 6.3.5 are new. The asymptotic behavior of solutions is studied by Lim [169]. Theorems 6.4.1 and 6.4.2 belong to Folynska and Werbowski [63]. Driver [56] studied asymptotic behavior of solutions of a system with small delay. For related work, see [9], [52], [57], [97], [126], [184], [191], [236], [238], [239], and [282].

References

1. Anderson, C. H., Asymptotic oscillation results for solutions to first order nonlinear differential difference equations of advanced type, J. Math. Anal. Appl., $\underline{24}$(1968), 430-439.

2. Angelova, D. C. and Bainov, D. D., On some oscillating properties of the solutions of a class of functional differential equations, Tsukuba J. Math., $\underline{5}$(1981), 39-46.

3. Angelova, D. C. and Bainov, D. D., On the oscillation of solutions to second order functional differential equations, Boll. Un. Mat. Ital. B(6) $\underline{1}$(1982), 797-807.

4. Angelova, D. C. and Bainov, D. D., On the oscillation of solutions to second order functional differential equations, Mathematics Reports Toyame University, $\underline{5}$(1983), 1-14.

5. Arino, O. and Seguier, P., Existence of oscillatory solution of certain differential equations with delay, Lect. Notes Math. $\underline{730}$(1979), 46-64.

6. Arino, O., Gyori, I., and Jawhari, A., Oscillation criteria in delay equations, J. Differential Equations, $\underline{53}$(1984), 115-123.

7. Atkinson, F. V., On second order nonlinear oscillation, Pacific J. Math., $\underline{5}$(1955), 643-647.

8. Ausari, J. S., A theorem on asymptotic behavior of functional equations and its application, J. Comput. Appl. Math., $\underline{2}$(1976), 35-39.

9. Banks, H. T. and Manitius, A., Projection series for retarded functional differential equations with applications to optimal control problems, J. Differential Equations, $\underline{18}$(1975), 296-332.

10. Barrett, J. H., Oscillation theory of ordinary linear differential equations, Advances in Math., $\underline{3}$(1969), 415-509.

11. Barwell, V. K., On the asymptotic behavior of the solution of a differential difference equation, Utililas. Math., $\underline{6}$(1974), 189-194.

12. Bernoulli, J., Meditations, Dechordis vibrantibis, Comment. Acad. Sci. Imper. Petropol., $\underline{3}$(1728), 13-28. Collected Works, 3, 198 p.

13. Birkoff, G. and Kotin, K., Asymptotic behavior of first order linear differential delay equations, J. Math. Anal. Appl., $\underline{13}$(1966), 8-18.

14. Birkoff, G. and Kotin, K., Integrodifferential delay equations of positive type, J. Differential Equations, $\underline{2}$(1966), 320-327.

15. Bobisud, L. E., Oscillation of second order differential equations with retarded argument, J. Math. Anal. Appl., $\underline{43}$(1973), 201-206.

16. Bogar, G. A., Oscillations of nth order differential equations with retarded argument, SIAM J. Math. Anal., $\underline{5}$(1974), 473-481.

17. Bradley, J. S., Oscillation theorems for a second order delay equation, J. Differential Equations, $\underline{8}$(1970), 397-403.

18. Brands, J. J. A. M., Oscillation theoreams for second order functional differential equations, J. Math. Anal. Appl., $\underline{63}$(1978), 54-64.

19. Brauer, F., Decay rates for solutions of a class of differential difference equations, SIAM J. Math. Anal., $\underline{10}$(1979), 783-788.

20. Buchanan, J., Bounds on the growth of a class of oscillatory solutions of $y'(x) = my(x - d(x))$ with bounded delays, SIAM J. Appl. Math., $\underline{27}$ (1974), 539-543.

21. Butler, G. J., Oscillation criteria for second order nonlinear ordinary differential equations, Colloquia Mathematica Sociatatie Janos Bolyai 30, Qualitative theor of differential equations, Ed. Szeged. 1979, 93-109.

22. Burkowski, F., Nonlinear oscillation of a second order sublinear functional differential equation, SIAM J. Appl. Math., $\underline{21}$(1971), 486-490.

23. Burkowski, F., Oscillation theorems for a second order nonlinear functional differential equation, J. Math. Anal. Appl., $\underline{33}$(1971), 258-262.

24. Burkowski, F. and Ponzo, P. J., The oscillatory behavior of a first order nonlinear differential equation with delay, Canad. Math. Bull., $\underline{17}$(2) (1974), 185-188.

25. Burton, T. and Grimmer, R., Oscillatory solution of $x''(t) + a(t)f(x(g(t))) = 0$, _Delay and Functional Differential Equations and their Applications_, 335-343, Academic Press, New York and London, 1972.

26. Burton, T. A. and Haddock, J. R., On solution tending to zero for the equation $x''(t) + a(t)x(t - r(t)) = 0$, Arch. Math. (Basel), $\underline{27}$(1976), 48-51.

27. Burton, T. A. and Haddock, J. R., On the delay differential equations $x'(t) + a(t)f(x(t - r(t))) = 0$ and $x''(t) + a(t)f(x(t - r(t))) = 0$, J. Math. Anal. Appl., $\underline{54}$(1976), 37-48.

28. Butlewski, Z., Un theoreme de l'oscillation, Ann. Polon. Math., $\underline{24}$ (1951), 95-110.

29. Bykov, Y. V. and Merzlyakova, G. D., On the oscillatoriness of solutions of nonlinear differential equations with a deviating argument. Differencial'nye Uravnenija, $\underline{10}$(1974), 210-220.

30. Bykov, Y. V. and Matekaev, A. I., Oscillation properties of a class of first order differential equations, Studies in integro-differential equations, 20-28, 256-257, "Ilim" Frunze, 1979.

31. Chen, Lu-san, Note on asymptotic behavior of solutions of a second order nonlinear differential equation, J. Math. Anal. Appl., $\underline{57}$(1977), 481-485.

32. Chen, Lu-san, Some oscillation theorem for differential equation with functional arguments, J. Math. Anal. Appl., $\underline{58}$(1977), 83-87.

33. Chen, Lu-san, Some nonoscillation theorem for the higher order non-linear functional differential equations, Ann. Mat. Pura. Appl., (4)117 (1978), 41-53.

34. Chen, Lu-san and Yeh, Chen-chin, Oscillations of nth order functional differential equations with perturbations. Rend. Istit. Mat. Univ. Trieste 11(1979), N. 1-2, (1980), 83-90.

35. Chen, Lu-san and Yeh, Chen-chin, On oscillation of solutions of higher order differential retarded inequalities, Houston. J. Math., 5(1979), 37-40.

36. Chen, Lu-san and Yeh, Chen-chin, Asymptotic nature of nonoscillatory solutions of nonlinear deviating differential equations, Ann. Mat. Pura and Appl., 121(1979), 199-205.

37. Chen, Lu-san and Yu, Fong-ming , On oscillation and asymptotic behavior of the solution of nth order retarded differential equations, Math. Japan 26(1981), N. 1, 145-151.

38. Chen, Lu-san and Yeh, Chen-chin, Necessary and sufficient conditions for asymptotic decay of oscillations in delayed functional equations, Proc. Roy. Soc. Edinburgh Sect. A(91) (1981/1982), 135-145.

39. Chen, Ming-po, Yeh, Chen-chin and Yu, Cheng-shu, Asymptotic behavior of nonoscillatory solution of nonlinear differential equation with retarded arguments, J. Math. Anal. Appl., 59(1977), 211-215.

40. Chiou, Kuo-liang, Oscillation and nonoscillation theorem for second order functional differential equations, J. Math. Anal. Appl., 45(1974), 382-403.

41. Cooke, K. L., Asymptotic theory for the delay differential equation $u'(t) = -au(t - r(u(t)))$, J. Math. Anal. Appl., 19(1967), 160-173.

42. Cooke, K. L. and Yorke, J., Equations modelling population growth, economic growth and gonorrhea epidemiology, NRL-MRC Conference, Washington, D.C., 1971, Academic Press, New York and London, 1972, 35-53.

43. Cooke, K. L. and Yorke, J., Some equations modeling growth processes and gonorrhea epidemics, Math. Biosci., 16(1973), 75-101.

44. Dahiya, R. S. and Singh. B., On oscillatory behavior of even order delay equations, J. Math. Anal. Appl., 42(1973), 183-190.

45. Dahiya, R. S., Nonoscillation of arbitrary order retarded differential equations of nonhomogeneous type, Bull. Austral. Math. Soc., 10(1974), 453-458.

46. Dahiya, R. S. and Singh, B., On the oscillation of a second order delay equation, J. Math. Anal. Appl., 48(1974), 610-617.

47. Dahiya, R. S., Oscillation generating delay terms in even order retarded equations, J. Math. Anal. Appl., 49(1975), 158-165.

48. Dahiya, R. S., Comparative oscillation of second order retarded differential equations, Indian J. Pure and Appl. Math., 6(1975), 282-287.

49. Dahiya, R. S., Oscillation theorems for nonlinear delay differential equations, Bull. Fac. Sci. Ibaraki. Univ. Ser. A N., 14(1982), 13-18.

50. Dahiya, R. S., Oscillation and asymptotic behavior of bounded solutions of functional differential equations with delay, J. Math. Anal. Appl., $\underline{91}$ (1983), 276-286.

51. Dahiya, R. S., Kusano, T. and Naito, M., On almost oscillation of functional differential equations with deviating arguments, J. Math. Anal. Appl., $\underline{98}$(1984), 332-340.

52. Delfour, M. C. and Mitter, S. K., Hereditary differential systems with constant delays, J. Differential Equations, $\underline{18}$(1975), 18-28.

53. Domshlak, Y. I., Comparison theorems of Sturm type for first and second order differential equations with sign variable deviations of the argument, (Russian). Ukrain. Mat. Zh., $\underline{34}$(1982), 158-163, 267.

54. Driver, R. D., Some harmless delays, Delay and Functional Differential Equations and their Applications, Academic Press, New York, 1972, 103-119.

55. Driver, R. D., Sasser, D. W. and Slater, M. L., The equation $x'(t) = ax(t) + bx(t - \tau)$ with "Small" delay, Amer. Math. Monthly, $\underline{80}$ (1973), 990-995.

56. Driver, R. D., Linear differential systems with small delays, J. Differential Equations, $\underline{21}$(1976), 148-166.

57. Driver, R. D., Ordinary and delay differential equations, Springer-Verlag, Berlin and New York, 1977.

58. Elbert, A., Comparison theorem for first order nonlinear differential equations with delay, Studia Sci. Math. Hungar., $\underline{11}$(1976), 259-267.

59. El'sgol'ts, L. E. and Norkin, S. B., Introduction to the theory and application of differential equations with deviating arguments, Academic Press, 1973.

60. Erbe, L., Oscillation criteria for second order nonlinear delay equations, Canad. Math. Bull., $\underline{16}$(1973), 49-56.

61. Erbe, L., Oscillation and asymptotic behavior of solutions of third order differential delay equations, SIAM J. Math. Anal., $\underline{7}$(1976), 491-500.

62. Fite, W. B., Properties of the solutions of certain functional differential equations, Trans. Amer. Math. Soc., $\underline{22}$(1921), 311-319.

63. Folynska, I. and Werbowski, J., On the oscillatory behavior of solutions of systems of differential equations with deviating arguments, Qualitative theory of differential equations, V. I, II Ed. Szeged, 1979, 243-256, Colloq. Math. Soc. Janos. Bolyai 30 New York, 1981.

64. Foster, K., Criteria for oscillation and growth of nonoscillatory solutions of forced differential equations of even order, J. Differential Equations, $\underline{20}$(1976), 115-132.

65. Foster, K. E. and Grimmer, R. C., Nonoscillatory solutions of higher order differential equations, J. Math. Anal. Appl., $\underline{71}$(1979), 1-17.

66. Fox, L., Mayers, D. F., Ockendon, J. R. and Tayler, A. B., On a functional differential equation, J. Inst. Math. Appl., $\underline{8}$(1971), 271-307.

67. Garner, B. J., Oscillatory criteria for a general second order functional equation, SIAM J. Appl. Math., $\underline{29}$(1975), 690-698.

68. Gollwitzer, H. E., On nonlinear oscillation for a second order delay equation, J. Math. Anal. Appl., $\underline{26}$(1969), 385-389.

69. Gopalsamy, K., Oscillation in linear systems of differential difference equations, Bull. Austral. Math. Soc., $\underline{29}$(1984), 377-388.

70. Grace, S. R. and Lalli, B. S., Oscillation theorems for n^{th} order delay differential equations, Nonlinear Analysis and Applications (St. Johns Nfld. 1981) 343-355. Lecture Notes in Pure and Appl. Math. $\underline{80}$, New York, 1982.

71. Grace, S. R. and Lalli, B. S., An oscillation criterion for n^{th} order nonlinear differential equations with functional arguments, Canad. Math. Bull., $\underline{26}$(1983), 35-40.

72. Grace, S. R. and Lalli, B. S., On oscillation of solutions of nth order delay differential equations, J. Math. Anal. Appl., $\underline{91}$(1983), 328-339.

73. Grace, S. R. and Lalli, B. S., Oscillation theorems for n^{th} order delay differential equations, J. Math. Anal. Appl., $\underline{91}$(1983), 352-366.

74. Grace, S. R. and Lalli, B. S., Oscillation theorems for n^{th} order nonlinear functional differential equations, J. Math. Anal. Appl., $\underline{94}$ (1983), 509-524.

75. Grace, S. R. and Lalli, B. S., On oscillation of solutions of higher order forced functional differential inequalities, J. Math. Anal. Appl., $\underline{94}$(1983), 525-535.

76. Grace, S. R., Oscillation theorem for n^{th} order differential equation with deviating arguments, J. Math. Anal. Appl., $\underline{101}$(1984), 268-296.

77. Grace, S. R. and Lalli, B. S., An oscillation criterion for n^{th} order nonlinear differential equations with retarded arguments, Czech. Math. J., $\underline{34}$(109)(1984), 18-21.

78. Grace, S. R. and Lalli, B. S., Oscillation theorems for n^{th} order nonlinear differential equations with deviating arguments, Proc. Amer. Math. Soc., $\underline{90}$(1984), 65-70.

79. Graef, J. R. and Spikes, P. W., Asymptotic properties of solutions of functional differential equations of arbitrary order, J. Math. Anal. Appl., $\underline{60}$(1977), 339-348.

80. Graef, J. R., Oscillation, nonoscillation and growth of solutions of nonlinear functional differential equations of arbitrary order, J. Math. Anal. Appl., $\underline{60}$(1977), 398-409.

81. Graef, J. R., Katamura, Y., Kusano, T., and Spikes, P. W., On the nonoscillation of perturbed functional differential equations, Pacific J. Math., $\underline{83}$(1979), 365-373.

82. Graef, J. R., Grammatikopoulos, M. K. and Spikes, P. W., Oscillatory behavior of differential equations with deviating arguments, C. R. Acad. Bulgare Sci., $\underline{33}$(1980), 1443-1446.

83. Graef, J. R., Grammatikopoulos, M. K. and Spikes, P. W., Asymptotic and oscillatory behavior of superlinear differential equations with deviating arguments, J. Math. Anal. Appl., $\underline{75}$(1980), 134-148.

84. Graef, J. R., Grammatikopoulos, M. K. and Spikes, P. W., Oscillatory and asymptotic properties of solutions of generalized Thomas-Fermi equations with deviating arguments, J. Math. Anal. Appl, $\underline{84}$ (1981), 519-529.

85. Graef, J. R., Grammatikopoulos, M. K. and Spikes, P. W., Classification of solutions of functional differential equations of arbitrary order, Bull. Inst. Math. Acad. Sinica, $\underline{9}$(1981), 517-532.

86. Graef, J. R., Kusano, T., Onose, H. and Spikes, P. W., On the asymptotic behavior of oscillatory solution of functional differential equations, Funkcialaj Ekvacioj, $\underline{26}$(1983), 11-16.

87. Graef, J. R., Nonoscillation of higher order functional differential equations, J. Math. Anal. Appl., $\underline{92}$(1983), 524-532.

88. Graef, J. R., Grammatikopoulos, M. K. and Spikes, P. W., On the decay of oscillatory solutions of a forced higher order functional differential equations, Math. Nachr., $\underline{117}$(1984), 141-153.

89. Graef, J. R. and Seal, E. H., An oscillation theorem for n^{th} order linear functional differential equations, J. Math. Anal. Appl., $\underline{99}$(1984), 458-471.

90. Grammatikopoulos, M. K., Oscillatory and asymptotic behavior of differential equations with deviating arguments, Hiroshima Math. J., $\underline{6}$(1976), 31-53.

91. Grammatikopoulos, M. K., Oscillation and asymptotic results for strongly nonlinear retarded differential equations, Sixth Balkan Congress. Summaries. Varna. (1977), 201.

92. Grammatikopoulos, M. K., Sficas, Y. G. and Staikos, V. A., Oscillatory properties of strongly superlinear differential equations with deviating arguments, J. Math. Anal. Appl., $\underline{67}$(1979), 171-187.

93. Grimmer, R., Oscillation criteria and growth of nonoscillatory solutions of even order ordinary and delay differential equations, Trans. Amer. Math. Soc., $\underline{198}$(1974), 215-228.

94. Gustafson, G. B., Bounded oscillations of linear and nonlinear delay differential equations of even order, J. Math. Anal. Appl., $\underline{46}$(1974), 175-189.

95. Gyori, I., On the oscillatory behavior of solutions of certain nonlinear and linear delay differential equations, Nonlinear Analysis, $\underline{8}$(1984), 429-439.

96. Haddock, J. R., On the asymptotic behavior of solutions of $x'(t) = -a(t)f(x(t - r(t)))$, SIAM Math. Anal., $\underline{5}$(1974), 569-573.

97. Halanay, A. and Yorke, J., Some new results and problems in the theory of differential delay equations, SIAM Rev., $\underline{13}$(1971), 55-80.

98. Hale, J. K., Functional Differential Equations, Springer-Verlag, New York, 1971.

99. Heard, M. L., Asymptotic behavior of solutions of the functional differential equations $x'(t) = ax(t) + bx(t^{\alpha})$, $\alpha > 1$, J. Math. Anal. Appl., $\underline{44}$ (1973), 745-757.

100. Hino, Y., On oscillation of the solution of second order functional differential equations, Funkcial. Ekvac., $\underline{17}$(1974), 94-105.

101. Hunt, B. R. and Yorke, J. A., When all solutions of $x'(t) = -\Sigma q_i(t)x(t - \tau_i(t))$ oscillate, J. Differential Equations, $\underline{53}$(1984), 139-145.

102. Ivanov, A. F., Kitamura, Y., Kusano, T. and Shevelo, V. N., Oscillatory solutions of functional differential equations generated by deviation of arguments of mixed type, Hiroshima Math. J., $\underline{12}$(1982), 645-655.

103. Ivanov, A. F. and Shevelo, V. N., Oscillation and asymptotic behavior of solutions of first order functional differential equations, Ukrain. Mat. Zh., $\underline{33}$(1981), 745-751, 859.

104. Johnson, G. W. and Yan, Jurang, Oscillation and nonoscillation criteria for nonlinear forced functional differential equations of nth order, Chinese Ann. Math. (to appear).

105. Kamenskii, G. A., On the general theory of equations with deviating argument, Dokl. Akad. Nauk, SSSR, $\underline{120}$(1958), 697-700.

106. Kartsatos, A. G. and Manougian, M. N., Further results on oscillation of functional differential equations, J. Math. Anal. Appl., $\underline{53}$ (1976), 28-37.

107. Kartsatos, A. G., Oscillation and existence of unique positive solutions for nonlinear n^{th} order equations with forcing term, Hiroshima Math. J., $\underline{6}$(1976), 1-6.

108. Kartsatos, A. G., Recent results on oscillation of solutions of forced and perturbed nonlinear differential equations of even order, Stability of Dynamical Systems Theory and Applications, Ed. Graef, J. R., V. 28(1977) Chapter III.

109. Kartsatos, A. G., Oscillation and nonoscillation for perturbed differential equations, Hiroshima Math. J., $\underline{8}$(1978), 1-10.

110. Kartsatos, A. G. and Toro, J., Oscillation and asymptotic behavior of forced nonlinear equations, SIAM J. Math. Anal., $\underline{10}$(1979), 86-95.

111. Kato, T. and Mcleod, J. B., The functional differential equation $y'(x) = ay(\lambda x) + by(x)$, Bull. Amer. Math. Soc., $\underline{77}$(1971), 891-937.

112. Katsuo, T., On global properties of solution of a certain delay differential equation, J. London Math. Soc., (2) $\underline{28}$(1983), 519-530.

113. Kiguradze, E. T., On the oscillation of solutions of the equation $\frac{d^m u}{dt^m} + a(t) u^m \operatorname{sign} u = 0$, Mat. Sb., $\underline{65}$(1964), 172-187.

114. Kitamura, Y. and Kusano, T., Vanishing oscillation of solutions of a class of differential systems with retarded argument, Atti. Acad. Naz. Lincei. Rend. Cl. Sci., Fis. Mat. Natur., $\underline{62}$(1977), 325-334.

115. Kitamura, Y. and Kusano, T., Oscillation of a class of nonlinear differential system with general deviating arguments, Nonlinear Analysis, $\underline{2}$(1978), 537-551.

116. Kitamura, Y., Kusano, T. and Naito, M., Asymptotic properties of solution of nth order differential equations with deviating argument, Proc. Japan Acad. Ser. A. Math. Sci. 54(1978), 13-16.

117. Kitamura, Y. and Kusano, T., Asymptotic properties of solutions of two dimensional differential systems with deviating argument, Hiroshima Math. J. 21(1978), 305-326.

118. Kitamura, Y. and Kusano, T., On the oscillation of a class of nonlinear differential systems with deviating arguments, J. Math. Anal. Appl. 66(1978), 20-36.

119. Kitamura, Y. and Kusano, T., Oscillation of first order nonlinear differential equations with deviating arguments, Proc. Amer. Math. Soc. 78(1980), 64-69.

120. Koplatadze, R. G., On oscillatory solutions of second order delay differential inequalities, J. Math. Anal. Appl., 42(1973), 148-157.

121. Koplatadze, R. G., The oscillatory nature of the solutions of differential inequalities and second order differential equations with retarded argument, Math. Balkanica, 5(1975), 163-172.

122. Koplatadze, R. G. and Canturija, T. A., On oscillatory properties of differential equations with deviating arguments, Tbilisi Univ. Press. Tbilisi, 1977 (Russian).

123. Koplatadze, R. G., Monotone solutions of first order nonlinear differential equations with retarded arguments (Russian), Tbilisi. Gos. Univ. Inst. Priki. Mat. Trudy, 8(1980), 24-28, 163.

124. Koplatadze, R. G., Asymptotic behavior of solutions of second order linear differential equations with time-lag. Differentcial'nye Uravnenija, 16(1980), 1247-1250.

125. Koplatadze, R. G. and Canturija, T. A., On the oscillatory and monotone solutions of the first order differential equations with deviating arguments, Differentcial'nye Uravnenija, 18(1982), 1463-1465, 1472.

126. Kozakiewicz, E., On asymptotic behavior of positive solutions of two differential inequalities with retarded argument, Colloq. Math. Soc. Janos. Bolyai. 15, Differential Equations. Keszthely (Hungary) 1975, 309-319.

127. Kulenovic, M. R. and Grammatikopoulos, M. K., First order functional differential inequalities with oscillating coefficients, Nonlinear Analysis, 8(1984), 1043-1054.

128. Kung, G. C. T., Oscillation and nonoscillation of differential equations with a time-lag, SIAM J. Appl. Math., 21(1971), 207-213.

129. Kusano, T. and Onose, H., Nonlinear oscillation of a sublinear delay equation of arbitrary order, Proc. Amer. Math. Soc., 40(1973), 219-224.

130. Kusano, T., Oscillatory behavior of solutions of high order retarded differential equations, Proceedings of the C. Caratheodory International Symposium (Athens, 1973), 370-389. Greek Math. Soc. Athens (1974).

131. Kusano, T. and Onose, H., Oscillatory and asymptotic behavior of sublinear retarded differential equations, Hiroshima Math. J., $\underline{4}$ (1974), 343-355.

132. Kusano, T. and Onose, H., Oscillations of functional differential equations with retarded argument, J. Differential Equations, $\underline{15}$ (1974), 269-277.

133. Kusano, T. and Naito, M., On the oscillation of fourth-order nonlinear differential equations with deviating argument, Qualitative research on stability of solutions of functional differential equations (Proc. Sympos. Res. Inst. Math. Sci. Kyoto. Univ. Kyoto (1976)).

134. Kusano, T. and Naito, M., Nonlinear oscillation of second order differential equations with retarded argument, Ann. Mat. Pura Appl., $\underline{106}$(1976), 171-185.

135. Kusano, T. and Onose, H., Nonoscillation theorems for differential equations with deviating argument, Pacific J. Math., $\underline{63}$(1976), 185-192.

136. Kusano, T. and Onose, H., Asymptotic decay of oscillatory solutions of second order differential equations with forcing term, Proc. Amer. Math. Soc., $\underline{66}$(1977), 251-257.

137. Kusano, T. and Onose, H., Nonlinear oscillation of second order functional differential equations with advanced argument, J. Math. Soc. Japan, $\underline{29}$(1977), 541-558.

138. Kusano, T. and Naito, M., Nonlinear oscillation of fourth order differential equations with deviation argument, Rev. Roum. Math. Pura et Appl. T. XXIII. N. 9. (1978), 1351-1365.

139. Kusano, T., On strong oscillation of even order differential equations with advanced arguments, Hiroshima Math. J., $\underline{11}$(1981), 617-620.

140. Kusano, T. and Naito, M., Comparison theorems for functional differential equations with deviating arguments, J. Math. Soc. Japan, $\underline{33}$(1981), 509-532.

141. Kusano, T., On even order functional differential equations with advanced and retarded arguments, J. Differential Equations, $\underline{45}$(1982), 75-84.

142. Kusano, T., Oscillation of the even order linear differential equations with deviating arguments of mixed type, J. Math. Anal. Appl., $\underline{98}$ (1984), 341-347.

143. Kusano, T. and Singh, B., Positive solutions of functional differential equations with singular nonlinear terms, Nonlinear Analysis, $\underline{8}$ (1984), 1081-1090.

144. Ladas, G. and Lakshmikantham, V., On asymptotic behavior of functional differential systems, An. Sti. Univ. Jasi. See. 1a., $\underline{17}$(1971), 79-88.

145. Ladas, G., Oscillation and asymptotic behavior of solutions of differential equations with retarded argument, J. Differential Equations, $\underline{10}$(1971), 281-290.

146. Ladas, G., Lakshmikantham, V. and Papadakis, J. S., Oscillation of higher-order retarded differential equations generated by retarded argument, Delay and Functional Differential Equations and their Applications, Academic Press, New York and London, 1972, 219-231.

147. Ladas, G., Ladde, G. and Papadakis, J. S., Oscillation of functional differential equations generated by delays, J. Differential Equations, 12(1972), 385-395.

148. Ladas, G. and Lakshmikantham, V., Oscillations caused by retarded actions, Applicable Anal., 4(1974), 9-15.

149. Ladas, G., Oscillatory effects of retarded actions, J. Math. Anal. Appl., 60(1977), 410-416.

150. Ladas, G., Sharp conditions for oscillations caused by delays, Applicable Anal., 9(1979), 95-98.

151. Ladas, G. and Stavroulakis, I. P., On differential inequalities with several deviating arguments, Proceedings of the IXth International Conference on Nonlinear Oscillations held in Kiev, USSR, from August 30 to September 6, 1981.

152. Ladas, G. and Stavroulakis, I. P., Oscillations caused by several retarded and advanced arguments, J. Differential Equations, 44 (1982), 135-152.

153. Ladas, G. and Stavroulakis, I. P., On delay differential inequalities of high order, Canad. Math. Bull., 25(1982), 348-354.

154. Ladas, G., Sficas, Y. G. and Stavroulakis, I. P., Asymptotic behavior of solutions of retarded differential equations, Proc. Amer. Math. Soc., 88(1983), 247-253.

155. Ladas, G., Sficas, Y. G. and Stavroulakis, I. P., Functional differential inequalities and equations with oscillating coefficients, Trends in Theory and Practice of Nonlinear Differential Equations, Ed. Lakshmikantham, V. Marcel Dekker, New York, 1983, 277-284.

156. Ladas, G. and Stavroulakis, I. P., A necessary and sufficient condition for oscillation of first order delay differential equations, Amer. Math. Monthly, 90(1983), 637-640.

157. Ladas, G., Sficas, Y. G. and Stavroulakis, I. P., Necessary and sufficient conditions for oscillations of higher order delay differential equation, Transactions of the Amer. Math. Soc., 28 1984), 81-90.

158. Ladde, G. S., Oscillations of nonlinear functional differential equations generated by the retarded argument, Delay and Functional Differential Equations and Their Applications, Academic Press, New York, 1972, 355-365.

159. Ladde, G. S., Oscillations of nonlinear functional differential equations generated by retarded actions I, Rend. Circolo. Matematico (Italia) T. 22(1973), 67-76.

160. Ladde, G. S., Oscillations caused by retarded perturbations of first order linear ordinary differential equations, Atti. Acad. Naz. Lincei. Rend. Cl. Sci. Fis. Mat. Natur. 63(1977), 351-359.

161. Ladde, G. S., Class of functional equations with applications, Nonlinear Analysis, 2(1978), 259-261.

162. Ladde, G. S., Stability and oscillation in single species process with past memory, Int. J. System. Sci. $\underline{10}$(1979), 621-647.

163. Ladde, G. S. and Zhang, B. G., Oscillation and nonoscillation for system of two first order linear differential equations with delay, J. Math. Anal. Appl., $\underline{115}$(1986), 57-75.

164. Lalli, B. S., Oscillation theorem for certain second order delay equations, Atti. Acad. Naz. Lineei. Rend. Cl. Sci. Fis. Mat. Natur. (8)$\underline{56}$(1971), 487-494.

165. Leighton, W., A first course in ordinary differential equations (1981), Wadsworth.

166. Levin, T. J., Boundedness and oscillation of some Volterra and delay equations, Advances in Differential and Integral Equations, Madison (1968), 183-191.

167. Levitan, B. M., Some problems of the theory of almost periodic functions, Uspehi. Mat. Nauk, $\underline{2}$(1947), 133-192.

168. Lillo, J. C., Oscillatory solutions of $y'(x) = m(x)y(x - n(x))$, J. Differential Equations, $\underline{6}$(1969), 1-35.

169. Lim, Eng-Bin, Asymptotic behavior of solutions of functional differential equation $x'(t) = Ax(\lambda t) + Bx(t)$, $\lambda > 1$, J. Math. Anal. Appl., $\underline{55}$ (1976), 794-806.

170. Lim, Eng-Bin, Asymptotic bounds of solutions of the functional differential equations, SIAM J. Math. Anal., $\underline{9}$(1978), 915-920.

171. Liossatos, G. E., Some oscillation theorems for second order nonlinear differential equations with functional argument, Bull. Soc. Math. Grece., $\underline{11}$(1970), 61-65.

172. Liu, Tsai-Sheng, Oscillation of even order differential equations with deviating arguments, Pacific J. Math., $\underline{61}$(1975), 493-502.

173. Liu, Tsai-Sheng, Classification of solutions of high order differential equations and inequalities with deviating arguments, J. Differential Equations, $\underline{21}$(1976), 417-430.

174. Liu, Tsai-Sheng, Oscillation of solution of higher order functional differential equations and damped differential equations, Utilitas. Math , $\underline{11}$(1977), 193-214.

175. Lovelady, D. L., Asymptotic analysis of a second order nonlinear functional differential equations, Funkcialaj, Ekvacioj., $\underline{18}$(1975), 15-22.

176. Lovelady, D. L., Oscillation and a class of linear delay differential equations, Trans. Amer. Math. Soc., $\underline{226}$(1977), 345-364.

177. Macki, J. M. and Wong, J. S. W., Oscillation of solutions of second order nonlinear differential equations, Pacific J. Math., $\underline{24}$(1968), 111-118.

178. Mahfoud, W. E., Oscillation and asymptotic behavior of solutions of nth order nonlinear delay differential equations, J. Differential Equations, $\underline{24}$(1977), 75-98.

179. Mahfoud, W. E., Characterization of oscillation of solutions of the delay equation $x^{(n)}(t) + a(t)f(x(g(t)))x$, J. Differential Equations, $\underline{28}$ (1978), 437-451.

180. Mahfoud, W. E., Remarks on some oscillation theorems for n^{th} order differential equations with a retarded argument, J. Math. Anal. Appl., $\underline{62}$(1978), 68-80.

181. Mahfoud, W. E., Comparison theorems for delay differential equations, Pacific J. Math., $\underline{83}$(1979), 187-197.

182. Marusiak, P., Note on the Ladas's paper on oscillation and asymptotic behavior of solutions of differential equations with retarded argument, J. Differential Equations, $\underline{13}$(1973), 150-156.

183. Marusiak, P., On oscillation of solutions of differential inequalities with retarded argument, Cas. Pester. Mat., $\underline{104}$(1979), 281-294.

184. Marusiak, P. and Norkin, S. B., The oscillatory nature of systems with deviating argument, Prace Stud. Vysokej. Skoly Dograr. Spojov. Ziline. Ser. Mat. Fyz, $\underline{3}$(1981), 13-25.

185. McCalla, C., Zeros of the solutions of first order functional differential equations, SIAM J. Math. Anal. $\underline{9}$(1978), 845-847.

186. McCalla, C., Oscillatory and asymptotic behavior of second order functional differential equations, Nonlinear Analysis, $\underline{3}$(1979), 283-291.

187. Mishev, D. P. and Bainov, P. P., Asymptotic behavior of the non-oscillating solutions of functional differential equations of nth order, Mathematics Reports Toyama University. Toyama, Japan. $\underline{6}$(1983), 83-94.

188. Morgenthal, K., On the asymptotic behavior of solutions of a third order differential equation with after effect I, II, Beitrage Anal. N. $\underline{18}$(1981), 139-148, 149-156.

189. Myskis, A. D., Linear differential equations with retarded argument, Moscow, Nauka, 1972, (Russian).

190. Nababan, S. and Noussair, E. B., Oscillation criteria of second order nonlinear delay inequalities, Bull. Austral. Math. Soc., $\underline{14}$ (1976), 331-341.

191. Nababan, S., Oscillation criteria for a class of nonlinear vector delay differential equations, Ann. Mat. Pura Appl., $\underline{117}$(1978), 55-66.

192. Naito, M., Oscillations of differential inequalities with retarded arguments, Hiroshima Math. J., $\underline{5}$(1975), 187-192.

193. Naito, M., Positive solutions of nonlinear differential inequalities, Hiroshima Math. J., $\underline{9}$(1979), 769-785.

194. Naito, M., On strong oscillation of retarded differential equations, Hiroshima Math. J., $\underline{11}$(1981), 553-560.

195. Nasyhova, L. G., The existence of oscillating solutions of an nth order differential equation with retarded argument, Differencial'nye Uravneija, $\underline{11}$(1978), 120-127.

196. Nichiro, K., Kusano, T. and Naito, M., Nonoscillatory solutions of forced differential equations of the second order, J. Math. Anal. Appl., $\underline{90}$(1982), 323-342.

197. Norkin, S. B., <u>Differential equations of the second order with retarded argument</u> (1972), V. 31. Translations of Mathematical Monographs, AMS, Providence, R.I.

198. Norkin, S. B., The two sides solution of linear system with asymptotic constant coefficient and delay, Czech. Math. J., $\underline{33}$(1983), 58-69.

199. Ohriska, J., The argument delay and oscillatory properties of differential equation of nth order, Czech. Math. J., $\underline{29}$(1979), 268-283.

200. Ohriska, J., Sufficient conditions for the oscillation of nth order nonlinear delay differential equation, Czech. Math. J., $\underline{30}$(1980), 490-497.

201. Ohriska, J., Oscillation of nth order linear and nonlinear delay differential equations, Czech. Math. J., $\underline{32}$(1982), 271-274.

202. Ohriska, J., Oscillation of second order delay and ordinary differential equations, Czech. Math. J., $\underline{34}$(1984), 107-112.

203. Olah, R., Note on the oscillatory behavior of bounded solutions of higher order differential equation with retarded argument, Arch. Mat., $\underline{14}$(1978), 171-174.

204. Olah, R., Note on the oscillation of differential equations deviating argument, Casopis. Pest. Math., $\underline{107}$(1982), 380-387, 428.

205. Onose, H., An oscillation theorem for delayed differential equations of arbitrary order, Nanta. Math., $\underline{9}$(1976), 117-120.

206. Onose, H., A comparison method for forced oscillations, Monatsh. Math., $\underline{83}$(1977), 37-41.

207. Onose, H., Oscillation of a functional differential equation arising from an industrial problem, J. Austral. Math. Soc. Ser. A $\underline{26}$(1978), 323-329.

208. Onose, H., Forced oscillation for functional differential equations of fourth order, Bull. Fac. Sci. Ibaraki. Univ. Ser. A $\underline{11}$(1979), 57-63.

209. Onose, H., Nonlinear oscillation of functional differential equations with complicated deviating argument, Bull. Fac. Sci. Ibaraki. Univ. Ser. A $\underline{13}$(1981), 29-43.

210. Onose, H., Asymptotic behavior of nonoscillatory solutions of a high order functional differential equations, Bull. Austral. Math. Soc., $\underline{24}$(1981), 85-92.

211. Onose, H., Oscillation property of functional differential equations with complicated arguments, Math. Seminar Notes, $\underline{10}$(1982), 715-720.

212. Onose, H., Oscillatory properties of the first order differential inequalities with deviating argument, Funkcialaj Ekvacioj, $\underline{26}$(1983), 189-196.

213. Onose, H., Oscillatory properties of the first order nonlinear advanced and delayed differential inequalities, Nonlinear Analysis, $\underline{8}$ (1984), 171-180.

214. Pandofi, L., Some observations on the asymptotic behaviors of the solutions of the equation $x'(t) = A(t)x(\lambda t) + B(t)x(t)$, $\lambda > 0$, J. Math. Anal. Appl., $\underline{67}$(1979), 483-489.

215. Philos, Ch. G., An oscillatory and asymptotic classification of the solutions of differential equation with deviating arguments, Atti. Acad.

Naz. Lincei. Rend. Cl. Sci. Fis. Mat. Natur. (8)$\underline{63}$(1977)N. 3-4, 195-203 (1978).

216. Philos, Ch. G., Oscillatory and asymptotic behavior of all solutions differential equations with deviating arguments, Proc. Roy. Soc. Edinburgh. Sect. A $\underline{81}$(1978), 195-210.

217. Philos, Ch. G., Oscillations caused by delay, Au. Stiint. Univ. "Al. I. Cuza" Iasi Sect. Ia Mat (N.S)$\underline{24}$(1978), 71-76.

218. Philos, Ch. G., Bounded oscillations generated by retardations for differential equations of arbitrary order, Util. Math., $\underline{15}$(1979), 161-182.

219. Philos, Ch. G., A new criterion for the oscillatory and asymptotic behavior of delay differential equations, Bull. Acad. Polon. Sci. Ser. Sci. Math. $\underline{29}$(1981), 367-370.

220. Philos, Ch. G., Nonoscillation and damped oscillations for differential equations with deviating arguments, Math. Nachr., $\underline{106}$(1982), 109-119.

221. Philos, Ch. G., Sficas, Y. G. and Staikos, V. A., Some results on the asymptotic behavior of nonoscillatory solutions of differential equations with deviating arguments, J. Austral. Math. Soc. Ser. A $\underline{32}$(1982), 295-317.

222. Philos, Ch. G. and Staikos, V. A., A basic asymptotic criterion for differential equations with deviating arguments and its applications to the nonoscillation of linear ordinary equations, Nonlinear Analysis, $\underline{6}$(1982), 1095-1114.

223. Philos, Ch. G. and Sficas, Y. G., Oscillatory and asymptotic behavior of second and third order retarded differential equations, Czech. Math. J. $\underline{32}$(1982), 169-182.

224. Philos, Ch. G., Some comparison criteria in oscillation theory, J. Austral. Math. Soc. Ser. A, $\underline{36}$(1984), 176-186.

225. Sibgatullin, G. K., A comparison theorem for nonoscillating solutions of differential equations of order n (n $\geq$ 2) with lag, (Russian), Partial differential equations. Ryazan. Gos. Ped. Inst. Ryazan. 1980, 87-93.

226. Sficas, Y. G. and Staikos, V. A., The effect of retarded actions on nonlinear oscillations, Proc. Amer. Math. Soc., $\underline{46}$(1974), 259-264.

227. Sficas, Y. G., The effect of the delay on the oscillatory and asymptotic behavior of the nth order retarded differential equations, J. Math. Anal. Appl., $\underline{49}$(1975), 748-757.

228. Sficas, Y. G. and Staikos, V. A., Oscillations of differential equations with deviating arguments, Funkcial. Ekvacioj., $\underline{19}$(1976), 35-43.

229. Sficas, Y. G., Strongly monotone solutions of retarded differential equations, U.I.G. Tech. Rep. $\underline{103}$(1977), 1-14.

230. Sficas, Y. G. and Stavroulakis, I. P., On the oscillatory and asymptotic behavior of a class of differential equations with deviating arguments, SIAM J. Math. Anal., $\underline{9}$(1978), 956-966.

231. Sficas, Y. G., On the behavior of nonoscillatory solutions of differential equations with deviating argument, Nonlinear Analysis, $\underline{3}$(1979), 379-394.

232. Sharkovskii, A. V. and Shevelo, V. N., On the oscillation and asymptotic behavior of solution to differential equations with deviating arguments, Problems of the Asymptotic Theory of Nonlinear Oscillation, Naukova Dunka Kiev, 1977, 257-263.

233. Sharkovskii, A. V., Shevelo, V. N. and Ivanov, A. F., Some results of the qualitative investigation of first order functional differential equations, Functional differential systems and related topics II. Proceedings of the second international conference held in Blazejeuko, May 3-10, 1981. Ed. M. Kosielewicz. Higher College of Engineering, Institute of Mathematics and Physics, Zielona Gora, 1981.

234. Shevelo, V. N., On oscillation of solution to differential equations with deviating arguments, Naukova Dunka. Kiev, 1978.

235. Shevelo, V. N. and Varekh, N. V., On oscillation of solutions to first order differential equations with deviating arguments, Asymptotic methods in the theory of nonlinear oscillations (Proc. All-Union Conf. Asymptotic methods in Nonlinear Mech., Katsiveli, 1977), 247-262, 278. Naukova, Dunka. Kiev, 1979.

236. Shevelo, V. N. and Varekh, N. V., Some properties of solutions of systems of functional differential equations. Qualitative investigations of functional differential equations, 153-171, 187. Naukova Dumka Kiev. 1980.

237. Shevelo, V. N., Varekh, N. V. and Gricai, A. G., Investigation of oscillation properties of solutions of systems of differential and differential-difference equations (survey), An investigation of differential and differential-difference equations, 157-183, 192. Akad. Nauk. Ukrain SSR. Inst. Mat. Kiev, 1980.

238. Shevelo, V. N. and Varekh, N. V., Some specific theorems for systems of differential-difference equations, "Functional Differential and Difference Equations," 110-125, 132. Akad. Nauk. Ukrain. SSR. Inst. Mat. Kiev, 1981.

239. Shevelo, V. N., Varekh, N. V. and Gritsai, A. G., Oscillatory properties of solutions of systems of differential equations with retarded argument, Akad. Nauk. Ukrain SSR. Inst. Mat. Preprint. 1982, N. 2, 48 pp.

240. Shreve, W. E., Oscillation in first order nonlinear retarded argument differential equations, Proc. Amer. Math. Soc., 43(1973), N. 2, 565-568.

241. Singh, B., Nonoscillation of forced fourth order retarded equation, SIAM J. Appl. Math., 28(1975), 265-269.

242. Singh, B., Forced oscillations in general ordinary differential equations with deviating arguments, Hiroshima Math. J., 6(1976), 7-14.

243. Singh, B., Forced nonoscillation in second order functional equation, Hiroshima Math. J., 7(1977), 657-665.

244. Singh, B., Vanishing nonoscillations of Lienard type retarded equations, Hiroshima Math. J., 7(1977), N. 1, 1-8.

245. Singh, B., Necessary and sufficient condition for maintaining oscillations and nonoscillations in general functional equations and their asymptotic properties, SIAM J. Math. Anal., 10(1979), N. 1, 18-31.

246. Singh, B., Necessary and sufficient condition for eventual decay of oscillation in general functional equations with delays, Hiroshima Math. J., 10(1980), 1-10.

247. Singh, B. and Kusano, T., On asymptotic limits of nonoscillation in functional equations with retarded arguments, Hiroshima Math. J., 10(1980), 557-565.

248. Singh, B. and Kusano, T., Asymptotic behavior of oscillatory solutions of a differential equations with deviating arguments, J. Math. Anal. Appl., 83(1981), 395-407.

249. Singh, B., Oscillation in second order functional equations with deviating arguments, Internat. J. Math. Math Sci., 4(1981), 137-146.

250. Singh, B., Nonlinear oscillations in disconjugate forced functional equations with deviating arguments, Internat. J. Math. Math Sci., 6(1983), 101-109.

251. Singh, G., On quick oscillations in functional equations with deviating arguments, J. Math. Anal. Appl., 101(1984), 598-610.

252. Smith, H. L., Bounded oscillation in a class of functional differential equations, J. Math. Anal. Appl., 56(1976), 223-232.

253. Staikos, V. A. and Sficas, Y. G., Oscillatory and asymptotic characterization of the solutions of differential equations with deviating arguments, J. London Math. Soc., 10(1975), 39-47.

254. Staikos, V. A. and Sficas, Y. G., Forced oscillations for differential equations of arbitrary order, J. Differential Equations, 17(1975), 1-11.

255. Staikos, V. A. and Philos, Ch. G., On the asymptotic behavior of nonoscillatory solutions of differential equations with deviating arguments, Hiroshima Math. J., 7(1977), 9-31.

256. Staikos, V. A. and Stavroulakis, I. P., Bounded oscillations under the effect of retardation for differential equations of arbitrary order, Proc. R.S.E., 77(1977), 129-136.

257. Staikos, V. A. and Philos, Ch. G. Nonoscillatory phenomena and damped oscillations, Nonlinear Analysis, 2(1978), 197-210.

258. Staikos, V. A., Basic results on oscillation for differential equations with deviating arguments, Hiroshima Math. J., 10(1980), 495-515.

259. Stavroulakis, I. P., Nonlinear delay differential inequalities, Nonlinear Analysis, 6(1982), 389-396.

260. Swanson, C. A., Comparison and oscillation theory of linear differential equations, New York and London, Acad. Press, 1968.

261. Szlaba, U., Some properties of solutions of differential equations with retarded arguments, Demonstr. Math., 9(1976), 563-571.

262. Szlaba, U., Note on "On oscillatory behavior of even order delay equations" by R. S. Dahiya and B. Singh, J. Math. Anal. Appl., 63 (1978), 313-318.

263. Terry, R. D. and Wong, P. K., Oscillatory properties of a fourth order delay differential equation, Funkcial. Ekvac., $\underline{15}$(1972), 209-220; $\underline{16}$(1973), 213-224.

264. Terry, R. D., Oscillatory properties of a delay differential equation of even order, Pacific J. Math., $\underline{52}$(1974), 269-282.

265. Terry, R. D., Delay differential equations of odd order satisfying property pk, J. Austral. Math. Soc., $\underline{20}$(1975), 451-467.

266. Terry, R. D., Some oscillation criteria for delay differential equations even order, SIAM J. Appl. Math., $\underline{28}$(1975), 319-334.

267. Terry, R. D., Oscillatory and asymptotic properties of homogeneous and nonhomogeneous delay differential equations of even order, J. Austral. Math. Soc. Ser. A, $\underline{22}$(1976), 282-304.

268. Terry, R. D., Positive solution of $D_n[r(t)D^n y(t)] = p(t)y(t - \tau(t))$, Funkcial. Ekvacioj. $\underline{20}$(1977), 49-60.

269. Terry, R. D., An application of Lyapunov's direct method to the study of oscillations of a delay differential equation of even order, J. Austral. Math. Soc. Ser. A, $\underline{25}$(1978), 201-209.

270. Teufel, H., Second order almost linear functional equations oscillation, Proc. Amer. Math. Soc. $\underline{35}$(1972), 117-119.

271. Teufel, H., A note on second order differential inequalities and functional differential equations, Pacific J. Math., $\underline{41}$(1972), 537-541.

272. Tomaras, A., Oscillations of an equation relevant to an industrial problem, Bull. Austral. Math. Soc., $\underline{12}$(1975), 425-431.

273. Tomaras, A., Oscillatory behavior of an equation arising from an industrial problem, Bull. Austral. Math. Soc., $\underline{13}$(1975), 255-260.

274. Tomaras, A., Oscillations of higher order retarded differential equations caused by delays, Rev. Roum. Math. Pures et Appl., $\underline{20}$(1975), 1163-1172.

275. Tomaras, A., Oscillations of a first order functional differential equation, Bull. Austral. Math. Soc., $\underline{17}$(1977), 91-95.

276. Tomaras, A., Oscillatory behavior of first order delay differential equations, J. Austral. Math. Soc., $\underline{19}$(1978), 183-190.

277. Tramov, M. I., Conditions for oscillatory solutions of first order differential equations with a delayed argument, Izvestiya Vysshikh Uchebnykh Zavedenii Matematika, $\underline{19}$(1975), 92-96.

278. Tramov, M. I., The oscillatory nature of solutions of differential equations with deviating argument, Differentcial'nye Uravnenija $\underline{18}$(1982), 245-253, 364.

279. Travis, C. C., Oscillation theorems for second order differential equations with functional arguments, Proc. Amer. Math. Soc., $\underline{31}$(1972), 199-202.

280. True, E. D., A comparison theorem for certain functional differential equations, Proc. Amer. Math. Soc., $\underline{47}$(1975), 127-132.

281. Varekh, V. N. and Shevelo, V. N., Some properties of solutions of systems of differential equations with retarded argument, Ukrain Mat. Zh. $\underline{34}$(1982), 1-8, 131.

282. Ved, Yu. A., Criterion for the existence of asymptotically constant solution of differential systems with deviating argument, Izv. Akad. Nauk. Kirgiz. SSR. (1982), 9-11.

283. Waltman, P., A note on an oscillation criterion for an equation with a functional argument, Canad. Math. Bull. 11(1968), 593-595.

284. Werbowski, J., Note on the asymptotic behavior of the solutions of differential equations with deviating arguments, Math. Slovaca, 28 (1978), 181-188.

285. Werbowski, J., On the asymptotic behavior of solutions of second order differential equations with deviating argument, Fasc. Math., 111(1979), 125-131.

286. Werbowski, J., Oscillations of nonlinear differential equations caused by deviating arguments, Funkcial. Ekvac., 25(1982), 295-301.

287. Willet, D. W., Classification of second order linear differential equations with respect to oscillation, Advances in Math., 3(1969), 594-623.

288. Winston, E., Comparison theorems for scalar delay differential equations, J. Math. Anal. Appl., 29(1970), 455-463.

289. Wong, J. S. W., On second order nonlinear oscillation, Funkcial. Ekvac., 11(1968), 207-234.

290. Wong, J. S. W., On the generalized Emden-Fowler equation, SIAM Rev., 17(1975), 339-360.

291. Yan, Jn-Rang, Oscillatory property of second order nonlinear differential equations with deviating argument, Kexue Tongbao 27(1982), 7-11.

292. Yan, Jn-Rang, Oscillatory behavior of forced second order nonlinear functional differential equations, Kexue Tongbao, 27(1982), 921-925.

293. Yan, Jn-Rang, Oscillatory properties of solutions of second order damped nonlinear differential equations, Acta, Mathematical Applicatae Sinica. 6(1983), 251-256.

294. Yeh, Cheh-Chih, An oscillation criterion for second order nonlinear differential equations with functional arguments, J. Math. Anal. Appl., 76(1980), 72-76.

295. Yeh, Cheh-Chih, Oscillations of nth order retarded differential equations, Publ. Inst. Math. (Boegrad)(N.S.)29(43)(1981), 293-297.

296. Yorke, J. A., Selected topics in differential delay equations, Lecture Notes in Math., 243(1971), 16-28.

297. Yoshizawa, T., Oscillatory property of second order differential equations, Tohoku Math. J., 22(1970), 619-634.

298. Zhang, Binggen, On the oscillation of the solutions for second order functional differential equations, J. of Shandong College of Oceanology, N. 1 (1980).

299. Zhang, Binggen, A survey of oscillation of solution to the differential equations with deviating arguments, Reprint. To present on the second functional differential equations Conference of all China, Oct. 8-10, 1981, Hefei.

300. Zhang, Binggen, Oscillation and nonoscillation for second order functional differential equations, Chinese Annals of Math., $\underline{2}$(1) 1981.

301. Zhang, Binggen, Oscillation of the solution of the first order advanced type differential equations, Science Exploration, N. 3 (1982).

302. Zhang, Binggen and Ding, Yidong, Influence of the advanced argument on oscillation of solutions, A Monthly Journal of Science $\underline{27}$(1982).

303. Zhang, B. G., Ding, Y. D., Feng, R. L., Wu, D. and Wang, O. S., Some new results about oscillation of solutions of functional differential equations, J. of Shandong College of Oceanology, $\underline{12}$(1982).

304. Zhang, B. G., A survey of the oscillation of solution to first order differential equation with deviating arguments, Proceeding of VIth Nonlinear Analysis International Conference, held at Arlington, Texas, June 18-22, 1984.

305. Zhang, B. G. and Ladde, G. S., On the oscillation to the even order delay differential equation, J. Math. Anal. Appl. (in press).

306. Zhang, B. G., Oscillation of first order nonlinear differential equations with deviating argument, Kexue Tongbao (in press).

307. Zhang, B. G., Oscillation behavior of solution of the first order functional differential equations, Funkcial, Ekvac, $\underline{28}$(1985), 93-101.

308. Zhang, B. G., On the oscillation of solution to the first order linear differential equations with deviating arguments, Acta Mathematica Sinica, $\underline{28}$(1985), 637-643.

309. Zhang, B. G., Oscillation for a kind of second order differential equations with deviating arguments, Kexue Tongbao, 1985, N. 23.

Index

DATE DUE

'88